**7 Day**

**University of Plymouth Library**

Subject to status this item may be renewed
via your Voyager account

**http://voyager.plymouth.ac.uk**

Exeter  tel: (01392) 475049
Exmouth  tel: (01395) 255331
Plymouth  tel: (01752) 232323

# IRON
# Nutritional and physiological significance

The Report of the British Nutrition Foundation's Task Force

The British Nutrition Foundation

Published by Chapman & Hall
for the British Nutrition Foundation

## CHAPMAN & HALL

London · Glasgow · Weinheim · New York · Tokyo · Melbourne · Madras

Published by Chapman & Hall, 2–6 Boundary Row, London SE1 8HN, UK

Chapman & Hall, 2–6 Boundary Row, London SE1 8HN, UK

Blackie Academic & Professional, Wester Cleddens Road, Bishopbriggs, Glasgow G64 2NZ, UK

Chapman & Hall GmbH, Pappelallee 3, 69469 Weinheim, Germany

Chapman & Hall USA, 115 Fifth Avenue, New York, NY 10003, USA

Chapman & Hall Japan, ITP-Japan, Kyowa Building, 3F, 2-2-1 Hirakawacho, Chiyoda-ku, Tokyo 102, Japan

Chapman & Hall, 102 Dodds Street, South Melbourne, Victoria 3205, Australia

Chapman & Hall India, R. Seshadri, 32 Second Main Road, CIT East, Madras 600 035, India

First edition 1995
© 1995 The British Nutrition Foundation

Typeset in 10 on 12pt Helvetica by Columns of Reading Ltd
Printed in Great Britain by T.J. Press (Padstow) Ltd, Padstow, Cornwall

ISBN 0 412 73170 3

A catalogue record for this book is available from the British Library
Library of Congress Catalog Card Number: 95-70860

∞ Printed on acid-free paper, manufactured in accordance with ANSI/NISO Z39.48-1992 (Permanence of Paper)

# CONTENTS

# MEMBERSHIP OF THE BNF IRON TASK FORCE

*Chairman*
**Professor Dame B. Clayton**
16 Chetwynd Drive
Bassett
Southampton

*Members*
**Professor P. Aggett**
Institute of Food Research
Colney Lane
Norwich

**Dr J. Brock**
Dept of Immunology
Western Infirmary
Glasgow

**Dr P. Evans**
Department of Public Health
University of Glasgow
2 Lilybank Gardens
Glasgow

**Dr S. Fairweather-Tait**
Institute of Food Research
Colney Lane
Norwich

**Dr R. Lansdown**
The Child-to-Child Trust
Institute of Education
20 Bedford Way
London

**Dr M. Lawson**
Institute of Child Health
30 Guilford Street
London

**Dr M. Nelson**
King's College London
Campden Hill Road
London

**Professor T. Peters**
Dept of Clinical Biochemistry
King's College Hospital
Bessemer Road
London

**Professor M. Pippard**
Dept of Haematology
Ninewells Hospital and Medical
School
Dundee

**Professor T.A.B. Sanders**
King's College London
Campden Hill Road
London

**Professor D. Thurnham**
Dept of Biological/Biomedical
Science
University of Ulster
Coleraine
Co. Londonderry

**Dr R. Ward**
Dept of Clinical Biochemistry
King's College Hospital
Bessemer Road
London

**Dr B.A. Wharton**
Director-General
British Nutrition Foundation

**Dr M. Worwood**
University Hospital of Wales
Cardiff

*Observers*
**Dr D.H. Buss**
Ministry of Agriculture, Fisheries
and Food
Nobel House
London

**Dr G. Sarna**
Medical Research Council
20 Park Crescent
London

**Dr M. Wiseman**
Department of Health
Skipton House
London

*Corresponding members*
**Professor J. Cook**
University of Kansas Medical
Center
Kansas 66103
USA

**Professor L. Hallberg**
University of Göteborg
Sahlgren Hospital
S-413 45 Göteborg
Sweden

**Professor C. Hershko**
Shaare Zedek Medical Centre
Jerusalem
Israel

*Secretariat*
**Dr Margaret Ashwell**
Science and Publications
Director
British Nutrition Foundation
(Editor of the Task Force Report
until March 1995)

**Ms Ursula Arens**
Senior Nutrition Scientist
British Nutrition Foundation
(Secretary to the Task Force
from August 1994;
Editor of the Task Force Report
from March 1995)

**Ms Catherine Rushton**
British Nutrition Foundation
(Assistant Editor from March
1995)

**Ms Hilary Groom**
Nutrition Scientist
British Nutrition Foundation
(Secretary to the Task Force
until August 1994)

**Ms Sandra Rodrigues**
PA to the Science Group
British Nutrition Foundation

This Report is the collective responsibility of all the members of the Task Force. Authors of the first draft of each chapter are given below.

1   Iron biochemistry
    *Dr S. Fairweather-Tait*
2   Iron absorption
    *Dr S. Fairweather-Tait*
3   Dietary sources and intakes of iron
    *Dr D.H. Buss*
4   Mechanisms of cellular iron homeostasis
    *Dr M. Worwood*
5   Measurement of iron status
    *Dr M. Worwood*
6   Iron as a pro-oxidant
    *Professor D. Thurnham*
7   Iron and anaemia
    *Professor M. Pippard*
8   Iron overload and toxicity
    *Dr R. Ward*
9   Effect of iron on work performance and thermogenesis
    *Professor M. Pippard*
10  Iron in infection and immunity
    *Dr J. Brock*
11  Iron and mental and motor behaviour in children
    *Dr R. Lansdown and Dr B.A. Wharton*
12  Iron and coronary heart disease
    *Dr M. Nelson*
13  Iron and cancer
    *Dr M. Nelson*
14  Iron, the brain and neurodegeneration
    *Dr P. Evans*
15  Iron in infancy and childhood
    *Dr M. Lawson*
16  Iron in adolescence
    *Dr S. Fairweather-Tait*
17  Iron and women in the reproductive years
    *Professor P. Aggett*
18  Iron status in older people
    *Professor D. Thurnham*
19  Iron status of vegetarians
    *Professor T.A.B. Sanders*
20  Public health considerations
    *Dr M. Wiseman*

# TERMS OF REFERENCE

The Task Force was invited by the Council of the British Nutrition Foundation to:

1. Review the present state of knowledge of:
   (a) the dietary sources and intakes of iron;
   (b) the absorption and metabolic and physiological functions of iron;
   (c) iron status in health and disease;
   (d) iron status and requirements of different population subgroups;
   (e) the risks of excess intakes of iron.
2. Prepare a report and, should it see fit, to draw conclusions, make recommendations and identify areas for future research.

# FOREWORD

The British Nutrition Foundation organizes independent 'Task Forces' to review, analyse and report in depth upon specific areas of interest and importance in the field of human nutrition.

These expert committees consist of acknowledged specialists and operate completely independently of the Foundation.

The Iron Task Force has reviewed and discussed much published information. This report summarizes the deliberations and findings of the Task Force and gives its conclusions and recommendations.

I am most grateful to the members of the Task Force who have contributed their time and expertise so generously. My sincere thanks also go to the Secretariat for their excellent support.

Professor Dame Barbara Clayton
Chairman of the Task Force

# ABBREVIATIONS USED IN THE TEXT

| | |
|---|---|
| AAP | American Academy of Paediatrics |
| ADP | adenosine diphosphate |
| ALA | 5-aminolevulinic acid |
| ATP | adenosine triphosphate |
| BMI | Body Mass Index |
| CDC | Centers for Disease Control and Prevention (USA) |
| CHD | coronary heart disease |
| CNS | central nervous system |
| COMA | Committee on Medical Aspects of Food Policy |
| CSF | cerebrospinal fluid |
| DH | Department of Health |
| DNA | deoxyribonucleic acid |
| DRV | Dietary Reference Values |
| EAR | Estimated Average Requirement |
| ELISA | enzyme-linked immunosorbent assays |
| EPO | erythropoietin |
| EPP | erythrocyte protoporphyrin |
| ESPGAN | European Society for Paediatric Gastroenterology and Nutrition |
| ESR | erythrocyte sedimentation rate |
| Fe(II) | ferrous iron |
| Fe(III) | ferric iron |
| FEP | free erythrocyte protoporphyrin |
| Ft | ferritin |
| $G_6PD$ | glucose-6-phosphate dehydrogenase |
| HC | haemochromatosis |
| Hb | haemoglobin |
| HLA | human leucocyte antigen |
| ICSH | International Committee for Standardization in Haematology |
| ID | iron deficiency (with or without anaemia) |
| IDA | iron deficiency anaemia |

| | |
|---|---|
| IQ | intelligence quotient |
| IRE | iron response element |
| IRF | iron regulatory factor |
| IRP | iron repressor protein |
| LAMMA | laser microprobe mass analyser |
| LRNI | Lower Reference Nutrient Intake |
| MAFF | Ministry of Agriculture, Fisheries and Food |
| MCHC | mean cell haemoglobin concentration |
| MCV | mean cell volume |
| MI | myocardial infarction |
| MRI | magnetic resonance imaging |
| MWt | molecular weight |
| NAA | neutron activation analysis |
| NADH | (reduced) nicotinamide adenine dinucleotide |
| NHANES | National Health and Nutrition Examination Survey |
| NK | natural killer cells |
| PIXE | proton-induced X-ray emission |
| PMN | polymorphonuclear leucocytes |
| PUFA | polyunsaturated fatty acids |
| RDW | red cell distribution width |
| ROS | reactive oxygen species |
| RNA | ribonucleic acid |
| RNI | Reference Nutrient Intake |
| SOD | superoxide dismutase |
| $T\frac{1}{2}$ | half-life |
| TBARS | thiobarbituric acid-reacting substances |
| TIBC | total iron binding capacity |
| Tf | transferrin |
| TfR | transferrin receptor |
| UIBC | unsaturated iron binding capacity |
| WHO | World Health Organization |
| ZPP | zinc protoporphyrin |

# 1

# IRON BIOCHEMISTRY

## 1.1 FUNCTION AND CHEMISTRY

Iron (atomic weight 55.85, atomic number 26), the second most abundant metal in the earth's crust, exists in two valency states: ferrous (FeII) and ferric (FeIII). Over 500 million years ago large amounts of reduced iron (FeII) would have been present in the low-oxygen environment, but now iron exists almost exclusively in the less soluble oxidized state (FeIII). This has greatly reduced its accessibility to many forms of life, including humans, and this is suggested to be one of the main contributory factors in the aetiology of iron deficiency anaemia, one of the most common nutritional deficiency disorders in the world (FAO/WHO, 1988).

The chemistry of iron is complex, primarily because of its dual valency and reactivity with oxygen. Ionic iron is an active promoter of free-radical reactions, and is toxic to living cells. Biological systems have therefore developed several ways of limiting the entry of iron into the body and converting any absorbed iron into a bound 'safe' form. In mammals serum iron is bound to transferrin, and most body iron is present as iron porphyrin complexes (haemoglobin, myoglobin and haem-containing enzymes). Iron is stored as ferritin and haemosiderin. The distribution of iron in the body is illustrated in Figure 1.1.

The nutritional need for iron in living organisms is derived from the central role that it plays in the energy metabolism of living cells. Iron is a transition metal and can take part in redox processes by undergoing reversible valency changes, such as reduction by an organic substrate and re-oxidation by oxygen. It can also bind oxygen either on its own or as part of a complex.

## 1.2 IRON COMPOUNDS IN THE BODY

Proteins of iron transport (transferrin) and storage (ferritin and haemosiderin) are discussed in Chapter 4. Functionally important forms of iron in the body are haemoglobin, myoglobin, cytochromes, iron–sulphur proteins, iron enzymes and lactoferrin.

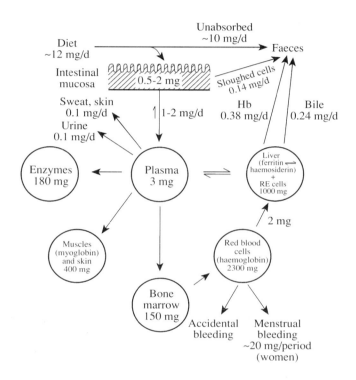

**Figure 1.1** Iron distribution (mg) and metabolism within the body. RE: Reticuloendothelial cells

### 1.2.1 Haemoglobin (Hb)

About two thirds of body iron is present in haemoglobin in red blood cells, where it is essential in the transport of oxygen. The haemoglobin molecule consists of four subunits, each of them a polypeptide chain bound to a haem molecule, and has a molecular weight of 65 000 and an iron content of 0.34%. Iron is stabilized in the ferrous state which allows it to be reversibly bound to oxygen. The synthesis of haem and its attachment to globin take place in the bone marrow in the late stages of the development of the red blood cell. Iron is carried to the bone marrow as ferric iron bound to transferrin: it is released within the red cell precursors, reduced to the ferrous form and transferred to protoporphyrin (Chapter 7).

### 1.2.2   Myoglobin

Myoglobin, the red pigment of muscle, is a single peptide homologue of haemoglobin with a molecular weight of 17 000. Its function is to store oxygen delivered to the tissues by haemoglobin for utilization during muscle contraction. This protein accounts for 5–10% of total body iron.

### 1.2.3   Cytochromes

Cytochromes, the electron-transport enzymes, are located in the mitochondria as well as in other cellular membranes. They are able to undergo reversible oxidation by way of changes in the oxidation state of iron. Cytochromes a, b and c are present in all aerobic cells, within the cristae of mitochondria, and are essential for the oxidative production of cellular energy in the form of ATP. Cytochrome c has a molecular weight of 13 000 and, like myoglobin, is made up of a single peptide chain and one haem group, containing an atom of iron. Extra-mitochondrial cytochromes include cytochrome P-450, located within microsomal membranes of the liver, which is involved in oxidative degradation of drugs and endogenous substrates, e.g. steroids.

### 1.2.4   Iron–sulphur proteins

The iron–sulphur proteins contain iron and acid-labile sulphur in equimolar amounts, and are involved in electron transport by undergoing reversible $Fe(II)–Fe(III)$ transitions. They include flavoproteins found in the mitochondria, such as NADH dehydrogenase and succinic dehydrogenase, and aconitase.

### 1.2.5   Iron enzymes

The haem enzymes catalase and peroxidase contain four haem groups, each with one iron atom; they are widely distributed in the body but are particularly abundant in red blood cells and the liver. They function in the reduction of hydrogen peroxide produced in the body. Other non-haem iron enzymes include aconitase and the flavoprotein, xanthine oxidase.

### 1.2.6   Lactoferrin

The glycoprotein lactoferrin (molecular weight 80 000) is a cationic iron carrier, very similar to transferrin, that is present in high concentrations in human breast milk (1 mg/ml). It binds two atoms of ferric iron per molecule and is found in neutrophilic granulocytes and on mucosal surfaces as part of their protective coat. Lactoferrin is believed to participate in the defence of the breast-fed infant against infection by depriving bacteria of the iron needed for growth, and by donating iron to generate reactive oxygen radicals to enhance the microbicidal mechanisms of phagocytes.

## 1.3   INTERACTIONS BETWEEN IRON AND OTHER MICRONUTRIENTS

There are a number of micronutrients that affect iron metabolism either before or after absorption from the gut (Table 1.1). Uptake of iron may be modified by other minerals, such as calcium and nickel. Interactions occur in the gastrointestinal tract, primarily with calcium, zinc, manganese and vitamin C (Chapter 2). Subsequent physiological utilization of the element may be impaired with riboflavin, vitamin A or copper deficiencies.

**Table 1.1**   Interactions between iron and other micronutrients

| Site of interaction | Nutrients involved |
| --- | --- |
| Food chain (soil–water–plant–animal) | Calcium, nickel |
| Lumen of gastrointestinal tract (uptake across brush border) | Calcium, zinc, manganese, copper, cobalt |
| Serosal transport (transport from mucosal cell to the circulation) | Riboflavin, calcium |
| Erythropoiesis | Vitamin A, riboflavin, copper |

# 2

# IRON ABSORPTION

## 2.1 INTRODUCTION

The capacity of the body to excrete iron is extremely limited (McCance and Widdowson, 1937), therefore the absorptive process plays the major role in the maintenance of iron homeostasis In general, only a small proportion of dietary iron is absorbed, and the amount is quite variable both between and within individuals (Kuhn *et al.*, 1968). It is significantly influenced by a number of factors, both diet- and host-related.

Measurements of iron intake are of limited value in assessing the nutritional value of a diet without some indication of iron bioavailability. This is defined as the proportion of the total intake that is potentially available for absorption and normal body functions, as discussed in the Report of the Panel on Dietary Reference Values (DH, 1991). Most absorbed iron is incorporated into red blood cells. The amount depends on the iron status and erythropoietic activity of the individual, and values of 80–95% have been reported.

## 2.2 MEASUREMENTS OF IRON AVAILABILITY FROM FOODS

The various techniques that have been developed to assess the availability of iron from foods can be broadly subdivided into *in vitro* and *in vivo* techniques (Table 2.1). The method adopted for any study must consider:

- resources available (e.g. financial, skill-base, and equipment);
- question(s) to be answered (e.g. iron bioavailability from specific foods, whole diets, or effects of processing techniques);
- ethical considerations (e.g. use of radioisotopes in subjects under investigation).

### 2.2.1 *In vitro* methods

One approach is to measure iron that is 'available for absorption' using *in vitro* methods, by determining ionizable iron (Narasinga Rao and Prabhavathi,

**Table 2.1** Techniques used to study iron bioavailability (source: Fairweather-Tait, 1992a)

| *In vitro* methods | *In vivo* methods |
|---|---|
| Measurement of soluble/dialysable iron Intestinal vesicle preparations Everted small intestinal rings/sacs | Perfusion experiments with ligated loops Rate of repletion following depletion Chemical balance Plasma appearance Whole body counting ($^{59}$Fe) Radio- or stable isotopic balance Haemoglobin incorporation of isotopes |

1978), or dialysable iron using equilibrium dialysis (Miller *et al.*, 1981) or continuous flow dialysis (Minihane *et al.*, 1993). These methods avoid the need to understand and control all the physiological factors that affect the efficiency with which iron is absorbed and should, in theory, provide a consistent and reproducible means of assessing the effect of dietary variables on absorbable iron. Clearly it is not possible to simulate the physiological factors that account for major variations in the absorption of iron. Thus the objective of the *in vitro* methods is to rank individual foods in order of available iron content and to predict correctly the relative effects of enhancers and inhibitors of absorption. *In vitro* methods are less expensive and generally require fewer resources than *in vivo* techniques, and are thus worthy of further development. However, caution must be applied in the interpretation of results from such studies (Valdez *et al.*, 1992; Miller and Berner, 1989).

### 2.2.2 *In vivo* methods

#### (a) Isotopic iron incorporation into haemoglobin

Incorporation of iron into haemoglobin is probably the only true method of determining bioavailability since it is a direct measure of iron utilization. Iron repletion studies are useful when considering the value of different iron sources in treating iron deficiency anaemia, but they are lengthy and require strict dietary control. Furthermore, they are not necessarily appropriate for the study of iron replete individuals, whose efficiency of absorption is much lower than that of iron deficient subjects. The method of choice is to use radioisotopes ($^{55}$Fe and $^{59}$Fe) to label extrinsically iron in individual foods or in meals, and to determine the incorporation of the dose into haemoglobin 14 days post-dosing (described by Bothwell et al., 1979).

The marked effect of body iron stores on absorption requires a method for correcting individual absorption values to a common reference point. The most widely applied technique is to use one isotope to label the food iron and a different one to label a reference dose (3 mg of iron as ferrous ascorbate). Absorption/haemoglobin incorporation of both isotopes is measured, and food absorption corrected to a mean reference value of 40% in each subject by multiplying by 40/R, where R is the reference dose absorption. The value of 40% is taken to represent the amount of iron that is absorbed by someone with virtually zero iron stores but with a 'normal' haemoglobin concentration (Magnusson et al., 1981). An alternative approach, suggested by Cook et al. (1991), is based on the inverse relationship between serum ferritin concentrations and iron absorption (Cook et al., 1974; Walters et al., 1975; Magnusson et al., 1981). Dietary absorption is corrected to a value corresponding to a serum ferritin of 40 μg/l (the overall mean of all the volunteers in the study) from the following equation:

$$\text{Log } A_c = \text{Log } A_o + \text{Log } F_o - \text{Log } 40$$

where $A_c$ is corrected dietary absorption, $A_o$ is observed absorption and $F_o$ is observed serum ferritin. Absorption can be predicted in groups of subjects with different levels of body stores (i.e. different serum ferritin values) but if this technique is to be employed the determination of serum ferritin must be carried out very carefully, and preferably more than once during the course of a study.

When determining isotopic incorporation into haemoglobin to measure iron absorption, an assumption has to be made as to the percentage of absorbed iron that is incorporated into the red blood cells. This is usually taken to be 80% (Bothwell et al., 1979) but it is possible to measure it accurately by injecting a known (small) dose of radiolabelled iron into the blood at the same time as giving the oral dose of iron labelled with a different isotope (Brise and Hallberg, 1962).

In studies of iron absorption multiple dose design is preferred, if possible, since it overcomes to some extent the problem of intra-subject variability in efficiency of absorption (Kuhn et al., 1968).

Where there are ethical constraints regarding the use of radioisotopes, alternative methods have been developed using stable isotopes (Janghorbani et al., 1986). These are more appropriate for work with infants (Fairweather-Tait et al., 1995; Fomon et al., 1989) where the doses needed to achieve a measurable enrichment in the blood are much lower than with adults (Barrett et al., 1992). However, a very important consideration when employing stable isotopes of iron for bioavailability studies is the validity of extrinsic labels, and the effects of adding non-negligible quantities of iron to produce labelled foods/meals (discussed by Sandstrom et al., 1993).

### (b) Whole body counting/faecal monitoring

Other in vivo techniques involve the measurement of iron absorption and/or retention in the body. The method of choice is to administer a meal labelled with the radioisotope $^{59}$Fe and then measure retention in the body by means of whole body counting. Where there is no access to a suitable counter, isotopic retention can be determined from faecal monitoring, which, unlike whole body counting, is also suitable for use with other isotopes of iron, both radio- and stable. As with the haemoglobin incorporation technique, the use of stable isotopes to label food (endogenous) iron requires further critical evaluation.

### (c) Plasma appearance

Plasma appearance of an oral isotope (radio- or stable) can be used to quantify iron absorption (Whittaker et al., 1991). An intravenous (iv) dose of a different iron isotope is given at the same time as the oral dose and the area under the curve (AUC) of the plasma enrichment of both isotopes is measured for at least 6 hours post-administration. For practical reasons blood sampling is usually discontinued before isotopic enrichment has returned to baseline values; therefore an extrapolation area is calculated from the enrichment of the last sample time and an estimate of the elimination rate constant. The smaller the contribution of the extrapolation area to the total area, the more accurate the estimation of total area. Absorption from the oral dose is calculated as:

$$\% \text{ absorption} = \frac{\text{AUC (oral)}}{\text{AUC (iv)}} \times \frac{\text{dose (iv)}}{\text{dose (oral)}} \times 100$$

### 2.3 SINGLE MEALS VS WHOLE DIET

A very important aspect of iron bioavailability studies that should be considered is the extent to which the single meal approach (i.e. the measurement of

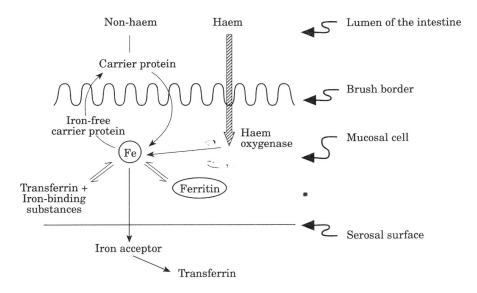

**Figure 2.1**  Mechanism of iron absorption.

iron absorption from a labelled single meal in a previously fasted subject) represents absorption from the diet as a whole. Cook *et al.* (1991) found that when subjects consumed their normal diets, there was good agreement between dietary absorption (6.4%) and representative single meals fed in the laboratory (6.1%). When the diet was modified to promote iron absorption, dietary absorption increased only slightly (8.0%) and remained significantly lower than it was from single meals (13.5%). With a diet selected to inhibit iron absorption, the decrease from single meals was similarly exaggerated. These results indicate the presence of short-term compensatory adaptive mechanisms whereby bioavailability data obtained from single test meals may be an over- or under-estimate for the same meals when taken in the context of the whole diet.

## 2.4  DIETARY FACTORS AFFECTING IRON BIOAVAILABILITY

### 2.4.1  Physico-chemical form

Iron in foods exists in two main forms:

- haem iron – in meat as part of haemoglobin and myoglobin;
- non-haem iron – in cereals, vegetables, meat and other foods.

Iron compounds can be added deliberately to fortify foods, or adventitiously via contamination from metal objects or soil. Iron supplements are another source of dietary iron but, as with the fortification of foods, the efficiency with which the iron is absorbed depends to a great extent on the physico-chemical form of the iron. The homeostatic control of iron absorption mediated via the intestinal mucosal cells is of special importance when considering the efficiency of absorption of high intakes of iron, such as from supplements.

Haem and non-haem iron are absorbed by different pathways (Figure 2.1) with different degrees of efficiency depending upon the chemical form, other dietary constituents and the level of iron stores in the individual (Hallberg, 1981).

It is generally agreed that 20–30% of haem iron is absorbed and that this is a constant figure, being relatively unaffected by other dietary or physiological variables (FAO/WHO, 1988). On the other hand, a large number of dietary variables that enhance (Table 2.2) or inhibit (Table 2.3) non-haem iron absorption have been identified (see reviews by Fairweather-Tait, 1992b; Hallberg, 1981). The various mechanisms whereby dietary substances affect iron absorption include:

- chemical reactions in the chyme such as chelation or changes in iron valency;
- effects on intestinal or mucosal function;
- competition with other minerals for transport protein.

### 2.4.2  Other dietary constituents

- Ligands, such as citric and ascorbic acid, fructose and amino acids, form soluble monomeric complexes with iron thus preventing precipitation and polymerization, and thereby promoting absorption.
- Other chelating compounds, including polyphenols (containing alkyl groups), phosphates,

**Table 2.2**   Dietary constituents that enhance the absorption of non-haem iron

| Enhancing food | Degree of effect | Active substance(s) | Reference |
|---|---|---|---|
| Guava, paw paw | +++ | Ascorbic and citric acids | Ballot *et al.* (1987) |
| Beef, lamb, pork, liver, chicken, fish | +++ | Cysteine-containing peptides | Cook and Monsen (1976b) Taylor *et al.* (1986) |
| Orange, pear, apple, pineapple juices | +++/++ | Ascorbic and citric acids | Rossander *et al.* (1979) Hallberg *et al.* (1986) Ballot *et al.* (1987) |
| Cauliflower | ++ | Ascorbic acid | Hallberg *et al.* (1986) |
| Beer | ++ | Ethanol, lactic acid | Derman *et al.* (1980) |
| Sauerkraut | ++ | Lactic acid | Gillooly *et al.* (1983) |
| Plum, rhubarb, banana, mango, pear, cantaloup | ++/+ | Ascorbic and citric acids | Ballot *et al.* (1987) |
| Carrot, potato, beetroot, pumpkin, broccoli, cauliflower, tomato, cabbage, turnip | ++/+ | Citric, malic and tartaric acids | Gillooly *et al.* (1983) |
| Salad (lettuce, tomato, green pepper, cucumber) | + | Ascorbic acid | Hallberg *et al.* (1986) |
| Wine | + | Ethanol | Hallberg and Rossander (1982) |
| Rice miso | + | | Macfarlane *et al.* (1990) |
| Soy sauce | + | Fermentation products | Baynes *et al.* (1990) |
| | + | Cysteine | Martinez-Torres *et al.* (1981) |
| | + | Glutathione | Layrisse *et al.* (1984) |

carbonates and oxalates have an adverse effect on bioavailability. Their inhibitory effect is usually due to the formation of large polymers.

- Meat, fish and poultry promote non-haem iron absorption. The mechanism is not yet known, but probably the formation of iron complexes with amino acids such as cysteine or peptides counteract luminal factors that inhibit absorption.
- Reducing agents such as ascorbic acid change the valency of iron from Fe(III) to Fe(II), which increases absorption. Fe(II) is more soluble than Fe(III) at pH values greater than 3, as found in the duodenum.
- Associated anions affect iron absorption; for example, ferric chloride is more soluble than ferric phosphate (an important constituent of vegetables), even at low pH.
- Competition between similar cations for uptake into the intestinal mucosal cells has been described between copper, zinc, manganese

and cobalt. The mechanisms for these interactions have not yet been established.
- Dietary constituents that alter gut secretions and transit time affect the bioavailability of iron. For example, alcohol and meat promote gastric acid production, lowering the pH of the proximal small intestine and increasing the solubilization of iron.
- Calcium reduces iron absorption. Hallberg *et al.* (1991) reported a marked inhibition of iron absorption from wheat rolls in the presence of calcium, but Turnlund *et al.* (1990) found that milk had no effect on iron absorption from cereal-based diets. In view of the high prevalence of iron deficiency in young women, Hallberg *et al.* (1992) recommended a reduction in the intake of dairy products with principal iron-providing meals, but the further implications of such a recommendation, in particular the effects on calcium nutrition, warrant careful consideration. Sokoll and Dawson-Hughes (1992)

**Table 2.3** Dietary constituents that inhibit the absorption of non-haem iron

| Inhibitory food | Degree of effect | Active substance(s) | Reference |
|---|---|---|---|
| Wheat bran | +++ | Phytate | Bjorn-Rassmussen (1974); |
| Tea | +++ | Polyphenols | Disler et al. (1975)<br>Hallberg and Rossander (1982) |
| Nuts | +++ | Phytate, polyphenols | Macfarlane et al. (1988) |
| Legumes | +++ | Phytate, polyphenols | Lynch et al. (1984) |
| Soya protein | +++ | Phytate | Cook et al. (1981)<br>Lynch et al. (1985, 1994) |
| Oats | +++ | Phytate | Rossander-Hulthen et al. (1990)<br>Brune et al (1989) |
| Oregano | +++ | Polyphenols | Brune et al. (1989) |
| Leafy vegetable<br>(*Leucaema glauca*) | +++ | Polyphenols | Tuntawiroon et al. (1991) |
| Coffee | +++/++ | Polyphenols | Hallberg and Rossander (1982)<br>Morck et al. (1983) |
| Maize (tortilla,<br>corn meal, bran) | +++/++ | Phytate | Acosta et al. (1984)<br>Hurrell et al. (1988)<br>Siegenberg et al. (1991) |
| Milk chocolate | ++ | Phytate, calcium,<br>polyphenols | Rossander et al. (1979) |
| Milk, cheese | ++ | Calcium plus phosphate | Deehr et al. (1990)<br>Gleerup et al. (1995)<br>Monsen and Cook (1976) |
| Rice | ++/+ | Phytate | Tuntawiroon et al. (1990) |
| Eggs | + | Phosphoprotein, albumin | Rossander et al. (1979)<br>Monsen and Cook (1979)<br>Hurrell et al. (1988) |
| Spinach | + | Polyphenols, oxalic acid | Brune et al. (1989) |
|  | + | Oxalic acid | Gillooly et al. (1983) |
|  | + | EDTA | Cook and Monsen (1976a) |

showed that calcium supplements (1000 mg/day for 12 weeks) did not reduce iron stores in premenopausal women, but further studies are needed to investigate the longer-term effects of high calcium intakes on iron status.

### 2.4.3 Iron dose

Not only does the form of iron affect its bioavailability, but also the quantity of iron has an effect.

There is a negative relationship between iron dose and percentage absorption but, provided that the iron is in an assimilable form, there is a progressive rise in the actual amount absorbed; the upper limit to iron absorption is determined by host factors such as iron 'status'. Acute iron poisoning can occur, but usually only when the physiological intestinal mucosal homeostasis is overwhelmed by a massive iron overload. A high oral intake over a prolonged period may contribute to chronic iron overload, but usually only if

**Table 2.4**   Host-related factors that affect non-haem iron absorption

| Variable | Effect on absorption | Reference |
|---|---|---|
| Size of body iron stores<br>• Low<br>• Normal, high | <br>Marked effect (inverse)<br>Minor effect (inverse) | Baynes *et al.* (1987) |
| Rate of erythropoiesis | Positive correlation | Skikne and Cook (1992) |
| Physiological state | Increased absorption<br>in pregnancy | Whittaker *et al.* (1991)<br>Barrett *et al.* (1994) |
| Iron content of mucosal cells | Exposure to iron reduces<br>subsequent absorption | O'Neil-Cutting and Crosby (1987)<br>Fairweather-Tait and Minski (1986) |
| High altitude, hypoxia | Increased absorption | Skikne and Baynes (1994) |
| Secretion of gastric juice | Positive correlation | Bezwoda *et al.* (1978) |
| GI secretions (bile,<br>pancreatic secretions, mucus) | Increased absorption in<br>presence of amino acids,<br>peptides, ascorbic acid<br>and mucoproteins | Bothwell *et al.* (1979) |

there is some underlying disturbance of iron metabolism or erythropoiesis (Chapter 8). The classic example of such iron overload, resulting in 'bronze diabetes', was first observed in South Africa and associated with the consumption of large quantities of local beer, brewed in iron pots, although this is now thought to be, in part, secondary to a genetic abnormality (Chapter 8). There are other documented cases of iron overload in subjects taking medicinal iron over many years but it is not known whether the subjects also carried the gene for haemochromatosis.

## 2.5   PHYSIOLOGICAL FACTORS AFFECTING ABSORPTION

### 2.5.1   Systemic factors

The amount of iron that is absorbed is markedly influenced by the iron content in the body. However, the efficiency of absorption is affected by a number of factors (Table 2.4):

- the level of iron to which the intestinal mucosal cells have been previously exposed (short-term control) (Fairweather-Tait, 1986; O'Neil-Cutting and Crosby, 1987);
- body iron stores, as measured by serum ferritin concentrations (long-term control) (Cook *et al.*, 1974);
- rate of erythropoiesis (Bothwell *et al.*, 1979);
- tissue hypoxia.

### 2.5.2   Physiological state

The absorption of iron is known to increase under conditions in which tissue iron is reduced, such as during growth and pregnancy. During the latter half of pregnancy the efficiency of iron absorption is increased from both the diet (Apte and Iyengar, 1970) and from inorganic iron (Whittaker *et al.*, 1991). Reports concerning the size of the increase vary, depending upon several factors including quantity and form of iron administered, iron status of the individual, method of measuring absorption, and stage of pregnancy. Recently Whittaker *et al.* (1991) used stable isotopes of iron to measure absorption in pregnant women from 5 mg iron as ferrous sulphate. They observed increases between 12, 24 and 36 weeks gestation from a mean of 7.6% to 21.1% and 37.4% respectively. Absorption was still elevated (26.3%) 12 weeks post-delivery.

### 2.5.3   Other physiological factors

Other physiological factors that affect iron absorption and utilization include:

- Gastric juice. Hydrochloric acid plays a key role in the release of iron from food during peptic digestion. Achlorhydria results in a reduced absorption of non-haem iron, although haem iron is unaffected. Gastric acid output is affect-

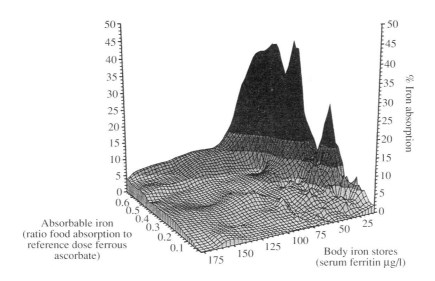

**Figure 2.2**   Effect of dietary and physiological factors on iron absorption.

ed by dietary constituents as well as factors unrelated to diet such as genetic predisposition and stress.

- Stomach emptying. The longer the food stays in the acidic environment of the stomach, the greater the proportion of iron that will be solubilized. Patients who have undergone partial gastrectomy have impaired iron absorption; this may be due to the partial loss of the reservoir function of the stomach and consequent accelerated progress of the food bolus through the upper gastrointestinal tract. As yet there is no evidence that small intestinal transit time has any effect on the efficiency of iron absorption (Fairweather-Tait and Wright, 1991).

- Pancreatic secretions. Pancreatin *per se* has no effect on iron absorption, but bicarbonate will promote the formation of unavailable iron hydroxide polymers. However, the overall effect of pancreatic juice may be to enhance iron absorption by releasing amino acids and polypeptides from foods which can then act as absorption promoting ligands.

- Biliary and other intestinal secretions. Animal studies indicate an enhancing effect of bile on iron absorption (Wheby *et al.*, 1962). It has been suggested that this is due to the ascorbic acid in bile, but *in vitro* studies have demonstrated the formation of mucoprotein ligands rather than iron–ascorbic acid complexes (Jacobs and Miles, 1970). Studies in rats have shown that fasting increases the quantity and iron-binding properties of the mucus layer (via a change in sialic acid content), which results in increased iron transport (Quarterman, 1987).

## 2.6 RELATIONSHIP BETWEEN DIETARY FACTORS AND PHYSIOLOGICAL FACTORS

Although the variations in non-haem iron absorption between (and within) individuals are very wide, depending on the dietary source of iron and accompanying dietary constituents, figures have been agreed for the purposes of deriving dietary reference values/recommended allowances. In diets containing generous levels of meat, poultry, fish and/or high in foods containing high amounts of ascorbic acid (promoters of iron absorption), as found for example in the UK, mean absorption from the whole diet (haem and non-haem iron) by individuals with low iron stores is taken to be 15% (DH, 1991; FAO/WHO, 1988).

Recently, Cook *et al.* (1991) examined the nutritional relevance of absorption studies investigating dietary factors believed to modify iron absorption. Iron absorption was found to be generally higher from test meals than the diet as a whole because of the experimental conditions used (e.g. fasting of subjects). The conclusion from the study was that for mixed Western diets, the bioavailability of non-haem iron is less important than absorption studies with single meals would suggest.

The relationship between absorbability of iron (as determined by dietary factors) and level of iron stores (the major physiological determinant), on iron absorption is illustrated in Figure 2.2. Individual data have been plotted from a series of iron absorption studies (Beard *et al.*, 1988b; Guindi *et al.*, 1988; Lynch *et al.*, 1984; Morck *et al.*, 1983) in which the absorbability of iron is expressed as the ratio of absorption of food iron to the reference iron salt (3 mg ferrous ascorbate) and iron stores as plasma ferritin concentration. Although a greater number of data points would undoubtedly have smoothed out the surface plot, it is evident from Figure 2.2 that dietary factors are only important in subjects with low iron stores. When serum ferritin values exceed 25 $\mu$g/l, there are only small differences in the absorption of iron from foods containing iron of very different absorbability.

## 2.7    MECHANISMS OF ABSORPTION

The precise details of intestinal iron absorption remain unclear, both in terms of the specific biochemical mechanisms of transport as well as their regulation. Iron absorption consists of three steps:

- uptake into the intestinal mucosal cell;
- movement through the intestinal cell;
- release from the cell to the circulation.

### 2.7.1    Iron uptake from the intestine to the mucosal cell

There are at least three pathways of mucosal iron uptake:

- Ionic iron is taken across the brush border membrane on a carrier protein by means of receptor-mediated endocytosis. The carrier was originally believed to be transferrin, but this is unlikely since repeated studies have failed to identify transferrin receptors in the intestinal brush border membrane. Results from *in vitro* studies suggest that membrane lipids may play a role in iron uptake into the brush border membrane (Simpson and Peters, 1987) and also indicate that there may be more than one path of transport of Fe(III) (Raja *et al.*, 1988).
- Iron from haemoglobin is absorbed by a pathway distinct from that of ionic iron since it is largely unaffected by dietary factors known to influence non-haem iron absorption. The intestinal haem receptor has been partially characterized from brush border membranes (Grasbeck *et al.*, 1979), indicating that the haem moiety is absorbed intact. Once inside the cell the iron is released from haem by haem oxygenase and enters a common pool within the cell.
- Iron from lactoferrin may enter the mucosal cell by a specific pathway since receptors for lactoferrin have been isolated in brush border membranes (Cox *et al.*, 1979), but their role in iron absorption needs further clarification.

### 2.7.2    Movement of iron within the intestinal cell

The pathway of iron within the mucosal cell is not fully understood. Ionic iron cannot exist in the cell, so upon release from the carrier protein it is immediately bound to a protein already present in the cell, either transferrin, ferritin or other iron-binding proteins. Although the transferrin gene is not expressed in the intestine, transferrin can enter the cell from the plasma via transferrin receptors that occur in the basal and lateral membranes of intestinal cells. Whether such intestinal transferrin has any role in iron absorption remains to be determined.

The connection between mucosal cell ferritin and iron absorption has been recognized for some time (Granick, 1949). Oral iron can induce apoferritin synthesis (Charlton *et al.*, 1965), but the ease with which iron can be mobilized and subsequently transferred from ferritin to the circulation is not known.

### 2.7.3    Transfer of iron from the cell to the circulation

The presence of transferrin receptors in the baso-lateral membrane of intestinal cells and their demonstrated increase in iron deficiency might suggest a role for mucosal transferrin in iron absorption (Banerjee *et al.*, 1986). However, several observations make this very unlikely.

- The inappropriate pH for the release of transferrin-bound iron at the baso-lateral membrane.
- The fact that the transcapillary exchange rate of transferrin between the mucosal extravascular space and the circulation is too slow to account for the rapid transfer of iron to the portal blood (Morgan, 1980).
- The observation that iron absorption in hypo-transferrinaemic mice is intact (Craven *et al.*, 1987).

Knowledge about iron transfer to the circulation is very limited and it is not known whether any intermediate acceptor is involved before binding to plasma transferrin in the portal circulation.

The most active sites of iron absorption are the duodenum and upper jejunum. Most iron taken up into the mucosal cells is transferred to the blood almost immediately, but after this rapid absorption phase, transfer continues at a much slower rate for up to 24 hours. The latter may be iron that has been temporarily stored within the cell as ferritin. Not all the iron that has entered the mucosa from the gastrointestinal tract is transferred to the plasma. It may be retained within the epithelial cells and eventually discarded largely in the form of ferritin, when the cells exfoliate.

## 2.8    REGULATION OF IRON ABSORPTION

### 2.8.1    Possible controlling factors

Iron absorption is directly related to both body stores and the amount of iron to which the intestinal mucosal cells have been exposed (Bothwell *et al.*, 1979), but mechanisms whereby information reaches intestinal absorptive cells to alter the efficiency of absorption and so maintain body homeostasis are unknown. Intestinal mucosal iron concentration has been implicated as the major regulating factor (Conrad and Crosby, 1963). The technique of isogeneic intestinal transplantation of iron-loaded and iron-deficient intestine into iron-deficient rats is a means of isolating the effects of body iron stores from the effects of intestinal mucosal iron concentration on iron absorption (Adams *et al.*, 1991); using this method, intestinal mucosal iron uptake was shown to regulate the uptake and transfer of iron in the intestine. On the other hand, the primary control of iron absorption may be via a circulating humoral mediator related to body stores (MacDermott and Greenberger, 1969); in experimental findings the serosal transfer of iron is the rate-limiting step in iron absorption (Huebers, 1986). Iron absorption may also be related to internal iron exchange, i.e. the relative size of intestinal and total exchangeable iron, as a function of plasma iron turnover (Cavill *et al.*, 1975).

### 2.8.2    Possible mechanisms

Iron absorption is regulated by two pathways, primary control being exerted according to the level of body stores, and 'fine tuning' according to the amounts of iron to which the intestinal mucosal cells are exposed.

A delayed response (several days) in the increase in iron absorption when animals are fed an iron-deficient diet is often described. This may be related to:

- a decrease in the amount of ferritin in the mucosal cell or a change in the partition of iron in the cell between ferritin and other iron-binding sites;
- an increase in the amount of iron carrier proteins.

The most likely explanation, however, is that the lag in iron absorption is related to the time taken for new cells, with a raised capacity to absorb iron, to become functional absorptive cells (Fairweather-Tait *et al.*, 1985). This strongly suggests that programming of mucosal cells takes place according to the state of body iron stores and previous exposure to iron. The mechanism may involve transferrin and/or ferritin receptors, since these are based on the serosal surface and would receive information about the amount of iron passing into circulation and also plasma ferritin levels, i.e. levels of body iron stores.

Although body iron content is the principal factor in the regulation of iron absorption, there are other physiological variables (Bothwell *et al.*, 1979). If the rate of erythropoiesis is stimulated by blood loss or acute haemolysis, iron uptake by the bone marrow is increased, plasma iron and transferrin saturation fall, and the efficiency of iron absorption is increased. Conversely, if erythropoiesis is inhibited by hypertransfusion, starvation or descent from high altitude to sea level, then iron absorption falls. There are, however, chronic dyserythropoietic anaemias such as those due to thalassaemia and sideroblastic anaemia in which transferrin saturation is normal, or even elevated. Such disease states are associated with a high efficiency of absorption, resulting in an accumulation of body stores of iron. On balance, it appears that erythropoiesis *per se* does not play a major role in the regulation of iron absorption in normal healthy individuals.

### 2.8.3    Genetic factors

In subjects with the iron-loading disorder, idiopathic haemochromatosis, genetic factors are responsible for the accumulation of excess iron from the diet. Current evidence suggests that there may be a defect in the mucosal and reticuloendothelial handling of iron. Inappropriately high rates of iron absorption occur from food (haem and non-haem iron) and from iron salts (Milder *et al.*, 1978; Bezwoda *et al.*, 1976). The fact that such enhanced absorption occurs in the presence of normal or high iron stores indicates a primary inherited abnormality in the regulation of iron balance. Some limited homeostatic regulation takes place since the increased absorption gradually decreases as body iron stores enlarge (Williams *et al.*, 1966).

## 2.9    EXAMPLES OF DIETS CONTAINING IRON OF LOW, MEDIUM AND HIGH BIOAVAILABILITY

Diets can be separated into three broad categories of 'low', 'intermediate' and 'high' bioavailability, with mean absorption (in individuals with very low iron stores but normal Hb concentrations) from the mixture of haem and non-haem iron of approximately 5, 10 and 15% respectively (FAO/WHO, 1988).

Low bioavailability diets (5% of iron absorbed)

contain cereals and root vegetables with negligible quantities of meat, fish or ascorbic acid-rich foods. Such diets contain a preponderance of foods that inhibit iron absorption (maize, beans, whole-grain flour) and are dominant in many developing countries, particularly among lower socio-economic groups.

Intermediate bioavailability diets (10% of iron absorbed) consist mainly of cereals and root vegetables but containing some ascorbic acid-rich foods and meat. A high bioavailability diet can be reduced to intermediate levels by the regular consumption of inhibitors of iron absorption, such as tea, coffee, cereal fibre, beans and/or high calcium foods with main meals.

High bioavailability diets (15% of iron absorbed) contain generous quantities of meat, poultry and fish. They also contain foods with high amounts of ascorbic acid such as citrus fruits and some vegetables. This is the type of diet often consumed by people in developed countries.

# 3

# DIETARY SOURCES AND INTAKES OF IRON

## 3.1 DIETARY SOURCES OF IRON IN BRITAIN

### 3.1.1 Food sources

The most important sources of iron are those foods which are not only rich in iron but also eaten in significant quantities, and from which iron is reasonably well absorbed (Chapter 2). The amounts of iron vary widely between foods, as indicated in Table 3.1. While milks (including human milk) and milk products, sugars and fats contain little, liver and shellfish which concentrate iron in their tissues are rich sources. However, the latter foods are not popular in Britain and the major sources in practice are the widely eaten cereal products, which provided nearly half the average household intake of iron in 1993 (Table 3.2). Meats and meat products provided a further 18% on average, but because about half of this is haem iron it would be better absorbed than the iron in cereals. Vegetables also provided about 16% on average of the total household iron intake. The intake provided 95% of the UK Reference Nutrient Intake (RNI), after allowance for meals not taken from the domestic food supply and for wastage of food within the home. The UK Dietary Reference Values (DRVs) are based on the assumption that a diverse diet (including meat, poultry, fish and reasonable amounts of vitamin C) is consumed and that iron absorption will be about 15%. Many dietary and physiological factors affect the bioavailability of iron (Chapter 2; DH, 1991) and may lower or raise the required intake. It is clear that the DRVs for iron must be interpreted with caution.

### 3.1.2 Fortified foods

Most of the naturally occurring iron in cereals is in the seed coat, so the white flour used in bread and pasta, and white rice, might be expected to contain comparatively little iron. The reason why cereal products are an important source of iron in the UK is that most wheat flours and many breakfast cereals have iron added to them. In the UK, all wheat flour other than wholemeal flour has been

**Table 3.1** Iron content per 100 g and per typical serving of selected foods (sources: Holland et al., 1991; Crawley, 1988)

| Foods | Iron (mg per 100 g) | Iron (mg/serving) |
|---|---|---|
| **Meats** | | |
| Beef, mince | 3.1 | 2.5–5 |
| Sirloin, roast | 1.6 | 1.5– 3 |
| Sausage, pork, grilled | 1.5 | 1.8 |
| Lamb chop, grilled | 1.9 | 1.9 |
| Chicken, light meat | 0.5 | 0.7 |
| Chicken, dark meat | 1.0 | 1.4 |
| Liver, calves | 8.0 | 7.3 |
| **Fish** | | |
| Cod, baked | 0.4 | 0.5 |
| Fish fingers, fried | 0.7 | 0.6 |
| Mackerel, smoked | 1.2 | 1.8 |
| Sardines | 2.9 | 2.9 |
| Tuna | 1.0 | 0.5 |
| Cockles | 26 | 6.5 |
| Mussels, boiled | 7.7 | 3.1 |
| **Eggs** | | |
| Whole | 1.9 | 1.0 |
| Yolk only | 6.1 | |
| **Cereals** | | |
| Bread, white | 1.6 | 1.0 |
| Bread, wholemeal | 2.7 | 1.9 |
| Rice, boiled | 0.2 | 0.3 |
| Pasta, boiled | 0.5 | 1.1 |
| Corn flakes | 6.7 | 2.7 |
| All-Bran | 12.0 | 6.0 |
| Porridge | 0.5 | 1.0 |
| **Vegetables** | | |
| Baked beans | 1.4 | 1.9 |
| Peas, frozen and boiled | 1.6 | 1.0 |
| Potatoes, boiled | 0.4 | 0.7 |
| Chips | 0.9 | 1.6 |
| Spinach, boiled | 1.6 | 1.4 |
| **Other foods** | | |
| Bananas | 0.3 | 0.3 |
| Chocolate, plain | 2.4 | 1.2 |
| Curry powder | 58 | 1.7 |
| Liquorice | 8.1 | |
| Milk | 0.06 | 0.06 |
| Human milk | 0.05 | |
| Peanuts | 2.5 | 1.0 |
| Wine, red | 0.9 | 1.1 |
| Wine, white | 0.5 | 0.6 |

Table 3.2   Main contributors to the average household intake of iron in Britain, 1993 (source: MAFF, 1994)

| Foods | Average household intake of iron | |
|---|---|---|
| | mg/person/day | % of total |
| Meat and meat products | **1.8** | **18.2** |
| of which carcase meats | 0.6 | 5.9 |
| meat products | 1.0 | 9.8 |
| Fish | **0.2** | **2.3** |
| Eggs | **0.3** | **2.6** |
| Cereals and cereal products | **4.8** | **48.3** |
| of which white bread | 0.8 | 7.9 |
| other bread | 1.3 | 13.1 |
| breakfast cereals | 1.4 | 14.1 |
| other cereals | 1.3 | 13.1 |
| Vegetables | **1.6** | **16.3** |
| of which potatoes | 0.4 | 3.7 |
| green vegetables | 0.2 | 2.7 |
| All other foods | **1.3** | **12.2** |
| TOTAL | **9.9** | **100** |

**Estimated Average Requirements**
Males    19–50 years  6.7
Females 19–50 years 11.4

required by law since the 1950s to contain at least 1.65 mg iron per 100 g (as well as added calcium, thiamin and niacin); the regulations also specify the particle size and solubility of the added iron to ensure that it is as bioavailable as possible (Statutory Instrument, 1984). The fortification of breakfast cereals is voluntary but most now contain about 6 mg per 100 g so that one serving would provide about 2 mg of iron. In addition, infant formulas, many other infant foods and some dietetic foods are also fortified. Flour is not fortified in most other countries, and breakfast cereals, which are not so commonly eaten elsewhere, might also not be fortified.

## 3.2   DIETARY INTAKES OF THE GENERAL BRITISH POPULATION

Intakes of iron from household food have been falling for many years, mainly because bread consumption has steadily declined from 250 g per person per day in the 1940s to about 100 g per day in 1993. There has also been a slight decline in the contribution from red meats, not only because their consumption has fallen but also because they now contain less iron as a result of changing husbandry practices. On the other hand, the contributions from poultry and from breakfast cereals have risen. Although on average the intake of iron from household food in Britain has fallen since 1974, the nutrient density of the household diet (expressed as mg iron/1000 kcal) has increased slightly (Figure 3.1).

There are only small regional and income differences in iron intake within Britain, but intakes per person tend to be proportionately smaller in larger households (MAFF, 1992). Thus in 1991, the intake in single adult households was 12.0 mg per day, while in families with two adults, intakes were 11.4 mg per person per day when there were no children, 9.4 mg with one child, 9.1 mg with two children, 8.6 mg with three children and 7.9 mg with four or more children. These trends broadly paralleled the smaller nutrient needs of children compared with those of adults.

In 1980, when the household intake of iron was 11.0 mg per person per day, 1.3 mg (12%) was estimated to be haem iron, 8.6 mg (78%) was natural non-haem iron and 1.1 mg (10%) was derived from the fortification of flour and breakfast cereals (Bull and Buss, 1980). The implications of these proportions for the amount of iron which can be absorbed and utilized are discussed in detail in Chapter 2. However, since vitamin C increases absorption and the tannins in tea inhibit it, it is of interest that household intakes of vitamin C between 1974 and 1991 rose from 50 mg to 55 mg per day – with the increased popularity of fruit juices more than offsetting a reduced consumption from fresh vegetables – while tea consumption declined from 9 g to 6 g per day during the same years.

The amounts of iron in the household diet will be supplemented by foods eaten outside the home and by chocolate, neither of which was recorded in the National Food Survey until 1992. Some people, particularly pregnant women, might also take additional iron in the form of supplements. Most iron and

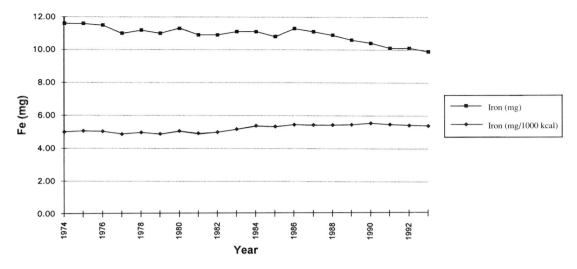

**Figure 3.1**   Iron intakes from 1974 to 1993. (Source: MAFF National Food survey.)

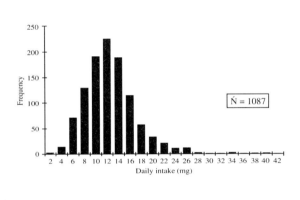

**(a)**

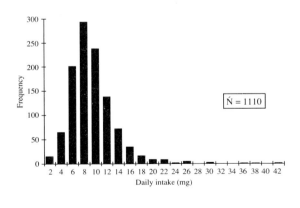

**(b)**

**Figure 3.2**   Daily iron intake (mg), from food sources only:
(a) for men; (b) for women. (Source: Gregory *et al.* 1990.)

multivitamin products provide around 12–15 mg per dose, which is close to the current RNI for women up to 50 years, i.e. 14.8 mg/day. All these items were included in the MAFF/DH 7-day dietary and nutrition survey of British adults in 1986–87, which showed total mean intakes of 14.1 mg iron per day for men and 12.3 mg per day for women (Gregory *et al.*, 1990; Figure 3.2). Intakes were skewed, with median intakes at 13.2 mg and 10.0 mg respectively, while the 2.5 centiles for men and for women were 6.5 mg and 4.7 mg. Iron supplements are more commonly taken by women than men, and provided 15% of the total intake for women compared with only 2% for men. The main dietary contributors were not dissimilar to those found in the National Food Survey, with 42% being derived from cereals, 23% from meats and 15% from vegetables. Women derived a slightly smaller proportion of their intake from meat and a slightly higher proportion from cereal products (especially from the higher fibre breakfast cereals).

These intakes compare with Estimated Average Requirements (EARs) for iron of 6.7 mg for men and 11.4 mg for the majority of women who are menstruating, and Lower Reference Nutrient Intakes (LRNIs) of 4.7 mg and 8.0 mg, respectively (DH, 1991). The potential for iron deficiency is greater in women, and Figure 3.2b shows that about one third of women in the survey had dietary intakes that were below the LRNI figure. Intakes were lower in younger men and women, although the age differences in dietary intakes in women were largely compensated for by higher intakes from iron supplements in the younger women. As in the National Food Survey, regional differences in iron intake were small, both for men and for women.

## 3.3   DIETARY INTAKES OF POPULATION SUBGROUPS

A number of other nationally representative surveys in this country, including those of older infants (Mills and Tyler, 1992), children between $1\frac{1}{2}$ and $4\frac{1}{2}$ years (Gregory *et al.*, 1995), schoolchildren (DH, 1989) and young adults (Bull, 1985), also indicated that some people in each group had very low intakes of iron. Women who were eating small amounts of food because they were watching their weight were at particular risk (Barber *et al.*, 1985) and 9–16% of pre-school children had iron intakes below the LRNI (Gregory *et al.*, 1995).

## 3.4   DIETARY INTAKES: INTERNATIONAL DATA

Dietary iron intakes in most other countries can be seen from FAO estimates of food supplies to be broadly similar to those in Britain (FAO, 1990). Thus, compared with an estimated supply of 14.2 mg iron per person per day in the UK, the food supplies in other European countries provide from 13.5 mg in Sweden and the Netherlands up to 19 mg in Portugal and Greece. The average supplies of iron in other broad areas of the world range between 12 mg and 19 mg per person per day, and although the values for individual countries may be higher or lower than this, the amount tends to reflect the total amount of food available.

# 4

# MECHANISMS OF CELLULAR IRON HOMEOSTASIS

## 4.1 PROTEINS OF IRON TRANSPORT

The study of iron homeostasis requires knowledge of the major forms of iron in the body, iron exchange between these components and iron intake and loss. This chapter considers the proteins of iron metabolism, extra- and intra-cellular iron exchange, and the regulation of synthesis and breakdown of the proteins of iron metabolism.

### 4.1.1 Transferrin

The iron-binding protein of the plasma is transferrin, which is very similar to lactoferrin found in granulocytes and in milk (reviewed by de Jong *et al.*, 1990). Both are monomeric glycoproteins with a

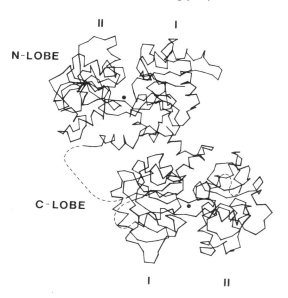

**Figure 4.1** Overall organization of the rabbit serum transferrin molecule. The polypeptide chain is folded into two lobes, each containing some 330 amino acids and a single iron-binding site. The shape of each lobe can be described by a prolate ellipsoid of approximate semiaxial dimensions 21 × 25 × 36 Å with the major axes of the N- and C-lobes running antiparallel to one another at an angle of 155°. In turn each lobe comprises two dissimilar domains, I and II. Broken lines denote poorly defined regions in the current model. (From Bailey *et al.*, 1988, with permission.)

molecular weight of about 80 kDa. The polypeptide chain of human transferrin has 679 amino acids organized in two homologous lobes – known as the N-terminal and C-terminal lobes. The protein contains about 6% carbohydrate on two branched heterosaccharide chains, both of which are in the C-terminal lobe (Figure 4.1). Each lobe has a single iron-binding site that requires both Fe(III) and an anion (usually carbonate or bicarbonate). The affinity of transferrin for iron is very high: at pH 7.4 in the presence of bicarbonate, the affinity constant for the binding of one iron atom is approximately $10^{20}M^{-1}$. However, iron can be released from both sites by lowering the pH to less than 5.5.

The plasma concentration of transferrin in adults is normally about 2.4 g/l and each mg of transferrin can bind 1.4 $\mu$g of iron. The protein is normally 20–40% saturated with iron.

### 4.1.2 Transferrin receptor

Iron is released from transferrin on acidification, but delivery of transferrin iron to cells (particularly to immature red cells for haemoglobin synthesis) takes place by interaction with specific receptors in the cell membrane (Huebers and Finch, 1987) followed by receptor-mediated endocytosis and by removal of iron and release of apotransferrin within the cell. The transferrin receptor (Figure 4.2) is a transmembrane glycoprotein consisting of two identical subunits of molecular mass 95 kDa joined by a disulphide bond (Trowbridge and Shackelford, 1986). The genes for both transferrin and its receptor are found on the long arm of chromosome 3. Although there are structural differences between the two iron-binding sites of the transferrin molecule, there do not appear to be significant differences in the way in which each site delivers iron to the red cell (Huebers and Finch, 1987). However, recent studies *in vitro* have shown that iron release from the transferrin receptor complex takes place more readily from the C-terminal site on the transferrin molecule than the N-terminal site (Bali and Aisen, 1991), thus explaining the high prevalence of N-terminal monoferric transferrin in

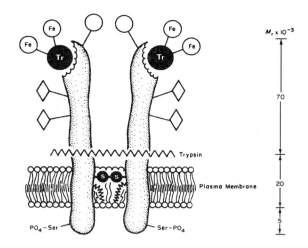

**Figure 4.2** Schematic representation of the human transferrin receptor. ◇ = high-mannose oligosaccharide; o = complex-type oligosaccharide; Tr = transferrin.
(From Trowbridge and Shackelford, 1986, with permission.)

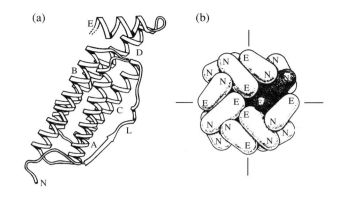

**Figure 4.3** Schematic drawing of a ferritin subunit showing the five helices A to E and the long inter-helix loop L. The loop L and the N-terminal residues N lie on the outside of the protein shell. Helix E runs from outside to inside. The C terminus is on the inside. (b) The ferritin molecule with 24 subunits surrounding the iron core (visible at centre, shown in black). N = N-terminus; E = N-terminal end of E helix. (From Harrison, 1986, with permission.)

the circulation. Furthermore iron release from receptor bound transferrin at pH 7.4 is slower than for transferrin in solution, whereas at pH 5.6 iron release is faster. Thus non-specific release of transferrin iron at the cell surface is prevented and release in the endosome at more acidic pH is ensured.

## 4.2   IRON STORAGE PROTEINS

### 4.2.1   Ferritin

Iron is stored in cells as ferritin which is a soluble, spherical protein enclosing a core of iron. Particularly high concentrations are present in the liver, spleen and bone marrow. Ferritin is also found in low concentrations in plasma and urine (Worwood, 1986). Human apoferritin (i.e. the molecule devoid of iron) has a molecular mass of about 480 kDa and is composed of 24 subunits whose molecular mass is about 19 kDa. The subunits are roughly cylindrical in shape and form a nearly spherical shell that encloses a central core containing up to 4500 atoms of iron in the form of ferric hydroxyphosphate (Figure 4.3). Normally ferritin in liver, spleen, heart and kidney contains about 1500 atoms Fe/molecule (Wagstaff *et al.*, 1982).

The amino acid sequences of human spleen and liver ferritin are known and X-ray crystallographic analysis at 2.8 Å resolution has demonstrated the arrangement of the subunits within the molecule, as well as channels between the subunits through which iron enters and leaves (reviewed by Theil, 1990).

Human ferritins are made up of two types of subunit in varying proportions. In liver and spleen ferritin, the 'L' subunit predominates. In the more acidic isoferritins found in the heart and in red cells, the 'H' subunit predominates. There is about 55% homology between the two subunit sequences. The gene for the L subunit is located on chromosome 19q13.3-q13.4 and the gene for the H subunit is found on chromosome 11q13 (Worwood, 1990).

The various isoferritins appear to have different functions. In iron-loaded tissues it is the L-rich isoferritins which predominate although H-rich isoferritins have the highest rates of iron uptake *in vitro* (Wagstaff *et al.*, 1982). The production of recombinant ferritins has demonstrated that although L subunits promote nucleation of the iron core (Levi *et al.*, 1992), ferroxidase activity is a property of the H subunit. H-rich isoferritins display inhibitory activity for haemopoiesis (Broxmeyer, 1992).

### 4.2.2   Haemosiderin

Haemosiderin is a degraded form of ferritin in which the protein shells have partly disintegrated, allowing the iron cores to aggregate (Richter, 1984). It is usually found in lysosomes and may be seen under the light microscope after tissue sections have been stained with potassium ferrocyanide in the presence of hydrochloric acid (Prussian blue or Perl's reaction). In a normal human with adequate iron status most of the storage iron is present as ferritin, but with increasing iron accumulation the proportion present as haemosiderin increases.

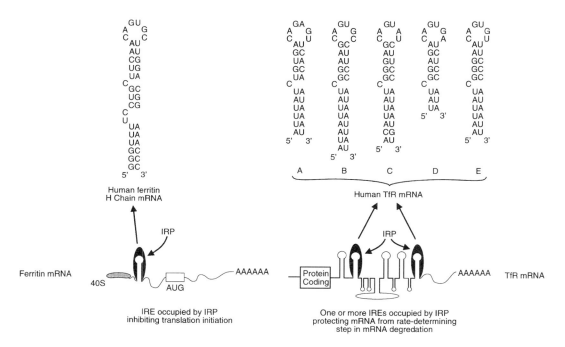

**Figure 4.6**    The interaction of the iron regulatory protein IRP (or IRE-BP) with 'stem-loop' structures on both ferritin and transferrin receptor (TfR) mRNA. Sequences for human H ferritin and human TfR mRNA are shown. (Redrawn from Klausner *et al.*, 1993.)

ALA synthase: the 'housekeeping' gene which lies at chromosome 3p21 and the erythroid-specific gene localized to Xp11.21 (Bishop, 1990; Cotter *et al.*, 1992). The ferrochelatase gene (the last stage of the pathway) is located at chromosome 18q.22 (Whitcombe *et al.*, 1994).

### 4.4.3    Haem catabolism

Haem catabolism is an enzymatic process mediated by haem oxygenase, which is located in the endoplasmic reticulum. The enzyme requires NADPH and molecular $O_2$. Enzyme activity (in the rat) is highest in the spleen and then in the bone marrow, liver, brain, kidney and lung. There is also activity in the intestinal mucosa. The cellular distribution of enzyme activity is consistent with the role of haem oxygenase in haemoglobin catabolism.

The enzyme is a monomeric protein with a molecular mass of approximately 33 kDa and exists as two isoenzymes (Schacter, 1988). Enzyme activity is stimulated by administration of haem (as haemoglobin or methaemalbumin) as well as many other agents. The inducible enzyme (HMOX1) is located at chromosome 22q12 and the constitutive enzyme (HMOX2) at 16p13.3 (Kutty *et al.*, 1994).

The immediate fate of the iron released from haem is unknown but iron is rapidly returned to the plasma as well as being incorporated into ferritin in the cells breaking down haem.

## 4.5    SYNCHRONIZED REGULATION OF SYNTHESIS OF TRANSPORT AND STORAGE PROTEINS

*In vitro*, ferritin can be shown to have all the qualifications required for an iron storage protein. Apoferritin will bind and oxidize Fe(II) and deposit Fe(III) within the protein. Release of iron may be affected by reducing agents. Studies in both animals and cultured cells show that apoferritin is synthesized in response to iron administration. This control is exercised at the level of translation (Kuhn, 1991). The 5' untranslated region of the ferritin mRNA contains a sequence which forms a 'stemloop' structure. This has been termed an 'iron response element' (IRE) (Figure 4.6). Cytoplasmic proteins which bind to this sequence and prevent translation have also been identified. In the presence of iron this repressor protein (iron regulatory protein; IRP) previously known as IRE-binding protein or iron regulatory factor, is unable to bind to the mRNA; polysomes form and translation proceeds. The iron regulatory protein has been purified and shown to be an iron–sulphur protein closely related to aconitase and encoded by a gene on chromosome 9 (and functioning as a cytosolic aconitase in its replete state). A model involving conformation changes which permit RNA binding has been proposed (Klausner *et al.*, 1993). There are now known to be two IRPs which bind specifically to the IREs. The functional significance of each protein is not yet known (Kim *et al.*, 1995).

A related mechanism operates in reverse for the transferrin receptor. Here there are stem-loop sequences in the 3' untranslated region and protein binding prevents degradation of mRNA. Hence iron deficiency enhances transferrin receptor synthesis. This translational regulation also applies to erythroid ALA synthase and aconitase.

Until recently control of transferrin synthesis had been considered in terms of transcriptional regulation involving not only iron but also developmental and tissue specific factors (Bowman *et al.*, 1988). Plasma transferrin levels increase in iron deficiency and decrease in iron overload but this latter change is not associated with a reduction in mRNA levels in the liver. Cox and Adrian (1993) have employed chimeric genes (human transferrin 5' regulatory region – chloramphenicol acetyl transferase transgene) to demonstrate post-transcription regulation by iron. They propose that IRP also binds to an IRE on transferrin mRNA. However, increased binding of IRP leads to an increase in transferrin mRNA translation rather than the decrease associated with binding to the ferritin IRE. The changes are also of a different magnitude, two-fold in the case of transferrin and up to 50-fold for ferritin. Thus post-transcriptional control by the IRE–IRP mechanism appears to be important for all three proteins regulating iron metabolism.

# 5

# MEASUREMENT OF IRON STATUS

## 5.1  IRON STATUS

Normal iron status implies both the presence of erythropoiesis which is not limited by iron and a small reserve of 'storage iron' to cope with normal physiological functions. The ability to survive the acute loss of blood (iron) which may result from injury is also an advantage. The limits of normality are difficult to define and some argue that physiological normality is an absence of storage iron (Sullivan, 1992a) but the extremes of iron deficiency anaemia and haemochromatosis are well understood.

Apart from too little or too much iron in the body there is also the possibility of maldistribution. An example is anaemia associated with inflammation or infection, where there is a partial failure of erythropoiesis and of iron release from the phagocytic cells in liver, spleen and bone marrow which results in accumulation of iron as ferritin and haemosiderin in these cells (Figure 5.1). Thus determination of iron status requires an estimate of the amount of haemoglobin iron (usually by measuring the haemoglobin concentration in the blood) and the

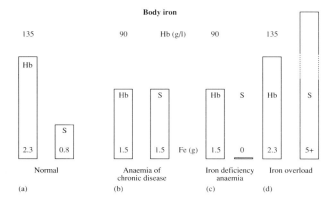

**Figure 5.1**   Body iron. Relationship between total haemoglobin (Hb) and storage iron (S): (a) normally; (b) in the anaemia of chronic disease; (c) in iron deficiency anaemia; and (d) in iron overload caused by genetic haemochromatosis. (The 'normal' example is for a 70 kg subject with a haemoglobin concentration of 135 g/l.)

level of storage iron. Occasionally, further investigations into iron loss, iron absorption and flow rates within the body are also required.

## 5.2  METHODS FOR ASSESSING IRON STATUS

The methods which have been used to assess iron status are summarized in Table 5.1. Some methods are not applicable to normal subjects but have value in the standardization of indirect methods.

### 5.2.1  Haemoglobin

The measurement of haemoglobin concentration depends on the conversion of haemoglobin to cyanmethaemoglobin and determination of the absorbance at 540 nm. An international standard is widely available for calibration and the measurement is included in the analysis provided by electronic blood cell counters in most haematology laboratories (Dacie and Lewis, 1991). Developmental changes in haemoglobin concentration have been reviewed by Yip (1994). Concentrations are higher in adult men than women (White *et al.*, 1993) but otherwise vary little with age in adult life. Haemoglobin concentrations increase with long-term exposure to high altitude (compensation for reduced oxygen supply) and are higher in cigarette smokers (White *et al.*, 1993; Yip, 1994) in compensation for reduced oxygen capacity due to the formation of carboxyhaemoglobin.

The investigation of anaemia begins with the blood count in order to distinguish between anaemia due to inadequate supply of iron or, more rarely, to $B_{12}$ or folate deficiency. Modern automated cell counters provide a rapid and sophisticated way of detecting the changes in red cells which accompany a reduced supply of iron to the bone marrow (Williams, 1990): a low mean cell volume (MCV), a low mean cell haemoglobin concentration (MCH) and a high red cell distribution width (RDW). Cell analysers can thus indicate the presence of microcytic anaemia which can be due to a reduced supply of iron to the bone marrow or to a deficit in

**Table 5.1** Assessment of body iron status

| Measurement | Representative reference range (adults) | Diagnostic use |
|---|---|---|
| **Functional iron** | | |
| Haemoglobin conc – Males | 130–180 g/l | Assessing the severity of IDA; response to a therapeutic trial of iron confirms IDA. Not applicable to assessment of iron overload. |
|       – Females | 120–160 g/l | |
| Red cell indices – MCV | 80–94 fl | |
|       – MCH | 27–32 pg | |
| **Tissue iron supply** | | |
| Serum iron | 10–30 µmol/l | Decreased saturation of transferrin, reduced red cell ferritin and increased zinc protoporphyrin, indicate impaired iron supply to the erythroid marrow. |
| Saturation of transferrin | 16–60% | |
| Red cell zinc protoporphyrin | <80 µmol/mol Hb (<70 µg/dl red cells) | Raised saturation of transferrin used to assess risk of tissue iron-loading (e.g. in haemochromatosis or iron-loading anaemias). |
| Red cell ferritin (basic) | 3–40 ag/cell | |
| Serum transferrin receptor | 2.8–8.5 mg/l | Particular value in identifying early iron deficiency and, in conjunction with a measure of iron stores, may be of value in distinguishing this from ACD. |
| **Iron stores** | | |
| Quantitative phlebotomy | <2 g iron | All measures are positively correlated with iron stores except TIBC which is negatively correlated. |
| Tissue biopsy iron | | |
| – liver (chemical assay) | 3–33 µmol/g dry wt | Quantitative phlebotomy, liver iron concentration, chelatable iron and MRI are of value only in iron overload. Bone marrow iron may be graded as absent, normal or increased, and is most commonly used to differentiate ACD from IDA. |
| – bone marrow (Prussian blue stain) | | |
| Serum ferritin | 15–300 µg/l | Serum ferritin is of value throughout the range of iron stores. |
| Urine chelatable iron (after 0.5 g I.M. desferrioxamine) Non-invasive methods (MRI etc.) | >2 mg/24 hr | |
| Serum TIBC/transferrin | 47–70 µmol/l | In IDA a raised TIBC is characteristic. |

Key:  IDA  = Iron deficiency anaemia
       ACD  = Anaemia of chronic disorders
       TIBC = Total iron-binding capacity.

haemoglobin synthesis such as in thalassaemia. Further tests are usually necessary to distinguish between simple iron deficiency (absence of storage iron) and a supply deficiency that is secondary to another disease process (Figure 5.2).

### 5.2.2 Quantitative phlebotomy

A direct way of measuring iron stores is by quantitative phlebotomy (removing up to 500 ml/week until anaemia develops). This gives a measure of the amount of iron available for haemoglobin synthesis. If blood is removed at a rate of 500 ml/week, most of the iron used for haemoglobin synthesis (250 mg Fe/week) is obtained from the stores (ferritin and haemosiderin) rather than by absorption (Torrance and Bothwell, 1980). Quantitative phlebotomy has been applied to validate the concept that serum ferritin concentrations in normal subjects reflect the level of available storage iron (Walters et al., 1973). It is also used to determine the initial level of storage iron during treatment of haemochromatosis.

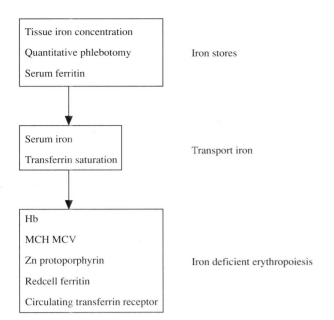

| Tissue iron concentration
Quantitative phlebotomy
Serum ferritin | Iron stores |

**Figure 5.2**   Indicators of iron status.

### 5.2.3   Tissue iron concentrations

The liver and bone marrow are important and relatively accessible storage sites and the amount of iron present in a tissue biopsy can be estimated either visually, using the Prussian blue reaction on tissue sections, or chemically. There is generally a good correlation between iron concentrations in liver and bone marrow (Gale et al., 1963). Methods for chemical and histological assessment of tissue iron concentration have been described in detail by Torrance and Bothwell (1980). Mean liver non-haem iron concentrations lie between 80 and 300 µg/g wet weight (Charlton et al., 1970). Chemical determination of liver iron concentrations is most widely applied for the demonstration of iron overload and allows the important distinction to be made between the relatively minor elevations of liver non-haem iron sometimes found in patients with cirrhosis of the liver, and iron overload associated with inherited haemochromatosis (Bassett et al., 1986). Estimation of iron concentration in bone marrow is, in contrast, usually by the histochemical method and is often used to detect iron deficiency. In particular, assessing marrow iron histologically distinguishes between 'true' iron deficiency and other chronic disorders in which there is impaired release of iron from reticuloendothelial cells.

### 5.2.4   Serum iron/total iron binding capacity and transferrin saturation

The International Committee for Standardization in Hematology (Expert Panel on Iron) has recom-

mended a reference method which is simple and reliable (ICSH, 1990). The simultaneous precipitation of serum proteins and the release of iron from transferrin in the presence of a reducing agent is followed by centrifugation to remove denatured protein and detection of the ferrous iron in the supernatant with a chromogen. The method avoids non-specific absorbance caused by interference with serum proteins and there is relatively little interference by haem iron or copper. Methods which do not require centrifugation have obvious advantages in terms of automation, and techniques in which iron is released from transferrin without precipitating serum proteins are also widely applied (Persijn et al., 1971). Procedures for measuring serum iron are available for most clinical chemistry autoanalysers, but there are concerns about accuracy (Tietz and Rinker, 1994).

Measurement of the serum iron concentration alone provides little useful clinical information because of the considerable variation from hour to hour and day to day in normal individuals. Transferrin iron is only 0.1% of the total body iron and the transferrin iron pool turns over 10–20 times each day. Changes in supply and demand due to infection, inflammation or injury (e.g. surgery) therefore cause rapid changes in serum iron concentration. Normal ranges for serum iron concentration are given in Table 5.2 and change little with age in adult life. Low concentrations are found in patients with iron deficiency anaemia and high concentrations in patients with iron overload (Bothwell et al., 1979). However, many hospital patients have a low serum iron concentration which is a response to inflammation, infection or surgery, and does not necessarily indicate an absence of storage iron. High concentrations are found in liver disease, hypoplastic anaemias, ineffective erythropoiesis and iron overload.

More information may be obtained by measuring both the serum iron concentration and the total iron binding capacity (TIBC), from which the percentage of transferrin saturation with iron may be calculated. The TIBC is a measurement of transferrin concentration and may be estimated by saturating the transferrin iron binding capacity with excess iron and removing the excess with solid magnesium carbonate, charcoal or an iron exchange resin. This is followed by determination of the iron content of the saturated serum (Bothwell et al., 1979). The unsaturated iron binding capacity (UIBC) may be determined by methods which detect remaining iron able to bind to chromogen, after adding an excess of iron to the serum. The problems inherent in these direct assays have been summarized by Bothwell et al. (1979). As for the serum iron determination, protocols for clinical chemistry analysers often include a method for

**Table 5.2**  Serum iron, total iron binding capacity and transferrin saturation during development

| Age | Number | Serum iron (µmol/l) Median | (95% range) | TIBC (µmol/l) Mean ± SD | % Saturation Median | 95% range | Reference |
|---|---|---|---|---|---|---|---|
| 0.5 months | 40* | 22 | 11–36 | 34 ± 8 | 68 | 30–99 | Saarinen and Siimes (1977a) |
| 1 months | 40* | 22 | 10–31 | 36 ± 8 | 63 | 35–94 | Saarinen and Siimes (1977a) |
| 2 months | 40* | 16 | 3–29 | 44 ± 10 | 34 | 21–63 | Saarinen and Siimes (1977a) |
| 4 months | 40* | 15 | 3–29 | 54 ± 7 | 27 | 7–53 | Saarinen and Siimes (1977a) |
| 6 months | 40* | 14 | 5–24 | 58 ± 9 | 23 | 10–43 | Saarinen and Siimes (1977a) |
| 0.5–2 years | 121 | 12.2 | 3.6† | – | 22 | 7† | Koerper and Dallman (1977) |
| 2–12 years | 164 | 12.5 | 3.6† | – | 25 | 7† | Koerper and Dallman (1977) |
| 12–18 years | 74 | 14.3 | 3.6† | – | 27 | 7† | Koerper and Dallman (1977) |
| Adult female | 517 | 16.1±7.4‡ | | 68.0 ± 12.6 | 24.6 | ± 11.8 | Jacobs et al. (1969) |
| Adult male | 499 | 18.0±6.3 | | 63.2 ± 9.1 | 29.1 | ± 11.0 | Jacobs et al. (1969) |

\* Approximate number, male and female infants who were not iron deficient
† Lower limit of 95% range, male and female
‡ mean ± SD

TIBC and doubts about accuracy apply as to serum iron (Tietz and Rinker, 1994). An alternative approach is to measure transferrin directly by immunological assay. Nephelometric methods are widely used and there is generally a good correlation between the chemical and immunological TIBC (Huebers et al., 1987). However, Bandi et al. (1985) reported great variability among a number of immunochemical assays for transferrin.

Normal ranges for both TIBC and transferrin saturation are given in Table 5.2. A transferrin saturation of 16% is usually considered to indicate an inadequate iron supply for erythropoiesis (Bainton and Finch, 1964). Of more significance is a raised TIBC (greater than 70 µmol/l), which is characteristic of a deficiency of storage iron. The measurement of TIBC or transferrin may sometimes lead to an apparent saturation of greater than 100% if there is non-transferrin iron present in the serum. Such non-transferrin iron may be ferritin if there is significant liver damage which releases ferritin into the blood. Some of the ferritin iron may be assayed in the determination of serum iron (ICSH, 1990) and other low molecular weight forms of iron may also be present in iron overloaded patients. These measurements of serum iron, which for many years formed the basis of much clinical investigation of iron metabolism, are now seen to provide an inadequate index of storage iron where they have been largely replaced by the assay of serum ferritin. In screening for idiopathic haemochromatosis, however, it is essential to measure both the serum iron concentration and the TIBC or transferrin concentration.

been considerable interest in its use in evaluating the iron supply to the bone marrow. The 'free' protoporphyrin concentration of red blood cells increases in iron deficiency. A widely used technique directly measures the fluorescence of zinc protoporphyrin (µmol/mol haem) in a haematofluorometer (Labbe and Rettmer, 1989). The small sample size (about 20 µl of venous or skin-puncture blood), simplicity, rapidity and reproducibility within a laboratory are advantages. Furthermore, the test has an interesting retrospective application. Because it takes some weeks for a significant proportion of the circulating red blood cells to be replaced with new cells, it is possible to make a diagnosis of iron deficiency anaemia some time after iron therapy has commenced. Chronic diseases that reduce serum iron concentration, but do not reduce iron stores, also increase protoporphyrin levels (Hastka et al., 1993).

The measurement of erythrocyte protoporphyrin levels as an indicator of iron deficiency has particular advantages in paediatric haematology and in large-scale surveys in which the small sample size and simplicity of the test are important. The normal range in adults is <80 µmol/mol haem. Mean values in normal women are slightly higher than in men (Yip, 1994). Garrett and Worwood (1994) found a mean of 44 (range 30–68) µmol/mol haem in women and 41 (29–64) in men. Mean concentrations vary relatively little with age but are slightly higher for children aged 1–3 years than the mean for adult blood (Deinard et al., 1983; Yip, 1994). In the general clinical laboratory, however, it provides less information about iron storage levels in anaemic patients than the serum ferritin assay.

### 5.2.5  Erythrocyte protoporphyrin

This assay has been performed for many years as a test for lead poisoning. More recently, there has

### 5.2.6  Serum ferritin

It was only after the development of a sensitive immunoradiometric assay (IRMA) that ferritin was detected in normal serum or plasma (Addison *et al.*, 1972). Reliable assays, both RIA (labelled ferritin) and IRMA (labelled antibody), have been described in detail (Worwood, 1980a). More recently radioactive labels have been supplanted by enzyme-linked immunoassays (ELISA) with colorimetric and fluorescent substrates or by antibodies with chemiluminescent labels. The solid phase may be a tube, bead, microtitre plate or (magnetic) particle. Numerous variations have been described and serum ferritin included in the latest (batch and random access) automated analysers for immunoassay. A simple enzyme-linked assay with standard reagents has been described and may have application as a reference method (Worwood *et al.*, 1991b) in evaluating automated methods.

The use of a reference ferritin preparation to calibrate the assay is recommended. The second WHO standard for the assay of serum ferritin is reagent 80/578 (obtainable from the National Institute of Biological Standards and Control, PO Box 1193, Potters Bar, Herts EN6 3QH, UK).

Serum ferritin concentrations are normally within the range 15–300 µg/l: values are lower in children than adults, and from puberty to middle age mean concentrations are higher in men than in women (Worwood, 1982). Mean concentrations and ranges throughout life are given in Tables 5.3 and 5.4.

The mother's iron status appears to have relatively little influence on cord serum concentrations,

mean values for which are in the range 100–200 µg/l. In a sample of Australian office workers, blood donation and alcohol intake influenced serum ferritin concentrations along with diet in women (Leggett *et al.*, 1990). The significant increase associated with alcohol consumption in both men and women has been confirmed in the *Health Survey for England* (White *et al.*, 1993). In this survey, ferritin levels also increased with increasing body mass index. In older men and women, mean concentrations are similar. In elderly, unselected patients high levels of ferritin are often associated with pathology (Touitou *et al.*, 1985). Good correlations have been found between serum ferritin concentrations and storage iron mobilized by phlebotomy, stainable iron in the bone marrow, and the concentration of both non-haem iron and ferritin in the bone marrow. This suggests a close relationship between the total amount of storage iron and serum ferritin concentrations in normal individuals, which was directly demonstrated by Walters *et al.* (1973). Serum ferritin concentrations are relatively stable in healthy persons (see later). In patients with iron deficiency anaemia, serum ferritin concentrations are less than 12–15 µg/l (depending on the assay) and a reduction in the level of reticuloendothelial stores is the only common cause of a low serum ferritin concentration. This is the key to the use of the serum ferritin assay in clinical practice (Worwood, 1982).

**Table 5.3**    Serum ferritin concentrations (µg/l) in infants, children and adolescents

| Number of children | Age | Population | Selection | Mean | Range | Reference |
|---|---|---|---|---|---|---|
| 46 | 0.5 months | Helsinki | Non-anaemic | 238 | 90–628 | Saarinen and Siimes (1978) |
| 46 | 1 months | Helsinki | Non-anaemic | 240 | 144–399 | Saarinen and Siimes (1978) |
| 47 | 2 months | Helsinki | Non-anaemic | 194 | 87–430 | Saarinen and Siimes (1978) |
| 40 | 4 months | Helsinki | Non-anaemic | 91 | 37–223 | Saarinen and Siimes (1978) |
| 514 | 0.5–15 years | San Francisco[f] | Non-anaemic | 30[a] | 7–142 | Siimes *et al.* (1974) |
| 323 | 5–11 years | Washington | Low income families | 21[a] | 10–45[d] | Cook *et al.* (1976) |
| 117 | 5–9 years | Nutrition Canada Survey | Random | 15[b] | 2–107[e] | Valberg *et al.* (1976) |
| 335 | 6–11 years | Denmark | Random, urban | 29[a] | 12–67[c] | Milman and Ibsen (1984) |
| 126 male 125 female | 12–18 years | Washington | Low income families | 23[a] 21[a] | 10–63[d] 6–48[d] | Cook *et al.* (1976) |
| 98 male 106 female | 10–19 years | Nutrition Canada Survey | Random | 18[b] 17[b] | 3–125[e] 2–116[e] | Valberg *et al.* (1976) |
| 269 male 305 female | 12–17 years | Denmark | Random, urban | 28[a] 25a | 11–68[c] 6–65[c] | Milman and Ibsen (1984) |

[a] median
[b] geometric mean
[c] 5–95% interval
[d] 10–90% interval

[e] 95% confidence interval
[f] there were no significant differences in median values for ages 6–11 months, 1–2, 2–3, 4–7, 8–10 and 11–15 years.

**Table 5.4**   Serum ferritin concentrations (µg/l) in adults

| Age (years) | Men | | | Women | | |
|---|---|---|---|---|---|---|
| | **No.** | **Mean** | **5–95%** | **No.** | **Mean** | **5–95%** |
| 18–24 | 107 | 80 | 15–223 | 96 | 30 | 5–73 |
| 25–34 | 211 | 108 | 21–291 | 226 | 38 | 5–95 |
| 35–44 | 202 | 120 | 21–328 | 221 | 38 | 5–108 |
| 45–54 | 166 | 139 | 21–395 | 177 | 60 | 5–217 |
| 55–64 | 140 | 143 | 22–349 | 162 | 74 | 12–199 |
| 65–74 | 127 | 140 | 12–374 | 138 | 91 | 7–321 |
| 75+ | 80 | 110 | 10–309 | 99 | 77 | 6–209 |
| Total | 1033 | 121 | 16–328 | 1119 | 56 | 5–170 |

Source: White *et al.*, 1993. Subjects being treated with drugs for iron deficiency (n = 26) were included.
For other surveys of populations in North America and Europe, see Cook *et al.* (1976), Finch *et al.* (1977); Jacobs and Worwood (1975); Millman *et al.* (1986); and Valberg *et al.* (1976).

Iron overload causes high concentrations of serum ferritin, but so may liver disease and some forms of cancer. High concentrations of serum ferritin can only be ascribed to iron overload after careful consideration of the clinical situation.

### 5.2.7   Red cell ferritin

The ferritin in the circulating erythrocyte is but a residue of that present in its nucleated precursors in the bone marrow. Normal erythroblasts contain ferritin which is immunologically more similar to heart (H subunit) than liver ferritin (L subunit) and mean concentrations are about 10 fg ferritin protein/cell ($10^{-15}$ g/cell) (Hodgetts *et al.*, 1986). Concentrations decline in late erythroblasts, decline further in reticulocytes and only about 10 ag/cell ($10^{-18}$ g/cell) remains in the erythrocyte (measured with antibodies to L ferritin) again with somewhat higher levels detected with antibodies to H type ferritin (Cazzola *et al.*, 1983). Red cell ferritin concentrations have been measured in many disorders of iron metabolism, usually with antibodies to L ferritin. In general, red cell ferritin levels reflect the iron supply to the erythroid marrow and tend to vary inversely with red cell protoporphyrin levels (Cazzola *et al.*, 1983). Thus in patients with rheumatoid arthritis and anaemia, low values of red cell ferritin are found in those with microcytosis and low serum iron concentrations regardless of the serum ferritin levels. Red cell ferritin levels do not therefore necessarily indicate levels of storage iron. High levels of red cell ferritin are also found in thalassaemia, megaloblastic anaemia and myelodysplastic syndromes, presumably indicating a disturbance of erythroid iron metabolism in these conditions.

Because it is necessary to have fresh blood in order to prepare red cells free of white cells (which have much higher ferritin levels), the assay of red cell ferritin has seen little routine application despite possible diagnostic advantages (Cazzola and Ascari, 1986), such as differentiation between hereditary haemochromatosis (higher concentration) and alcoholic liver disease (lower concentration).

### 5.2.8   Serum transferrin receptor

Soluble transferrin receptors are detectable in the circulation by immunoassay and appear to reflect the number of transferrin receptors on immature red cells and thus the level of bone marrow erythropoiesis (Cazzola and Beguin, 1992). The levels determined depend on the assay. A mean level of 5.6 mg/l has been reported by Flowers *et al.* (1989). The assay is potentially of considerable value as it provides an alternative to the very cumbersome ferrokinetic studies which were previously necessary. In normal subjects the serum transferrin receptor level also provides a sensitive indicator of functional iron deficiency in subjects with absent iron stores but who have not yet developed iron deficiency anaemia (Skikne *et al.*, 1990). The serum transferrin receptor level is not elevated in patients with acute infection, including hepatitis, in chronic liver disease and other patients with the anaemia of chronic disease if there is no co-existing iron deficiency. It is not yet clear what happens to receptor levels in cases where there is iron deficiency (Petterson *et al.* 1994; Zoli *et al.* 1994). In chronic disease the normal levels seem to reflect the lack of any increase in erythropoietic activity in the bone marrow (Ferguson *et al.*, 1992). The ability to distinguish the anaemia of chronic disease from iron deficiency anaemia makes the transferrin receptor assay a potentially valuable addition to haemoglobin and ferritin in clinical practice and in epidemiological surveys of iron status (Ferguson *et al.*, 1992). However, its usefulness in specific groups, e.g. adolescents, has recently been questioned (Kivivuori *et al.*, 1993).

### 5.2.9   Non invasive methods for determining tissue iron concentrations

Dual-energy computed tomography, magnetic susceptibility and magnetic resonance imaging are all being exploited but at present are only applicable to the detection of iron overload (Stark, 1991).

### 5.3   METHODOLOGICAL AND BIOLOGICAL VARIABILITY OF ASSAYS

The blood assays vary greatly in both methodological and biological stability. Haemoglobin concentrations are stable and the simple and well-standardized method of determination (Dacie and Lewis, 1991) ensures relatively low day-to-day variation in individuals (Table 5.5). Automated cell counters analyse at least 10 000 cells and thus reduce errors. The more complicated procedures involved in immunoassays mean higher methodological variation for ferritin assays (coefficient of variation, CV, of at least 5%) and this, coupled with some physiological variation, gives an overall CV for serum ferritin for an individual over a period of weeks of the order of 15%. There is, however, little evidence of any significant diurnal variation in serum ferritin concentration (Dawkins *et al.*, 1979). The serum iron determination is an example of extremes with reasonably low methodological variation coupled with extreme physiological variability giving an overall 'within subject' CV of approximately 30% when venous samples are taken at the same time of day. A diurnal rhythm has been reported with higher values in the morning than in late afternoon, when concentrations may fall to 50% of the morning value (Bothwell *et al.*, 1979). The circadian fluctuation is due largely to variation in the release of iron from the reticuloendothelial system to the plasma. Results from a number of studies of overall variability are given in Table 5.5 but it should be noted that the type of blood sample, length of study period and the statistical analysis vary from study to study. The somewhat higher variability for Hb and ferritin reported by Borel *et al.* (1991) may be due to their use of capillary blood and plasma. Pootrakul *et al.* (1983) have demonstrated that mean plasma ferritin concentration is slightly higher in capillary specimens than venous specimens and that within and between sample variation was approximately three times greater. Variability was less in capillary serum but still greater than venous serum.

These results have clear implications for the use of these assays in population studies (Dallman *et al.*, 1984; Looker *et al.*, 1990; Wiggers *et al.*, 1991) and in the assessment of individual patients (Borel *et al.*, 1991). For accurate diagnosis, either a multiparameter analysis is required or the assay of several samples.

### 5.4   APPLICATION OF BLOOD ASSAYS FOR DETERMINATION OF IRON STATUS

Two important applications need to be considered:

- population surveys
- evaluation of patients in hospital or clinics.

### 5.4.1   Use in population surveys

Factors other than variation in iron stores which might influence the analyses should always be considered. In older people there may be a higher incidence of chronic disease and in some parts of

**Table 5.5**   Overall variability of assays for iron status (within-subject, day-to-day CV for healthy subjects)

| Reference | Haemoglobin | Serum ferritin | Serum iron | TIBC | EP |
|---|---|---|---|---|---|
| Dawkins *et al.* (1979) | – | 15 (MF) | – | – | – |
| Gallagher *et al.* (1989) | 1.6 (F) | 15 (F) | – | – | – |
| Statland and Winkel (1977) | – | – | 29 (F) | – | – |
| Statland *et al.* (1976) | – | – | 27 (M) | – | – |
| Statland *et al.* (1977) | 3 (MF) | – | – | – | – |
| Pilon *et al.* (1981) | – | 15 (MF) | 29 (MF) | – | – |
| Romslo and Talstad (1988)* | – | 13 (MF) | 33 (MF) | 11 (MF) | 12 (MF) |
| Borel *et al.* (1991) | 4 (MF) | 14 (M) | 27 (M) | – | – |
|  |  | 26 (F) | 28 (F) | – | – |

* anaemic patients

the world infection may be prevalent. High zinc protoporphyrin levels or normal transferrin receptor levels in conjunction with normal or high serum ferritin levels make it possible to detect chronic disease in anaemic subjects. Blood donation may reduce iron stores significantly. There is no information on seasonal factors influencing most of these analyses although seasonal changes in red cell parameters have been reported (Kristal-Boneh et al., 1993). Only limited information on changes during the menstrual cycle is available (Kim et al., 1993).

The serum ferritin assay is the only method which can provide a semiquantitative indication of the levels of storage iron but its application is limited by both methodological and biological variation. Starvation, or even fasting for a short period, can cause elevation of the serum ferritin concentration (Lundberg et al., 1984) and vitamin C deficiency may reduce it (Chapman et al., 1982b). Two definitions are commonly used but sometimes confused. **Iron deficiency anaemia** refers to anaemia caused by the lack of iron in the body. **Iron deficient erythropoiesis** refers to an impairment of iron supply to the bone marrow in the absence of storage iron and may occur with haemoglobin levels still in the reference range. Overlap in laboratory values for iron stores between anaemic and non-anaemic individuals means that using haemoglobin measurements alone leads to an overestimate of the prevalence of iron deficiency anaemia. Similarly the use of only one indicator of storage iron levels or tissue iron supply – ferritin, transferrin saturation or erythrocyte protoporphyrin – may overestimate the number of individuals with lesser degrees of iron deficiency. The concepts are illustrated in a paper by Cook et al. (1986) in which they show that more precise figures for the prevalence of iron deficient erythropoiesis and iron deficiency anaemia may be obtained by using multiple criteria (abnormal values for any two of serum ferritin, transferrin saturation or erythrocyte

protoporphyrin). The 'multiple-test' approach has been further refined with the introduction of soluble transferrin receptor measurements.

Skikne et al. (1990) employed quantitative phlebotomy to evaluate the use of serum ferritin concentrations and soluble transferrin receptor concentrations to indicate levels of iron stores. The log of the ratio of these two measurements (receptor/ferritin as $\mu g/\mu g$) shows a linear relationship with iron stores (mg Fe/kg body weight). This combination makes it possible to evaluate iron stores and also the impairment of iron supply for production of functional iron proteins which develops as more and more body iron is lost.

The results of the application of the multiple test and the single test approaches are clearly contrasted in two papers (Cook et al., 1986; Hallberg et al., 1993a). The final confirmation about the correctness of these prevalence figures will require an examination of the response to iron supplementation (Garby et al., 1969).

In order to allow greater consistency in the comparison of iron status data from population surveys, the Task Force has suggested cut-off points for children (Table 5.6) and adults (Table 5.7).

### 5.4.2   Use in individual patients

In combination with measurement of blood haemoglobin, assays of zinc protoporphyrin (ZPP) and transferrin saturation can indicate iron deficient erythropoiesis. Furthermore a high TIBC provides good evidence that there is no storage iron. However, most adult hospital patients have anaemia which is secondary to infection, inflammation, malignant disease or surgery. If information about the amount of storage iron is required, it is generally necessary to assay serum ferritin or to assess tissue iron levels directly. Iron deficient erythropoiesis in the presence of normal or raised levels of storage iron suggests secondary anaemia, not

**Table 5.6**   Suggested cut-off points for differing iron status in young children from epidemiological studies (source: Oski et al. 1983)

| Factor | Iron sufficient | Iron depleted non-anaemic | Iron deficient erythropoiesis | Iron deficient anaemia |
|---|---|---|---|---|
| Hb g/l | $\geq 110$ | $\geq 110$ | $\geq 110$ | $< 110$ |
| Ferritin µg/l* | $\geq 12$ | $< 12$ | $< 12$ | $< 12$ |
| Transferrin saturation % | $\geq 10$ | $\geq 10$ | $< 10$ | $< 10$ |
| EPP µmol/mol haem | $< 100$ | $< 100$ | $\geq 100$ | $\geq 100$ |

* Others have adopted more stringent criteria, e.g. ferritin of 7 µg/l has been suggested by Siimes et al. (1974) and various surveys, including the DH nationally representative sample of children have adopted this figure.

**Table 5.7**    Suggested cut-off points for defining iron status in adults from epidemiological studies

| Factor | Iron overload | Iron sufficient | Iron depleted non-anaemia | Iron deficient erythropoiesis | Iron deficient anaemia |
|---|---|---|---|---|---|
| Hb[a] g/l | – | ⩾ 130 men ⩾ 120 women | ⩾ 130 men ⩾ 120 women | ⩾ 130 men ⩾ 120 women | < 130 men < 120 women |
| Ferritin[b] μg/l | > 300 men > 200 women | ⩾ 13 | < 13 | < 13 | < 13 |
| Transferrin saturation % | > 60% men > 50% women | > 16% | ⩾ 16% | < 16% | < 16 |
| EPP[c] μmol/mol haem | – | < 80 | < 80 | ⩾ 80 | ⩾ 80 |
| Transferrin receptor mg/l | – | < 8.5 | < 8.5 | > 8.5 | >8.5 |

[a] WHO (1972) recommended cut-off points for IDA are below 120 g/l for women and below 130 g/l for men, but the Health Survey of England (1991) uses 110 g/l as the lower cut-off point for men and women.
[b] The Adult Survey uses 13 μg/l ferritin (Ft) as the lowest cut-off point but defines less than 25 μg/l as low iron stores.
[c] EPP is usually measured as ZPP. If washed cells are used, the cut-off value should be 40 μmol/mol haem.

primary iron deficiency. Thus the serum ferritin assay should provide an excellent discriminator between iron deficiency anaemia and the anaemia of chronic disease. In practice, interpretation of serum ferritin is more complicated.

In the anaemia of chronic disease, the most important factor controlling serum ferritin concentrations is the level of storage iron. Serum ferritin levels are higher than those found in patients with similar levels of storage iron but without infection and inflammation (Worwood, 1980b). There is experimental evidence from studies of rat liver that the rapid drop in serum iron concentration which follows the induction of inflammation may be due to an increase in apoferritin synthesis which inhibits the release of iron to the plasma (Konijn and Hershko, 1977). Interleukin-I (IL-I) is the primary mediator of the acute-phase response which in iron metabolism is indicated by the drop in plasma iron concentration (Dinarello, 1984). There is direct evidence from studies of cultured human hepatoma cells that IL-1β (which also causes changes in protein synthesis that mimic the acute phase response in cultured hepatoma cells) directly enhances the rate of ferritin synthesis by control of translation (Rogers et al., 1990).

Many clinical studies have demonstrated that patients with the anaemia of chronic disease, and no stainable iron in the bone marrow, may have serum ferritin concentrations considerably in excess of 15 μg/l and there has been much debate (Witte, 1991) about the practical application of the serum ferritin assay in this situation. Values < 15 μg/l indicate the absence of storage iron and values > 100 μg/l indicate the presence of storage iron. It is the 'grey' area from 15 to 100 μg/l which is difficult to interpret. It would seem logical to combine the assay of serum ferritin with a measure of disease severity such as the ESR or C-reactive protein. Witte et al. (1986) have described such an

approach and claimed to be able to confirm or exclude iron deficiency (absence of storage iron in the bone marrow) in almost all patients with secondary anaemia. However these conclusions have been challenged (Coenen et al., 1991). As described earlier, measurements of soluble transferrin receptor concentration may provide a valuable diagnostic aid for this problem area.

In assessing the adequacy of iron stores for replenishing haemoglobin the degree of anaemia must also be considered. Thus a patient with haemoglobin concentration of 100 g/l may benefit from iron therapy if the serum ferritin concentration is below 100 μg/l (Cavill et al., 1986).

The other major influence confounding the use of the serum ferritin assay to determine iron stores is liver disease. The liver contains much of the storage iron in the body and any process that damages liver cells will release ferritin. It is also possible that liver damage may interfere with clearance of ferritin from the circulation. It was suggested by Prieto et al. (1975) that the ratio of serum ferritin to aspartate aminotransferase activity might provide a good index of liver iron concentration. Glycosylated ferritin concentrations might relate directly to storage iron concentrations, while non-glycosylated levels would relate to the degree of liver damage (Worwood, 1979). However, neither the ferritin : aspartate aminotransferase ratio (Valberg et al., 1978; Batey et al., 1978) nor the measurement of glycosylated ferritin levels (Worwood et al., 1980; Chapman et al., 1982a) has proved to be any more reliable than the simple assay of serum ferritin as an index of liver iron concentration. In patients with liver damage a low value always indicates absent iron stores; normal values indicate absent or normal levels but rule out iron overload; whereas high values may indicate either normal or high iron stores and further investigation may be necessary to distinguish these.

## 5.5    SPECIFIC APPLICATIONS

The general application of the ferritin assay for the determination of iron storage levels has been described. There are a number of common diagnostic applications where a more specific application of the various indicators of iron metabolism is justified.

### 5.5.1    Iron deficiency in infancy and childhood

The serum ferritin concentration is a less useful guide to iron deficiency in children than in adults partly because of the low concentrations generally found in children over 6 months of age. ZPP provides a useful indicator of iron-deficient erythropoiesis, although high values may indicate lead poisoning rather than iron deficiency. The small sample volume for ZPP determination is also an advantage in paediatric practice. Hershko *et al.* (1981) studied children in villages from the Golan Heights in Israel and concluded that erythrocyte protoporphyrin (EP) was a more reliable index of iron deficiency than serum ferritin and serum iron. They suggested that a significant incidence of chronic disease affected both ferritin and iron values. However, Zanella *et al.* (1989) did not find that EP was a better predictor of iron deficiency than ferritin.

### 5.5.2    Treatment of iron deficiency anaemia

Oral iron therapy at conventional doses (60 mg Fe, three times daily) has little immediate effect on serum ferritin levels which rise slowly as the haemoglobin concentration increases. With double doses there is a rapid rise of serum ferritin to within the normal range (within a few days) which probably does not represent the increase in storage iron (Wheby, 1980). Intravenous iron causes a rapid rise in ferritin concentrations, which may be above the normal range, and then gradually drop back to the normal range.

### 5.5.3    Screening blood donors for iron deficiency

Regular blood donation reduces storage iron levels and this has been demonstrated in a number of surveys of blood donors (Skikne *et al.*, 1984; Milman and Kirchoff, 1991). A well-known difficulty is that the conventional 'copper sulphate' screening test for anaemia (Mollison *et al.*, 1988) is somewhat inaccurate and donors may be deferred unnecessarily. It has been suggested that secondary screening using ZPP would provide an immediate confirmation of iron deficiency (Raftos *et al.*, 1983; Schifman and Rivers, 1987). Despite the availability of the serum ferritin assay during the last 20 years, there has been little attention to the fundamental relationship between storage iron levels and the ability to give blood. Screening blood donors by routinely assaying serum ferritin may make it possible to predict the development of iron deficiency anaemia, identify donors with high iron stores who may give blood more frequently than usually permitted, and also identify donors homozygous for genetic haemochromatosis who are beginning to develop iron overload (Worwood *et al.*, 1991a; Worwood and Darke 1993).

# 6

# IRON AS A PRO-OXIDANT

A disturbance in iron metabolism is a secondary effect of diseases. The profound hypoferraemia associated in mammals with neoplasms, infection and other conditions which stimulate inflammation has been recognized for nearly half a century (Weinberg, 1983). In addition there is currently considerable interest in the possibility that high levels of storage iron are associated with the risk of myocardial infarction (Sullivan, 1992a; Salonen *et al.*, 1992b) and cancer (Stevens *et al.*, 1988; Selby and Friedman, 1988). In vitro studies show that iron can promote lipid peroxidation and the breakdown of tissue integrity (Halliwell and Gutteridge, 1984). These considerations raise the possibility that, as a pro-oxidant, iron may play a part in the pathogenesis of a range of disorders in addition to its known effects in human iron overload disease.

## 6.1 TRANSITION METALS AS CATALYSTS OF OXIDATIVE TISSUE DAMAGE

### 6.1.1 Transition metals

A free radical is an atom or molecule containing one or more unpaired electrons. Iron belongs to the group of elements known as the transition metals which include copper, zinc, manganese, molybdenum etc. Most transition metals can catalyse free radical reactions and this gives them the ability to take part in single-electron (radical) transfers. As a consequence many of these metals are components of oxidation–reduction reactions.

The reactivity of a mixture of Fe(II) salts and hydrogen peroxide on many organic molecules was first observed by Fenton in 1894 who gave his name to the following reaction.

$$Fe^{2+} + H_2O_2 = Fe^{3+} + OH\cdot + OH^- \qquad \text{Reaction 1}$$

Both iron and copper will also catalyse *in vitro* the reaction of superoxide radicals and hydrogen peroxide, producing hydroxyl radicals (OH·). In both cases the lower valency form iron (II) and copper (I) are the more active. This reaction was first described by Haber and Weiss in 1934 and has become known as the Haber–Weiss reaction:

$$O_2^- + H_2O_2 \underset{\text{catalyst}}{\overset{\text{Iron-salt}}{=}} O_2 + OH\cdot + OH^- \qquad \text{Reaction 2}$$

In the absence of a transition metal, however, the rate constant is virtually zero.

### 6.1.2 Endogenous production of free radicals

There are several radicals derived from oxygen such as superoxide, hydroxyl radical and nitric oxide (NO). The peculiar nature of the oxygen molecule is such that when it attempts to oxidize another molecule by accepting two electrons, both must have antiparallel spin characteristics. This imposes certain restrictions on electron transfer with the net effect that the reaction is sluggish and electrons are accepted sequentially. The addition of a single electron to the ground state oxygen molecule produces the superoxide radical ($O_2^-$). Addition of one more electron will give the peroxide ion ($O_2^{2-}$) and as this is accompanied by two protons, the product is hydrogen peroxide. The addition of two further electrons and two protons produces water (Halliwell and Gutteridge, 1985a). *In vitro* studies on systems producing superoxide and/or hydrogen peroxide have been observed to inactivate and destroy many biological systems but the evidence suggests that neither of these forms are particularly damaging in themselves but react through the production of the hydroxyl radical (OH·). The hydroxyl radical, in contrast to the other oxygen products, reacts with extremely high rate constants with almost every type of molecule found in living cells but requires transition metal ions for its formation. It is believed that iron is the main metal ion responsible for the formation of these reactive oxygen species (ROS) *in vivo*.

### 6.1.3   Nitric oxide and iron regulatory protein

Endothelium-derived relaxing factor (EDRF), which plays a central role in vascular tone and platelet aggregation, was first identified in 1987 as nitric oxide (NO), the NO being released from vascular endothelial cells (Palmer et al., 1987). It is produced by endothelium, macrophages, neutrophils and brain synaptosomes and is a free radical which is formed enzymatically from a terminal guanidino-nitrogen of L-arginine by two enzymes. The generation of NO by the constitutive $Ca^{2+}$-dependent nitric oxide synthase (NOS) has an important role in the homeostasis of the endothelium and nervous system and possibly other tissues, and activity may be controlled by the modulation of Ca entry into the cell (Manzoni et al., 1992; Clark et al., 1993) as well as other mediators. In addition, the interaction of NO with cytoplasmic iron regulatory protein (IRP) may coordinate cellular iron traffic (Pantopoulos et al., 1994).

IRP is a cytoplasmic homologue of mitochondrial aconitase (Chapter 4). Iron status determines the activity of this protein since when iron availability is high, only aconitase activity is expressed. However, in the absence of iron, IRP exerts high affinity for several iron-responsive elements within the untranslated regions of mRNAs involved in iron uptake, storage and utilization in higher eukaryotic cells (Pantopoulos et al., 1994). As a consequence of binding, ferritin and haem synthesis are blocked while transferrin-receptor activity is protected. Nitric oxide is believed to interact with the iron–sulphur (Fe–S) clusters in the iron-regulating/aconitase protein, stimulating IRP activity – i.e. mimicking the role of iron starvation and promoting iron uptake (Weiss et al., 1993b).

In addition to the $Ca^{2+}$-dependent NOS, there is an inducible $Ca^{2+}$-independent NOS which is expressed in target tissues when stimulated by some cytokines and bacterial endotoxin. Synthesized by the latter pathway in vessel walls, liver, lung and neuronal tissue, NO triggers vasodilation (Moncada et al., 1991). It is lipophilic and easily migrates through membranes; consequently when the inducible enzyme is switched on, there is evidence to suggest that the activity of the $Ca^{2+}$-dependent, constitutive enzyme is switched off by blockage of $Ca^{2+}$ entry channels into cells (Clark et al., 1993; DeBelder et al., 1993).

The $Ca^{2+}$-independent NOS appears to be particularly important in activated macrophages. Nitric oxide can react with iron and this may have physiological importance in both the inhibition of DNA replication in tumour cells and the killing of parasites (James and Hibbs, 1990). Cellular enzymes containing iron–sulphur (Fe–S) groups appear to be inhibited by stimulated macrophages and may be the target of attack by NO. Tumour cells co-cultivated with activated macrophages also release a significant fraction of their intracellular iron in parallel with the development of inhibition of DNA synthesis and mitochondrial respiration. It has also been suggested that NO can interact with superoxide and produce hydroxyl radicals (Beckman et al., 1990) and excess iron may inactivate NO (Liew and Cox, 1991).

Nitric oxide produced by both constitutive and inducible NOS can regulate IRP activity in vitro inducing changes in cellular metabolism which promote iron uptake. Whether there is greater activity in specific tissues such as the macrophages of liver, spleen and bone marrow, which tend to take up iron during inflammation, has yet to be determined. When NO is produced by macrophages it may have the dual role of promoting iron uptake in specific host tissues while scavenging iron in non-host (e.g. tumour or parasite) tissues.

## 6.2   MEMBRANE STRUCTURE AND INTEGRITY

The major constituents of biological membranes are lipid and protein, the amount of protein roughly depending on the number of functions of the membrane. The dominant lipids are phospholipids based on glycerol. Phospholipids contain a number of saturated and polyunsaturated fatty acid side-chains. The fatty acid side-chains of membrane lipids in animal cells have unbranched carbon chains and the double bonds are in the cis configuration. The cis double bond prevents rotation of the groups attached to the carbon atoms, thus producing 'kinks' in the molecule. As the number of double bonds in the fatty acid molecule increases, so the melting point decreases. The proportion of polyunsaturated to saturated fatty acids in a membrane determines membrane fluidity, which in turn influences the rate of diffusion of lipid and protein within the membrane. Thus any factor which alters fatty acid double bond organization, such as peroxidation, potentially affects fluidity, structure and function of the membrane.

### 6.2.1   Oxidative damage

All tissue components are susceptible to free radical damage but as lipid peroxidation illustrates the mechanisms involved, this process will be described. Lipid peroxidation is the oxidative deterioration of polyunsaturated lipids, i.e. those which contain more than two carbon–carbon double covalent bonds. This process, otherwise known as rancidity, is a widely recognized problem in the

Lipid peroxidation

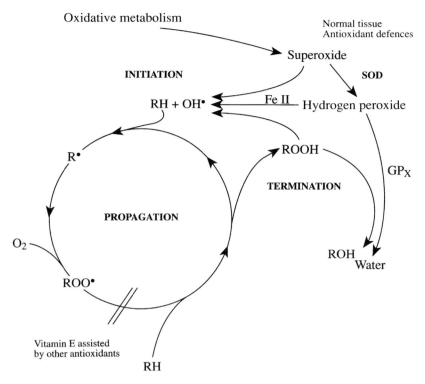

**Figure 6.1**   The autocatalytic chain of lipid peroxidation. Superoxide and hydrogen peroxide are normal metabolites of oxidative metabolism and are converted to water under the action of superoxide dismutase (SOD) and glutathione peroxidase (GPx). To initiate lipid peroxidation, ferrous iron (FeII) can react with these oxygen metabolites via the Haber–Weiss or Fenton reactions to generate hydroxyl radicals (OH•) which then react with lipid (RH) or any other organic molecule. Free radical generation is propagated by oxidation of the lipid radical (R•) to produce a peroxyl lipid radical (ROO•) and to lipid hydroperoxide (ROOH) and more R• by reaction with another hydrogen donor (RH). Termination takes place by interaction with vitamin E or via the action of glutathione peroxidase as shown.

storage of fats and oils (Dormandy, 1969).

The hydroxyl radical is a powerful stimulator of lipid peroxidation and thus of disruption in cell organization and structure (Figure 6.1). Peroxidation comprises three components or phases: initiation, propagation and termination. Initiation is brought about by the abstraction of a hydrogen atom from the methylene link (-CH$_2$-) between two double bonds in a polyunsaturated fatty acid. Since a hydrogen atom has only one electron, it leaves behind an unpaired electron on the methylene carbon. The carbon radical rapidly undergoes a molecular rearrangement to produce a conjugated diene which then easily reacts with an oxygen molecule to form a peroxyl radical R-OO•.

Peroxyl radicals can abstract a hydrogen atom from another lipid molecule which then undergoes the same process. This is described as the propagation phase and can be repeated many times with lipid molecules of other organic compounds. When peroxyl radicals abstract the hydrogen atom, they form a lipid hydroperoxide R-OOH. Lipid hydroperoxides behave like hydrogen peroxide; that is, they

are relatively stable but will react with transition metals to form alkoxyl radicals RO•. These react like the hydroxyl radical and are capable of initiating more lipid peroxidation.

### 6.2.2   Termination of oxidative damage

The propagation phase ends either when two radicals interact or when a radical scavenger such as vitamin E reacts with the lipid radical. Vitamin E is called a chain-breaking antioxidant: these are able to react easily with lipid radicals, donating a hydrogen atom and forming a relatively stable intermediate form. Chain-breaking antioxidants block further peroxidation and enable repair processes to take place.

Various products result from lipid peroxidation including pentane, ethane and a whole range of carbonyl-containing compounds, especially aldehydes such as nonenal and malonaldehyde. Aldehyde products are fairly stable and can therefore migrate and react with thiol and amino groups on proteins forming intra- and inter-molecular cross-links. Such cross-links affect protein function, causing aggregation of proteins, inhibiting membrane surface receptor molecules or enzyme activity, etc. and so contribute to declining function in radical-damaged tissues.

## 6.3   PROTECTION OF TISSUES FROM OXIDATIVE DAMAGE

Polyunsaturated fatty acids are present throughout the body at all levels of cellular structure; likewise, oxygen is of fundamental importance to life. Superoxide is not particularly reactive but its widespread production by many processes in the body (mitochondrial respiration, microsomal $P_{450}$ oxidation, in the carriage of oxygen by haemoglobin) makes it a potential source of radical stress. To control oxidation, all aerobic organisms have evolved enzymatic and other protective mechanisms (Fridovich, 1983).

### 6.3.1   Enzymatic defence systems

The discovery of superoxide dismutase activity in tissues led to the present interest in the superoxide theory of oxygen toxicity (Fridovich, 1983). The distribution of this enzyme in all aerobic tissues and at many levels of cellular structure, illustrates the importance which evolution has given to removing oxygen. Superoxide is dismutated to hydrogen peroxide:

$$O_2^- + O_2^- + 2H^+ = H_2O_2 + O_2 \qquad \text{Reaction 3}$$

Tissues also contain two other enzymes to catalyse the conversion of hydrogen peroxide to water, namely catalase, found principally in peroxisomes, and glutathione peroxidase ($GP_x$), which is more widely distributed. $GP_x$ also reacts with lipid hydroperoxides (R-OOH) produced during lipid peroxidation, converting them to harmless hydroxy-acids (ROH).

### 6.3.2   Tissue integrity

Oxidative stress is also minimized by the fact that cells possess structural integrity which protects tissue components that are susceptible to oxidative damage from autocatalytic oxidation (Dormandy, 1983). In particular, transition metal cations, notably ferrous iron, which must exist within the cell to participate in many oxidation–reduction reactions, are localized to specific sites and thus contact with 'foreign' components is minimized.

### 6.3.3   Binding of transition metals

Transition metals are abundant in the body and their availability is essential in the theory of oxygen toxicity. In the circulation, metals are bound to proteins (e.g. transferrin, caeruloplasmin, metallothionine and albumin) and the amounts of iron bound to transferrin in circulating fluids are restricted. Likewise, the availability of ionic iron, from metabolically active sites within cells, is tightly controlled by the structural integrity of tissues.

### 6.3.4   Chain-breaking or scavenging antioxidants

Accumulating evidence suggests that higher intakes of dietary components such as vitamins E and C, the carotenoids (Gey, 1993), bioflavonoids (Hertog et al., 1993; Hertog et al., 1995) etc. are associated with the reduced risk of certain chronic diseases. These components have various antioxidant functions involving scavenging radical products within the tissues and probably also in the gut (Thurnham, 1990a, 1992a). However, caution is needed before large amounts of these substances are consumed since there is uncertainty about the role of vitamin C during disease (Thurnham, 1994; 1995), and supplementation of smokers with β carotene has been associated with an increased rate of lung cancer (ATBC Study Group, 1994).

## 6.4   VITAMIN C AND IRON METABOLISM

Vitamin C (ascorbate) acts as an antioxidant in vitro and in vivo (Niki, 1991; Frei, 1991). Epidemiological studies indicate that, in standardized mortality rates for all causes of death, vitamin C intakes are strongly inversely related for males and weakly for females (Enstrom et al., 1992). However, in spite of a key role in protecting cells against oxidative damage, in the presence of Fe(III) or Cu(II), ascorbate can promote the generation of the same ROS (OH·, $O_2^-$, $H_2O_2$) and Fe(II) it is known to destroy (Stadtman, 1991). While the formation of Fe(II) in the gut improves the absorption of iron, the formation of Fe(II) or Cu(I) in vivo promotes damage to DNA and protein at metal binding sites on these macromolecules (Stadtman, 1991).

The levels of ascorbate range from 11 to 200 μmol/l in the plasma to millimolar levels in eye lens, brain, lung and the adrenals. There is little evidence that ascorbate behaves as a pro-oxidant under normal conditions, but the limiting factor may be the availability of metal ions. The total iron content of most tissues is quite high but it is tightly sequestered in protein complexes. Conditions which facilitate the release of iron may promote ascorbate-mediated radical damage, and it is likely that the ratio of iron to ascorbate will determine the ability of ascorbate to express pro-oxidant activity (Stadtman, 1991).

Plasma and leucocyte ascorbate levels are often reduced in infections (Bingol et al., 1975; Hume and Weyers, 1973), trauma (Crandon et al., 1961; Irvin et al., 1978), and the elderly (DHSS, 1979). In chronic diseases, plasma concentrations are lower than expected for dietary intakes (Cunningham et al., 1991). It has been suggested that changes accompanying trauma may facilitate the removal of ascorbate from the circulation to reduce the possibility of pro-oxidant activity (Thurnham, 1994). Leucocytes accumulate ascorbate to levels 25–80-fold above plasma (Evans et al., 1982) and in vitro studies with isolated human granulocytes showed that this is increased by antigen stimulation and by temperature up to 40°C (Moser and Weber, 1984).

Surgical trauma (Crandon et al., 1961; Irvin et al., 1978) and infection (Hume and Weyers, 1973) are followed immediately by a temporary fall in leucocyte ascorbate concentrations, but this is related to a post-operative leucocytosis; concentrations return to the pre-surgical level within about five days (Irvin et al., 1978; Vallance, 1986). Leucocytosis may provide the means of removing this potential pro-oxidant.

## 6.5 'OBLIGATORY' OXIDATIVE DAMAGE IN TISSUES

Experimental and human studies have also shown that many of the enzymes needed to remove superoxide and its products (SOD, $GP_x$, glucose-6-phosphate dehydrogenase and other pentose phosphate enzymes) are inducible when the level of oxidized stress is raised (Howarth et al., 1981) although a rapid change in oxidative metabolism can exceed this protection (Chow, 1976). Infection, inflammatory diseases or over-vigorous exercise in an untrained person may dramatically increase oxidative stresses. However, oxidative damage to DNA also occurs in the normal, healthy, unstressed individual. Ames et al. (1985) examined urinary DNA metabolites and suggested that a thousand molecules of thymine are oxidized in each of the body's $6 \times 10^{13}$ cells every day. The nucleus possesses repair mechanisms which are probably capable of restoring all this damage under most circumstances. However, repair processes are relatively slow and where replication occurs before repair, damage may become permanent or the cell may be destroyed (Thurnham, 1993). Products of lipid peroxidation in normal tissues (Avogaro et al., 1988; Moison et al., 1993) indicate that oxidative damage occurs in spite of antioxidant defences and is certainly exacerbated by a variety of diseases (Yagi, 1987).

## 6.6 SOURCES OF IRON IN BLOOD AND TISSUES

### 6.6.1 Transport and storage iron

Plasma transferrin is normally only about 30% saturated with iron and is therefore unlikely to release free iron into solution. In fact normal serum has powerful abilities to sequester iron and prevent lipid peroxidation (Stocks et al., 1974). In tissues, iron is stored bound to ferritin or, in an even more insoluble and unavailable form, as haemosiderin. Ferritin appears to prevent excessive intracellular accumulation of non-protein bound iron and apoferritin synthesis can be stimulated by loading cells with iron salts (Kuhn, 1991). Ferritin is rarely saturated with iron, and iron bound to the proteins of ferritin and transferrin or haemoglobin reacts slowly, if at all, with $H_2O_2$ (Halliwell and Gutteridge, 1985b). Some oxidative damage is probably a normal component of oxidative metabolism, so some iron and other transition metals must elude these protective mechanisms.

### 6.6.2 Functional iron in tissues

Low molecular weight iron complexes present in the human body under normal conditions (e.g. iron–citrate or iron–ATP) are probably the major catalysts of reactions 1 and 2 in vivo (Halliwell and Gutteridge, 1985b). Other protein complexes of iron within cells are also likely to be a source of ionic iron to catalyse the production of ROS, since their enzymatic functions require the availability of iron as a single electron carrier for oxidation–reduction reactions. Several mechanisms exist to minimize damage from these sources of iron.

## 6.7 PROTECTIVE RESPONSES TO UNBOUND IRON DURING TRAUMA

The body's rapid protective response to trauma provides strong indirect evidence for the potential oxidative reactivity of tissue contents. Infection, tumour cells, physical damage, etc. promote neutrophil activation which can itself result in further tissue destruction (Weiss, 1989). However, they also stimulate macrophages to produce cytokines, which in turn activate the acute phase response: a series of reactions taking place throughout the body (Koj, 1985). The reactions are protective, at

**Table 6.1**   Responses to trauma by acute phase proteins (adapted from Thurnham, 1990b and 1992b)

| Acute phase protein | Plasma concentration | Function | Response to trauma |
|---|---|---|---|
| Caeruloplasmin (g/l) | 0.155–0.592 | Fe(II) to Fe(III) | Increases 30–60% |
| Transferrin (g/l) | 1.9–2.58 | Binds Fe(III) in cells. Function in plasma unknown | Reduction up to 30% |
| Ferritin (µg/l) | 15–250 | Binds Fe(III) (in tissues) | Increase |
| Haptoglobin (g/l) | 0.70–3.79 | Binds haemoglobin | Increases 200–500% |
| Haemopexin (g/l)* | 0.52–1.14 | Binds haem | No change |
| Proteinase inhibitors (g/l) | 1.5–3.0 | Inhibits lysosomal enzymes, e.g. serine proteinase, elastase, collagenase and gelatinase | Increases 200–400% |

* Not an acute phase protein, but included because of its role in binding haem.

least in the short term, and tend to restore tissue integrity and reduce the effects of trauma. Integrity is threatened if lipid peroxidation occurs, as the resulting damage to cell membranes releases more transition metal, increasing the possibility of more damage.

### 6.7.1   Acute phase response to trauma

The body responds to trauma by altering (mainly increasing) the plasma concentration of several proteins involved in the sequestration and transport of iron (Table 6.1) (Koj, 1985; Thurnham, 1990b). These proteins can be regarded as protective antioxidants as they assist in scavenging ferrous iron and/or minimizing lipid peroxidation (Stocks *et al.*, 1974). One of the earliest responses (within 24 hours) is an increase in vascular endothelial cell permeability. This allows intermediately sized proteins like albumin and transferrin to move more readily into the extracellular fluid compartment. Such movements may facilitate the binding by transferrin of iron released from damaged tissues. Also within 24 hours, there is marked leucocytosis (Crandon *et al.*, 1961; Vallance, 1986) which may play a role in removing ascorbate from the circulation. Ascorbate is a potential pro-oxidant in the presence of iron (Stadtman, 1991) and if the leucocytosis is a means of scavenging ascorbate, this could account for the reports of low plasma ascorbate accompanying both acute disease (Hume and Weyers, 1973; Irvin *et al.*, 1978) and chronic disease (Sinclair *et al.*, 1991; Yoshioka *et al.*, 1984).

### 6.7.2   Prolonged response to trauma

A secondary response to trauma takes place in the liver. In general, acute phase proteins show stimulation of synthesis but there are some notable exceptions. Transferrin, albumin and retinol binding protein are negative acute phase proteins; that is, their concentrations in the plasma are lower during infection. Partly as a consequence, plasma concentrations of substances carried by these proteins, such as iron, zinc and retinol, fall in disease. Iron saturation of transferrin in disease is initially maintained but in chronic disease falls progressively (Bothwell *et al.*, 1979; Weinberg, 1992), so that the iron scavenging capacity of transferrin in plasma remains high, even though the absolute concentration falls and delivery of iron to tissues is compromised. This may contribute to the 'anaemia of chronic disease' (e.g. in rheumatoid arthritis and cancers) and also to the low haemoglobin levels of children in much of the Third World.

Serum ferritin is markedly increased in some infections such as malaria (Phillips *et al.*, 1986; Adelekan and Thurnham, 1990). Much of this increase reflects augmented synthesis of apoferritin within cells where it facilitates iron binding. It has been reported that ferritin iron can stimulate lipid peroxidation *in vitro* in direct proportion to the degree of iron loading (Gutteridge *et al.*, 1983). Ascorbate can itself reduce the iron in ferritin to release Fe(II). However, caeruloplasmin, which has the opposite effect of ascorbate on iron by promoting its oxidation to Fe(III), can block ferritin-dependent peroxidation *in vitro* (Samokyszyn *et al.*, 1989) and may function in a similar manner *in vivo*. Serum caeruloplasmin increases during trauma and promotes the conversion of ferrous to ferric iron, the latter being the form of iron which is taken up by transferrin and ferritin (Gutteridge, 1986). These processes will scavenge and remove any ferrous iron from the circulation and promote iron retention by macrophages of liver and spleen.

Other proteins with a role in this iron scavenging process include lactoferrin in exocrine fluids, and plasma haptoglobin and haemopexin, which bind haemoglobin and haem respectively, thus minimizing the risk of tissue damage originating from the iron- or haem-catalysed formation of free radicals.

Lactoferrin found in the granules of neutrophils is released at the site of inflammation, where it is presumably able to bind iron. Much of the lactoferrin may be scavenged by macrophages but it is also taken up from the plasma by liver cells (Young and Aisen, 1988). Proteinase inhibitors also assist in protecting tissue integrity by inhibiting lysosomal enzymes, released upon tissue damage.

## 6.8 CONSEQUENCES OF CHRONIC DISEASE ON IRON METABOLISM

If withdrawal of iron from the circulation as part of the acute phase response is an important component of the body's defence against disease, then:

1. The administration of iron in disease, or where the risk of disease is high, is contraindicated.
2. The anaemia associated with chronic disease may be a protective response to protect tissue integrity.
3. The consequence in a child exposed to continuous infection may be that iron is permanently trapped in the liver and other internal organs and results in a condition akin to iron overload.

### 6.8.1 Iron supplementation in areas where there is a high prevalence of infectious diseases and/or malaria

Keusch and Farthing (1986) reviewed reports of adverse reactions from supplemental iron administered in Third World circumstances and concluded that mild iron deficiency anaemia might give some protection against infection. Others have also drawn attention to the conflicting results following iron therapy, suggesting that parenteral rather than orally administered iron was associated with the greatest risk of increased infection in developing countries (Tomkins and Watson, 1989). The implication was that the cleanliness of the environment might influence infection rates from parenteral iron. However, increased morbidity has followed both oral (Murray et al., 1980) and parenteral iron in developing countries (McFarlane et al., 1970; Barry and Reeve, 1976) and there is strong circumstantial evidence that mortality risk may increase in severely malnourished children given iron.

Two studies of iron supplementation in malaria endemic areas support the conclusion that iron can increase morbidity. Oppenheimer et al. (1986) administered iron or placebo intramuscularly to infants under 6 months in Papua New Guinea. At 6 and 12 months, there was increased splenomegaly and malaria parasite slide positivity. Likewise, Smith et al. (1989), who administered oral iron or placebo

to Gambian infants (6 months to 5 years), found a significantly increased risk of fever associated with severe malaria parasitaemia and splenomegaly in children at 2 years in the iron group. However, in a further study in Papua New Guinea in which iron was given to school age children, there were no adverse effects (Harvey et al., 1989).

It is generally considered that iron promotes malaria replication in these studies but the evidence supporting a direct effect of iron on malaria growth is weak (Peto and Thompson, 1986). In contrast, the pro-oxidant theory of iron toxicity fits the evidence much better. The inflammatory response following malaria infection, plus the haemolysis, releases tissue iron which may exacerbate free radical damage in the tissues. Thus the higher the iron status, the greater the potential damage or disease severity. Oppenheimer (1989b) reported that infants with the highest haemoglobins at birth (thus a higher total body iron content) were significantly more likely to have malaria at follow-up and to be admitted to hospital with malaria. Thus a high iron status in a malaria endemic area appears to be a disadvantage.

The different susceptibility to the damaging effect of iron in relation to age, may be explained by the pro-oxidant hypothesis. To protect themselves against malaria, younger children depend to a much greater extent on cellular immune function than on acquired immunity (Reibnegger et al., 1987). They depend on the generation of ROS by macrophages and polymorphonucleocytes to destroy malaria parasites. Such inflammatory activity is invariably accompanied by tissue damage which may be exacerbated by normal or raised tissue iron status, potentially increasing the severity and duration of disease. This may account for why younger infants were vulnerable to the exacerbating effects of iron supplementation on malaria (Oppenheimer et al., 1986) whereas there were no adverse effects in older children (Harvey et al., 1989). The conflicting data on the role of iron in infection and immunity are discussed further in Chapter 10.

### 6.8.2 The 'anaemia of chronic disease' and of cancer

The haematological picture of anaemia in cancer and other chronic diseases is similar. The anaemia is initially normochromic and normocytic (though may become increasingly microcytic and hypochromic if prolonged) with an inappropriately low reticulocyte count, and is usually mild (Hb > 90 g/l). Serum iron and transferrin saturation are low and morphological examination of the bone marrow reveals erythroid precursors with a normal appearance and adequate or increased stainable macrophage iron stores (Cartwright, 1966; Miller *et al.*, 1990).

It had been assumed that depressed plasma iron levels in infection were a defensive mechanism by the host to 'starve' invading microorganisms of iron. Weinberg (1992) holds similar views about cancer cells, suggesting that the body withholds iron from the circulation to suppress their active growth. However, low circulating iron in infectious disease is probably only a temporary protection, for many microorganisms can produce specific high-affinity iron binding compounds (siderophores) that chelate Fe(III) into an assimilatable form, or have other means of competing effectively for the host's iron (Barclay, 1985). Similarly, once a cancer is established, mild impairment of iron supply is probably no obstacle to its growth.

A more likely reason for the anaemia is that it confers protection to the host from host tissue damage by the developing cancer or disease, as in both conditions there is stimulation of the acute phase response. Miller *et al.* (1990) have suggested that the depressed erythropoietin levels in cancer patients are inappropriately low by comparison with those of patients with uncomplicated iron deficiency anaemia, and that this may be the cause of anaemia in cancer. However, low erythropoietin levels may also be a feature of the acute phase response. If this is the case then it is probably in response to tumour necrosis factor (TNF) which in rats accelerates red blood cell clearance and reduces iron incorporation into red cells by 40% (Moldawar *et al.*, 1989) and not Interleukin 1 which, if anything, stimulates haemopoietic activity (Mochizuki *et al.*, 1987).

The potential damaging effects of iron therapy in inflammatory diseases are evident in rheumatoid arthritis. Blake and colleagues (Blake *et al.*, 1985) reported that of 11 patients who received an infusion of iron dextran for anaemia, two had anaphylactic reactions and the remaining nine had an exacerbation of synovitis. Small joints of the hands were maximally affected, though larger joints when previously inflamed also worsened; uninflamed joints were rarely affected. It is also relevant to note that iron absorption is suppressed in patients with active rheumatoid arthritis with and without iron deficiency (Weber *et al.*, 1988).

### 6.8.3  The iron trap hypothesis

Golden and colleagues (Golden and Golden, 1985; Golden and Ramdath, 1987) have suggested that free radical damage may be partly responsible for the clinical picture of kwashiokor. Such children are usually infected with various microorganisms, are anaemic and have high ferritin values; all the children who died had plasma ferritin greater than 250 μg/l. Earlier studies in India had also demonstrated that children with kwashiokor had high plasma ferritin values (Srikantia, 1958). Waterlow (1947) has shown that children dying with malnutrition had high levels of hepatic iron at $27.6 \pm 4.7$ (SD) μmol/g fat-free dry weight (n = 14) compared with normal values of approximately 3.5–12.5 μmol/g (Golden and Golden, 1985).

It is paradoxical that infants and children suffering continuous bouts of disease and diarrhoea, frequently for several months, and in areas of the world where the predominant iron status is of iron deficiency anaemia, acquired sufficient iron to appear to become overloaded. The effect of the acute phase response on iron metabolism offers an explanation. Infection with pathogenic organisms often precipitates and accompanies malnutrition (Jackson and Golden, 1983). Infection induces a redistribution of iron from the peripheral tissues and the circulation to the liver and spleen. During infection, the acute phase response is stimulated and transferrin in the circulation is reduced, lowering the level of iron carried in the plasma. Children in such conditions usually fail to grow. Raised levels of proteinase inhibitors are reported in rural Thai children (Schelp *et al.*, 1981) and Northrop-Clewes *et al.* (1994) found $\alpha_1$-antichymotrypsin to be almost continuously elevated throughout the first two years of life in rural Gambian infants. The overall effect is a trapping of iron in the liver, low transferrin iron in the circulation, and low haemoglobin but high plasma ferritin, particularly where malaria is a problem (Adelekan and Thurnham, 1990). In a malnourished child, exposed to continuous infection and losing weight, the liver will also acquire iron from tissue degradation due to the efficient iron retaining properties of the body. As a result of the iron redistribution, such children develop increased iron stores.

### 6.8.4  Plasma ferritin as a predictor of cardiovascular disease

Although the plasma ferritin concentration is usually interpreted principally as a marker of iron storage, raised ferritin levels may also be the consequence of an acute phase reaction in chronic inflammatory diseases. The latter may confound studies of the relationship between iron status and risk of cardiovascular disease. Elevated fibrinogen levels and white blood cell counts are both indicators of a stimulated acute phase response and are associated with an increased risk of heart disease (Yarnell *et al.*, 1991). Likewise, albumin is a negative acute phase protein and low albumin concentrations have been linked with the risk of both heart disease and cancer in several prospective studies (Phillips *et al.*, 1989; Kuller *et al.*, 1991; Gillum and Makuc, 1992). In a 10-year prospective study on British men, Phillips *et al.* (1989) reported that, even when they excluded persons dying in the first five years of the study, albumin was still the best predictor of mortality from cancer and cardiovascular disease in the remaining five years. Such results suggest that occult cancer and cardiovascular disease may stimulate an acute phase response. Northern Ireland has one of the highest heart disease rates in the world, and in a report on iron in the adult population Strain *et al.* (1990) suggested that the disproprotionately elevated serum ferritin relative to transferrin saturation was indicative of chronic inflammation in men and older women. Therefore the suggestion that high ferritin levels indicate high iron stores, and that the latter are causally related to the risk of heart disease and/or cancer, should be interpreted with caution at present because elevated ferritins may be the consequence of occult disease.

# 7

# IRON AND ANAEMIA

## 7.1 INTRODUCTION

The major part of the body's iron is normally found in circulating red blood cells in the form of haemoglobin, essential for oxygen transport to the tissues. A reduction in the blood haemoglobin concentration, or packed red cell volume/haematocrit, to below that which is normal for a given individual (i.e. not limited by nutritional deficit or disease) is defined as anaemia. Threshold criteria for haemoglobin concentration (Table 7.1) and haematocrit have been defined; in adults lower limits for haemoglobin concentration of 130 g/l for males and 120 g/l for females are frequently used to define anaemia (Cook and Skikne, 1989). However, it should be recognized that these limits are not absolute and refer only to the likelihood (around 95%) of anaemia being present; in some individuals haemoglobin concentrations above the threshold are still lower than those normally maintained, while in others individual haemoglobin concentrations may normally be below the threshold. Any anaemia, whether due to iron deficiency or secondary to other conditions (see below), is necessarily associated with changes in the total amount, or the distribution, of body iron. Iron deficiency is the commonest cause of anaemia but in other forms of anaemia iron stores are usually increased (the iron previously in haemoglobin being diverted to macrophage stores). Indeed, in some 'iron loading' anaemias, notably the more severe thalassaemia disorders (see below), the need for regular blood transfusions and/or increased dietary iron absorption can lead to a marked excess of storage iron which is toxic, resulting in potentially fatal tissue damage.

## 7.2 NORMAL RED CELL PRODUCTION

Circulating red blood cells are derived from nucleated precursor cells (erythroblasts) in the bone marrow. These morphologically recognizable erythroblasts themselves arise from more primitive precursors identified by *in vitro* colony cultures, and ultimately from a common haemopoietic stem cell.

**Table 7.1** Haemoglobin concentrations below which anaemia is likely to be present at sea level. (source: WHO, 1972.)

| Group | Subgroup | Hb (g/l) |
|-------|----------|----------|
| Children | 6 months to 6 years | 110 |
| | 6–14 years | 120 |
| Adults | Males | 130 |
| | Females (non-pregnant) | 120 |
| | Females (pregnant) | 110 |

The earliest of the precursors committed to erythroid differentiation is the erythroid burst forming unit (BFU-E) which in turn gives rise to erythroid colony forming units (CFU-E).

Control of red cell production depends on an interaction between the progenitor cells and other accessory cells in the microenvironment of the bone marrow, together with a variety of cytokines and growth factors that enhance or suppress erythroid proliferation and differentiation. The hormone erythropoietin, produced primarily in the kidney in response to hypoxia, is the principal stimulus to proliferation and differentiation of committed red cell precursors, where it acts to prevent apoptosis (cell death) (Krantz, 1991). Receptors for erythropoietin are found on late BFU-E but are most numerous on CFU-E. They decline in the morphologically recognizable erythroblasts where there are sequential peaks in expression of receptors for transferrin and in haemoglobin synthesis, the latter continuing until the final stages of red cell maturation.

## 7.3 HAEMOGLOBIN

Each haemoglobin molecule is made up of four protein chains (globin chains), each of which binds a haem molecule. Haem consists of a porphyrin ring with a central ferrous iron capable of reversibly binding oxygen.

**Table 7.2.** Haemoglobin tetramers at various stages of development (the sequential production of globin chains follows the order in which the corresponding genes are arranged (5' to 3') on chromosomes 16 and 11)

| Stage of development* | Haemoglobin | Globin chain constitution | | Haemoglobin tetramer |
| | | From $\alpha$-gene cluster (chromosome 16) | From $\beta$-gene cluster (chromosome 11) | |
| --- | --- | --- | --- | --- |
| Embryonic | Hb Gower I | $\zeta$ | $\epsilon$ | $\zeta_2\epsilon_2$ |
| | Hb Portland | $\zeta$ | $\gamma$ | $\zeta_2\gamma_2$ |
| | Hb Gower II | $\alpha$ | $\epsilon$ | $\alpha_2\epsilon_2$ |
| Fetal | HbF | $\alpha$ | $\gamma$ | $\alpha_2\gamma_2$ |
| Adult | HbA$_2$ 2.5% | $\alpha$ | $\delta$ | $\alpha_2\delta_2$ |
| | HbA 97% | $\alpha$ | $\beta$ | $\alpha_2\beta_2$ |

* The switch from embryonic to fetal haemoglobin occurs at 6–10 weeks gestation, and from fetal to adult haemoglobin at around the time of birth

### 7.3.1 Globin

In each haemoglobin molecule, two of the globin chains are derived from a cluster of related genes on chromosome 16 ($\alpha$-globin gene cluster) and two from a cluster of related genes on chromosome 11 ($\beta$-globin gene cluster). During development from embryonic to fetal, and from fetal to extra-uterine life, there is an ordered switch in activation of genes from each cluster. The resulting haemoglobin tetramers formed at the various stages of development are shown in Table 7.2. In fetal life, haemoglobin F ($\alpha_2\gamma_2$) dominates, its high oxygen affinity aiding oxygen uptake across the placenta at a relatively low oxygen tension. At around the time of birth there is a switch from $\gamma$ to $\beta$ globin chain production to form adult haemoglobin, HbA ($\alpha_2\beta_2$), which has a lower oxygen affinity and is more efficient in delivering oxygen to the tissues in extra-uterine life. A small amount of HbF (less than 1%) continues to be present in adult life, together with another minor adult component, HbA$_2$ ($\alpha_2\delta_2$).

### 7.3.2 Haem

The initial stages of porphyrin synthesis occur within the mitochondria of the erythroblasts. After further modifications by cytosolic enzymes, protoporphyrin re-enters the mitochondria for the incorporation of iron.

### 7.3.3 Iron

Iron is supplied bound to plasma transferrin and is taken up by the erythroblasts by a specific receptor-mediated endocytosis (Chapter 4). The number of transferrin receptors is increased in iron deficiency, stability of the messenger RNA for the transferrin receptor protein being enhanced by a reduced intracellular iron content. The transferrin receptors have a much greater affinity for diferric (fully saturated) transferrin than for monoferric transferrin (Huebers et al., 1985). Since the loading of transferrin iron binding sites with iron appears to be a random process (Huebers et al., 1984), the amount of diferric transferrin is dependent upon the overall saturation with iron of the plasma total iron binding capacity (TIBC): measurement of the percentage saturation of the TIBC is thus a better overall guide to the iron supply to the erythroblasts than the serum iron concentration, a saturation of less than 16% being unable to support normal amounts of erythropoiesis (section 5.2.4). Details of the mechanisms involved in the release of iron from transferrin, and the incorporation of iron into protoporphyrin, the final stage of haem synthesis, are described in Chapter 4. Haem is then combined with globin chains to form haemoglobin.

## 7.4 PATHOPHYSIOLOGY OF ANAEMIA

Anaemia may result from defects at any stage of red cell and haemoglobin production or, less commonly, when an increased rate of red cell destruction (haemolysis) exceeds the capacity of the bone marrow to mount a compensatory increase in production (Table 7.3). Changes in the relationships between red cell and plasma volumes may also result in a reduced haemoglobin concentration: such changes occur physiologically in pregnancy where red cell volume is increased less markedly than plasma volume, but also contribute to the 'sports anaemia' seen in endurance athletes (Chapter 9).

### 7.4.1 Abnormal red cell destruction (haemolytic anaemias)

In the haemolytic anaemias, iron released from red cells at the end of their shortened life span is usually

**Table 7.3**.   Pathophysiology of anaemia

**(a) Defective red cell production**

| Disorder | Nature of defect | Erythropoiesis |
| --- | --- | --- |
| (1) Defects in proliferation and/or maturation of red cell precursors | | |
| Aplastic anaemia | Stem cell defect/immune mediated | Hypoplastic |
| Red cell aplasia | Stem cell defect/immune mediated | Hypoplastic |
| Anaemia of chronic renal failure | Erythropoietin deficiency | Hypoplastic |
| Anaemia of chronic disorders | Erythropoietin deficiency/ inhibitory cytokines/ impaired iron supply | Hypoplastic |
| Myelodysplasia | Stem cell defect | Ineffective |
| Megaloblastic anaemias | Impaired DNA synthesis due to vitamin $B_{12}$ or folate deficiencies | Ineffective |
| Congenital dyserythropoietic anaemias | Stem cell defect | Ineffective* |
| (2) Defects in haemoglobin synthesis | | |
| Iron deficiency | Impaired haem synthesis | Hypoplastic (plus ineffective if severe) |
| Thalassaemia disorders | Impaired globin synthesis | Ineffective* |
| Sideroblastic anaemias | Impaired haem (porphyrin) synthesis | Ineffective* |
| Lead poisoning | Impaired haem synthesis | Hypoplastic |

**(b) Increased red cell destruction**

| Type of anaemia | Nature of defect |
| --- | --- |
| Inherited | Haemoglobinopathies (e.g. sickle cell disease; unstable haemoglobin variants; HbH disease) Red cell membrane defects (e.g. hereditary spherocytosis) Defects of red cell metabolism (e.g. glucose-6-phosphate dehydrogenase deficiency) |
| Acquired | Immune mediated (e.g. autoimmune) Non-immune (e.g. infection, drug-induced etc.) |

* Ineffective erythropoiesis may be accompanied by marked erythroid hyperplasia and risk of excessive iron absorption.

effectively recycled by phagocytic macrophages for new red cell production. There is erythroid hyperplasia of the bone marrow but delivery of mature red cells to the circulating blood is effective. Iron overload may develop if the haemolytic anaemia is severe enough to warrant regular blood transfusions. However, there is little risk of iron loading from excessive iron absorption, even in the more chronic haemolytic disorders, though in rare cases coincidental inheritance of a single gene for hereditary haemochromatosis may combine with the haematological defect to increase this risk (Mohler and Wheby, 1986). Indeed, where there is substantial intravascular haemolysis, free haemoglobin may be lost by glomerular filtration into the urine and iron deficiency may result.

### 7.4.2   Defects in red cell production

Defects of red cell production resulting from abnormal proliferation or maturation of red cell precursors include the megaloblastic anaemias caused by vitamin $B_{12}$ or folate deficiency, as well as a variety of anaemias resulting from intrinsic defects of red cell precursors. The anaemia of iron deficiency, though it is clearly associated with impaired haemoglobin formation within individual erythroblasts (see below), is predominantly a hypoplastic anaemia, with a lesser increase in the number of erythroblasts in the bone marrow than would be expected in a subject with no limitation of iron supply but the same degree of anaemia and erythropoietin stimulation. With increasing severity of the iron deficiency, the anaemia becomes associated with more florid evidence of impaired maturation with increasingly misshapen red cells. In the very common 'anaemia of chronic disorders' associated with many inflammatory and infective disorders, an inappropriately small rise in serum erythropoietin concentrations in response to the anaemia, impaired responsiveness of red cell precursors to erythropoietin, a mildly shortened red cell life span and a reduced iron supply are all contributory mechanisms (Means and Krantz, 1992).

### 7.4.3   Defects in haemoglobin synthesis

Where anaemia is accompanied by impaired haemoglobin synthesis, the resulting red cells tend to be small and poorly haemoglobinized, i.e. microcytic and hypochromic. Defects in haemoglobin production encompass impaired haem synthesis due to inadequate iron supply or defects in porphyrin synthesis, and defective globin chain synthesis. By far the commonest defect is impaired haem synthesis resulting from iron deficiency where the reduced iron supply to the mitochondria leads to an accumulation of free protoporphyrin in the red cells. However, as discussed above, impaired iron supply is also a feature of the common 'anaemia of chronic disorders', where it results from a malutilization, rather than an overall deficit of body iron; increased ferritin synthesis within stimulated macrophages and decreased iron release to circulating transferrin are probably central to the retention of iron within macrophage stores (section 10.5). Defects in porphyrin synthesis are characteristic of the much less common sideroblastic anaemias, where the lack of protoporphyrin results in the accumulation of iron in the mitochondria of developing erythroblasts, impaired mitochondrial function and early cell death within the bone marrow, i.e. ineffective erythropoiesis.

Ineffective erythropoiesis and/or shortened red cell survival are the predominant features of the anaemias of variable severity which result from inherited defects in the production of globin from one or more of the globin genes. These haemoglobinopathies, the thalassaemia disorders, are among the commonest of the single gene disorders (Weatherall et al., 1989) and in some populations well over 50% of people are carriers of at least one thalassaemia gene. In heterozygous β-thalassaemia (where one of the two β-globin genes is defective) there is a compensatory increase in δ-chain production and a measurable increase in the minor adult haemoglobin, $HbA_2$. Mild forms of α-thalassaemia (where one or two of the four α-globin genes is defective) are less easy to confirm without

gene mapping to identify the responsible gene deletions, but the unpaired globin which continues to be produced from the β-globin gene cluster may form identifiable inclusion bodies of $\gamma_2$ tetramers (Hb Barts) in fetal life, or $\beta_4$ tetramers (HbH) in adult life. In these mild or heterozygous forms of thalassaemia the anaemia is not symptomatic, and though it may be associated with severe hypochromia and microcytosis of red cells it requires no treatment. However, in individuals who have inherited thalassaemic defects of both β-globin genes, the intracellular precipitation of excess α chains causes damage to the developing erythroblasts and red cells, resulting in ineffective red cell production, shortened red cell survival and severe anaemia. Where three of the four α-globin genes are defective a haemolytic anaemia, HbH disease, of variable severity results, sometimes requiring regular blood transfusions. A total absence of functional α-globin genes leads to a failure to produce HbF in fetal life, a condition giving rise to Hb Barts $(\gamma_4)$ hydrops fetalis, and death *in utero* or soon after birth.

The anaemias associated with impaired haemoglobin production are all accompanied by increased erythropoietin concentrations in the plasma, though the erythropoietin response to anaemias is blunted in the 'anaemia of chronic disorders'. An impaired plasma iron supply, as in iron deficiency anaemia, appears to limit the proliferative response of the erythroblasts. However, in some patients with sideroblastic anaemias, and in severe β-thalassaemia disorders, the ineffective erythropoiesis may be associated with gross expansion of the erythroid bone marrow, sometimes resulting in distortion of the bones, and a huge increase in plasma iron turnover. The ineffective erythropoiesis is also associated with a marked increase in iron absorption from the gut, despite there being no overall deficit in body iron. This alone may eventually result in accumulation of excess body iron, though where the anaemia is severe enough to require blood transfusions the transfused haemoglobin is a major source of iron loading (section 7.7). Iron overload in the α-thalassaemia disorder, HbH disease, is not usually seen unless regular blood transfusions are required. These 'iron loading' anaemias, though less common than iron deficiency anaemias, are important to differentiate, since in both types the red cells are hypochromic and microcytic and mistaken iron therapy may contribute to ultimately fatal iron overload. Population-based approaches to reducing iron deficiency may also be complicated by the high frequency of thalassaemia genes in areas of past or current endemic malaria (*Plasmodium falciparum*) infection (e.g. Southern Europe, the Middle East and South-East Asia), areas where iron deficiency is also common (section 7.6.4).

## 7.5   PREVALENCE OF ANAEMIA

### 7.5.1   Global

A meta-analysis of 523 studies of the prevalence of anaemia (De Maeyer and Adiels-Tegman, 1985) used WHO threshold criteria to derive the mean percentage prevalence of anaemia by continent, by region (developed or developing), and by age and sex. An estimated 30% of the world's population (1.3 billion people) was found to be anaemic using this data, with the vast majority being in developing regions, particularly South Asia and Africa, where the prevalence of anaemia was 36% compared with 8% in developed regions. Prevalence was particularly high in pre-school children (51% in less developed and 10% in more developed regions) and in adult females (50% in less developed and 13% in more developed regions). By making the assumption that anaemia in adult males is likely to be due to causes other than iron deficiency (and this will not be completely true), a crude correction to obtain the total number of people affected by anaemia due to iron deficiency may be made: this very conservative estimate is approximately 500 million and takes no account of the additional, greater numbers of people who will have borderline iron status with depleted iron stores (Cook *et al.*, 1976).

### 7.5.2   Trends in developed regions

There is a low prevalence of iron deficiency anaemia in all age groups in developed regions, where anaemia is increasingly likely to be related to causes other than iron deficiency, even in pre-school children (Yip *et al.*, 1987a ; Tershakovec and Weller, 1991). Nevertheless, a number of recent studies in the UK have shown a high prevalence of iron deficiency among pre-school children. Those of Asian background were at even greater risk, most likely as a result of reduced iron availability in predominantly vegetarian diets (Chapters 15 and 19; Wharton, 1992b). It should be noted, however, that other causes of microcytic anaemia, including α- and β-thalassaemia traits, may also be present in such children (Earley *et al.*, 1990; Stevens, 1991) and these inherited conditions need to be excluded if the frequency of iron deficiency anaemia is not to be overestimated.

## 7.6   IRON DEFICIENCY ANAEMIA

### 7.6.1   Pathophysiology and diagnosis

Iron deficiency anaemia is the final and most obvious stage of a progressive 'negative iron balance', in which iron uptake from the gastrointestinal tract is insufficient to meet the need for an expanding volume of red blood cells (e.g. in growing infants or in pregnancy), or to keep pace with obligatory iron losses (predominantly through loss of gut enterocytes or desquamation of skin cells) or with pathological iron losses (e.g. haemorrhage into the gastrointestinal tract).

Under such circumstances of negative iron balance, the serum ferritin concentration falls as any storage iron is mobilized to circulating plasma transferrin to be delivered to iron-requiring tissues, where it is incorporated into functional iron compounds including red cell haemoglobin and essential tissue enzymes. As storage iron is exhausted, the saturation of the plasma transferrin with iron falls below the level (16% saturation) which is necessary to support normal amounts of erythropoiesis. At this point iron deficient erythropoiesis begins to develop, even though the total amount of circulating haemoglobin may still be above the threshold used to define anaemia. Such iron deficient erythropoiesis is associated with an increased concentration of serum transferrin receptors, derived from the increased expression on the cell membranes of iron deficient developing erythroblasts (Skikne *et al.*, 1990), as well as an accumulation of free erythrocyte protoporphyrin, and an increase in red cell heterogeneity as assessed by the red cell size distribution width, RDW (Bessman *et al.*, 1983). Finally, frank anaemia develops, associated with progressively more severe microcytosis (reduced mean cell volume, MCV) and hypochromia of the red cells on the blood film (Cook, 1982).

### 7.6.2   Causes

From the preceding section it will be clear that iron deficiency may arise as a result of increased iron losses, increased tissue iron requirements, reduced dietary iron content or bioavailability, or, more rarely, malabsorption of iron due to intrinsic gastrointestinal disease. Increased iron losses may be physiological (e.g. menstrual losses, iron losses to the fetus during pregnancy, iron lost in the milk during lactation), or pathological. Pathological causes of iron loss are dominated by gastrointestinal haemorrhage, though urinary losses (e.g. with chronic intravascular haemolysis in paroxysmal nocturnal haemoglobinuria) may occasionally be important. Gastrointestinal losses, unlike those in the urine, are often hidden. On a world scale, hookworm infestations are the commonest cause of gastrointestinal blood loss and iron deficiency anaemia. However, oesophagitis, gastritis, peptic ulcer disease, varices, neoplasia, angiodysplasia and inflammatory bowel disease may all give rise to iron deficiency anaemia (Beveridge *et al.*, 1965), sometimes as the first indication of a serious underlying disease. In infancy, occult gastrointestinal blood loss is commonly associated with feeding with unprocessed cow's milk (Fomon *et al.*, 1981), adding to the effects of the low iron content and poor availability of the iron in cow's milk.

A physiological increase in iron requirements is seen during periods of rapid growth in infants, children and adolescents, and iron deficiency is more common at these times. Pathological increases in iron requirements are seen in anaemias associated with erythroid hyperplasia, where although there may be an overall increase in body iron, and an increase in transferrin saturation with iron, the blood supply to the expanded marrow may be insufficient to satisfy the enormous demand for iron (Pootrakul *et al.*, 1984). With erythropoietin therapy in the anaemia associated with chronic renal failure, erythroid expansion may outstrip the mobilization of iron stores and give rise to iron deficient erythropoiesis (Eschbach *et al.*, 1987; Macdougall *et al.*, 1992).

The majority of the world's population eats a predominantly vegetarian diet with a relatively low bioavailability of iron compared with meat-containing diets. While such diets are unlikely to give rise to iron deficiency in adult males, any increased physiological blood losses or iron requirements are less likely to be met from such diets. The combination of physiological factors and poor dietary iron availability is probably the major contributor to the development of iron deficiency anaemia in the developing regions of the world.

### 7.6.3   Treatment

Treatment of iron deficiency anaemia in an individual patient is usually readily achieved with an oral preparation of a simple iron salt, preferably ferrous sulphate (200 mg thrice daily provides a total of 180 mg per day of elemental iron). Adverse gastrointestinal effects occur in a minority of patients: these are related to the amount of available iron in the preparation and can be ameliorated by reducing the dose or advising that the iron be taken with meals. High dose carbonyl iron (a form of elemental iron) was found in animal and human studies to have a greater safety margin than simple iron salts but with a similar profile of gastrointestinal side effects (Gordeuk *et al.*, 1987). Slow release preparations of iron are an unnecessary and expensive way of reducing the amount of available iron and thus adverse effects. There has been recent interest in the demonstration that, with repeated doses of iron supplements, the gastrointestinal mucosa

appears to show adaptation, with reduction in the proportion of iron retained. As a result, it has been suggested that intermittent use of iron supplements may be as effective as continuous use in correcting anaemia due to iron deficiency, with fewer adverse effects (Shultink *et al.*, 1995). However, others have shown no significant difference in iron absorption from a single oral iron dose with daily versus weekly iron supplements, with or without food, and further studies are required before embarking on large-scale clinical trials.

The use of parenteral iron is rarely needed and should be restricted to patients with proven iron deficiency who cannot tolerate oral preparations or in whom the rate of continuing blood loss is so great that oral iron cannot keep pace. Of the two preparations which have been used in the UK, iron dextran ('Imferon') and iron sorbitol ('Jectofer'), only the latter remains readily available because of production problems with the former. Iron sorbitol can be given only by intramuscular injection, whereas iron dextran has the advantage that it can also be given intravenously, sometimes as a total dose infusion, replenishing the calculated iron deficit with a single infusion over a few hours. Both preparations can have unpleasant adverse effects. Iron dextran may give rise to fever, flushing, nausea, vomiting, joint pains and lymphadenopathy, and severe anaphylactic responses have been seen occasionally. Iron sorbitol contains less tightly bound iron, and release of free iron is responsible for the adverse effects of metallic taste in the mouth, nausea, vomiting and malaise. With the use of recombinant erythropoietin therapy in the anaemia associated with chronic renal failure, there is a need for a readily available form of parenteral iron that can be given by intravenous injection, since oral iron therapy cannot always maintain an adequate iron supply in such patients, particularly during the initial phases of treatment and haemoglobin regeneration.

### 7.6.4   Prevention of iron deficiency anaemia

Interventions aimed at preventing iron deficiency and its attendant liabilities range from the highly directed screening of groups known to be at high risk and subsequent supplementation of iron deficient individuals, to more widespread supplementation and general iron fortification of staple foods. Where the prevalence of iron deficiency is relatively low, the former approach is likely to be favoured, while where iron deficiency is common, more universal measures may be required. However, even where the development of iron deficiency is known to be common (e.g. in the later stages of pregnancy), there may be little agreement about whether this confers any disadvantage and whether universal iron supplementation is appropriate or necessary (Chapter 17).

Strategies to prevent iron deficiency within a population (e.g. through iron fortification of foods) need to consider the relative frequency and severity of the conditions that are likely to benefit from, and those that may be harmed by, the intervention. For individuals with iron loading anaemias or hereditary haemochromatosis, there are potential hazards associated with increasing dietary iron intake among the whole population. It is much less clear whether there are any risks in the much more common asymptomatic heterozygous carrier states. Modest increases in serum ferritin have been reported in some subjects with heterozygous β-thalassaemia, but the disorder of erythropoiesis is mild and thus the risk of serious iron load is thought to be slight (Pippard and Wainscoat, 1987). There have been concerns in recent years about the potential risks to normal individuals of increasing iron intake. However, the normal downregulation of iron absorption as iron stores increase probably acts as an effective protection against any significant increase in uptake in normal individuals (Cook, 1990). Furthermore, the suggestion that an increase in body iron within the accepted normal limits is related to the development of ischaemic heart disease or colorectal cancer (reviewed in Chapters 12 and 13) is currently unconvincing.

## 7.7   IRON LOADING ANAEMIAS

In chronic anaemias that are refractory to simple treatment with nutritional supplements, there is a risk of progressive iron loading. There is no physiological mechanism for excretion of excess iron, and any iron administered to such patients – e.g. in the form of blood transfusions (one unit contains 200 mg of iron) or taken up by excessive iron absorption can lead to ultimately fatal storage iron accumulation. As with hereditary haemochromatosis (Chapter 8), the main target organs for iron induced tissue damage are the heart, endocrine glands and liver.

### 7.7.1   Diagnosis and prediction of risk

In patients requiring regular blood transfusions for the treatment of severe anaemias, whether due to ineffective erythropoiesis or to red cell aplasia or other forms of bone marrow failure, the risk of progressive iron accumulation is predictable. In patients with less severe anaemias associated with ineffective erythropoiesis, particularly in the thalassaemia disorders and the rarer sideroblastic

anaemias, the risk of iron overload from excessive iron absorption may be concealed until patients present with established organ damage. In these patients the degree of erythroid hyperplasia is currently the best long-term guide to the risk of iron loading (Pootrakul et al., 1988), which may occur even in patients where the anaemia is asymptomatic (Peto et al., 1983). Precise measures of erythropoietic activity, often including complex ferrokinetic assessments, have been difficult on a routine basis; it may be that the availability of measures of serum transferrin receptor concentrations (which reflect the overall size of the erythron as well as the adequacy of its iron supply) and of reticulocytes (reflecting the effectiveness of delivery of young red cells to the circulation) will allow a more readily available assessment of ineffective erythropoiesis (Cazzola and Beguin, 1992). As in patients with hereditary haemochromatosis, a persistent elevation in the percentage saturation of the plasma transferrin indicates a risk of iron deposition within parenchymal tissues. The increase in transferrin saturation precedes the increase in iron stores which may be assessed by measurement of serum ferritin followed by liver biopsy where appropriate (Chapters 5 and 8).

### 7.7.2   Treatment

The demonstration that further tissue damage could be prevented, and life expectancy enhanced, by removing the excess iron through regular phlebotomy in patients with hereditary haemochromatosis stimulated a search for effective alternatives for the iron loading anaemias. The iron chelating agent, desferrioxamine, was introduced in the early 1960s but only in the late 1970s was it realized that continuous infusions of the drug were required to keep pace with the rate of iron loading from blood transfusions and to make inroads on already established stores (Propper et al., 1976). Twelve-hour, overnight subcutaneous infusions used five or six times a week are now the standard therapy (Pippard, 1989). In the absence of iron chelation treatment, patients with transfusion dependent thalassaemia fail to go through puberty, develop cardiac complications in the second decade, and are usually dead by the end of the second decade of life. It is now clear that, with intensive desferrioxamine iron chelation therapy, cardiac problems are reduced and life expectancy enhanced (Brittenham et al., 1994; Olivieri et al., 1994). Adverse effects, including visual and auditory disturbances and growth retardation, have occurred in some patients, particularly those with relatively small amounts of iron overload given large doses of desferrioxamine (Porter and Huehns, 1989).

Desferrioxamine is expensive and it is inconvenient to administer, particularly in the developing countries where thalassaemia disorders are common. There has been an extensive search for an alternative, orally active iron chelating agent. A range of compounds, including polyanionic arylhydrazones and 3-hydroxypyridin-4-ones, have been investigated (Hershko, 1994). Of these, the hydroxypyridinones have shown the most promise and one member of the group, deferiprone, has undergone limited clinical studies: these have confirmed that it is an effective iron chelator but that its therapeutic margin is narrow, with several cases of neutropenia being described (Brittenham, 1992). Further work in this area continues and needs the continued backing of major pharmaceutical companies.

# 8

# IRON OVERLOAD AND TOXICITY

## 8.1 INTRODUCTION

Since humans are unable to excrete significant quantities of iron, it is essential that the uptake of iron is carefully controlled by the intestinal mucosa. In certain pathological conditions iron overload may result from an increased absorption of dietary iron, by parenteral administration of iron, or both. The magnitude, rate and distribution of iron accumulation will influence the onset and severity of complications and differ for the various pathological conditions, a number of which have an inherited genetic basis.

## 8.2 CHRONIC IRON OVERLOAD

### 8.2.1 Genetic haemochromatosis

In genetic haemochromatosis, an inborn error of iron metabolism leads to an inappropriate increase in the absorption of iron from the diet for the levels of iron stores present. It results from the homozygous state for an abnormal gene closely linked to the HLA-A locus on chromosome 6. Iron is progressively stored in all tissues over a period of many years, particularly in the liver and pancreas in which the iron concentration may reach 50–100 × normal. A liver biopsy from a patient with symptomatic untreated genetic haemochromatosis will reveal heavy iron loading, frequently associated with fibrosis on light microscopy, and on electron microscopy will show the characteristic paracrystalline arrays of ferritin, eventually precipitating as haemosiderin (Figure 8.1). In addition, the iron content of other organs, particularly the endocrine glands and cardiac muscle, is elevated. The excessive amounts of accumulated iron are related to the clinical manifestations of cirrhosis, diabetes mellitus, heart disease and endocrine dysfunction, as well as the characteristic bronze skin pigmentation. Less clearly related to the iron overload is an arthropathy, which has been reported in 25–50% of patients with haemochromatosis. Arthralgia has been recognized as an early symptom, which may occur before the development of cirrhosis or dia-

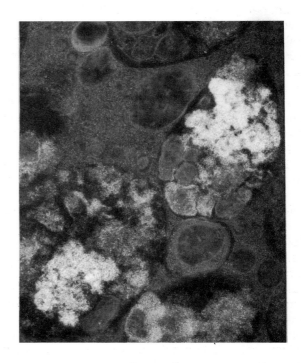

**Figure 8.1** Electron micrograph of iron loaded liver. In the cytosol, ferritin can be seen as spherical dots which form paracrystalline arrays in certain circumstances. Within the two secondary lysosomes, deposits of haemosiderin are observed. (Unstained section: magnification × 60 000.)

betes mellitus; genetic haemochromatosis should be suspected in patients presenting with arthralgia before the age of 40 years. Arthropathy usually does not regress after phlebotomy treatment.

Since the iron overload from excessive iron absorption takes place over many years, symptomatic genetic haemochromatosis does not usually develop until middle or later life. Expression is more common in men than women, presumably because of the protective effects of regular menstrual losses during the child bearing years. Once the disease is diagnosed, regular treatment by phlebotomy can remove the excessive iron stores, after which maintenance phlebotomy every few months will maintain iron stores at a normal level. Effective iron removal increases life expectancy, though in individuals presenting with symptomatic iron overload established tissue damage may not be reversible and there is a greatly increased (20–30%) risk of carcinoma of the liver in those with established cirrhosis (Bomford and Williams, 1976; Niederau *et al.*, 1985). As homozygotes are increasingly identified at a presymptomatic stage, through family or more general screening programmes, the pattern of morbidity and mortality is changing: individuals with iron overload in the absence of tissue damage have a normal life expectancy after phlebotomy treatment.

The cause of the high iron absorption in genetic haemochromatosis has not yet been identified, the defect being ascribed to the intestinal mucosa, the liver or the reticuloendothelial system by various authors. A primary defect within the intestinal mucosa could be caused by a failure to switch from neonatal to adult control of iron absorption (Srai *et al.,* 1984), a decrease in ferritin expression (Pietrangelo *et al.,* 1992), alteration in the brush border membrane and soluble iron binding proteins (Teichmann and Stremmel, 1990; Conrad *et al.,* 1990) or by some as yet unidentified factor.

Genetic haemochromatosis has a gene frequency of approximately one in 20 in the Northern European countries, such that approximately one in every 300 individuals is affected. That makes this the most commonly occurring genetic disease in Caucasians – even more common than cystic fibrosis. It is an inherited autosomal recessive disorder, tightly linked to the HLA-A locus and showing allelic association with HLA-A3 histocompatibility antigen complex on chromosome 6 (Simon *et al.*, 1976). The identification of the gene for genetic haemochromatosis appears to be imminent. Using highly polymorphic markers to chromosome 6p, a recent study of 34 individuals from three large families affected by genetic haemochromatosis showed that the haemochromatosis allele is proximal to D6S89 and distal to D6S105 markers (Jaz Winska *et al.,* 1993).

### 8.2.2   Neonatal haemochromatosis

Neonatal haemochromatosis is an uncommon and generally fatal disorder of infancy. It is characterized by hepatic disease which is generally present at birth, and by stainable iron in the tissues with a distribution resembling that seen in HLA-linked haemochromatosis. First degree relatives of such patients do not appear to be at risk of iron overload (Dalhj *et al.*, 1990) and in a number of cases the extrahepatic siderosis appears to be caused by hepatic injury at a time of life when tissue iron concentrations are high, rather than a primary defect in iron metabolism, e.g. excessive transport of iron from mother to fetus (Hoogstraten *et al.*, 1990).

### 8.2.3   Secondary haemochromatosis

The disorders of erythropoiesis that may be associated with iron overload as a result of increased iron absorption and/or regular blood transfusions are discussed in Chapter 7. Parenteral administration of iron by multiple blood transfusion, although ameliorating anaemia (e.g. thalassaemias) will cause iron loading of a variety of tissues.

### 8.2.4   Thalassaemias

Several forms of thalassaemia exist. The disease is found mainly in a belt extending through southern Europe, North Africa and the Middle East to India, Indonesia and the Far East.

The thalassaemias are characterized by various genetic defects of α or β globin chain synthesis which is caused by changes in regulatory sequences in the gene or by gene deletions that result in decreased or absence of α- or β-chains. These are referred to as α or β thalassaemias, respectively, and are discussed, along with the problems of iron overload, in Chapter 7.

### 8.2.5   Sub-Saharan iron overload ('Bantu-type siderosis')

Gross iron overload resulting from excess absorption of orally ingested iron is very rare but has been described in Bantu people in South Africa. A high intake of iron over a prolonged period occurred in Bantu people who consume large amounts of 'Kaffir beer', an alcoholic beverage fermented in iron pots. The histological pattern of hepatic iron deposition in such patients is different from that of patients with HLA-linked haemochromatosis, i.e. the iron is more prominent in the cells of the mononuclear-phagocyte system (Brink *et al.*, 1976). The pattern of expression in family studies suggests that such iron overload is related to a gene distinct from any HLA-linked gene but that expression is also dependent on the environmental exposure to the high oral iron intake, both factors being required (Gordeuk *et al.,* 1992).

### 8.2.6   Alcohol misuse

Various liver diseases, especially cirrhosis and portal–systemic shunting, may increase the liver iron content. Patients with alcoholic liver disease often show increased stores of iron in the liver which may be caused in part by alcohol increasing the intestinal mucosa permeability to iron, and in part by the high amount of iron present in certain wines and beers. The increased iron absorption may also be related to ineffective erythropoiesis associated with alcohol-related folate and sideroblastic abnormalities (Conrad and Barton, 1980). The extra amount of iron is usually moderate, and less than would be expected in patients of the same age with genetic haemochromatosis. Alcoholic subjects with excessive iron overload are almost always homozygous for the haemochromatosis gene and both iron accumulation and liver damage are accelerated by the alcohol misuse.

## 8.3   CHRONIC TOXICITY OF IRON IN IRON OVERLOAD

Extensive tissue damage often occurs in iron loaded tissues. In early studies, the pathogenesis of this lesion has been described in terms of increased levels of the iron storage proteins, ferritin and especially haemosiderin, and of iron catalysed lipid peroxidation.

Heterogeneity of haemosiderin, the iron storage protein, in genetic haemochromatosis and thalassaemias has been identified with respect to both the iron core and the associated protein shell (Mann *et al.,* 1988; Ward *et al.,* 1992, 1994b). Various mineralization products in addition to that of ferrihydrite have been identified, including amorphous ferric oxide and goethite, although it remains unresolved whether the presence of a certain form infers a greater toxicity to the iron loaded cell.

Initially, iron catalysed lipid peroxidation was an attractive hypothesis. The potential importance of iron in mediating oxidative damage by the production of radicals, via the Haber–Weiss reaction/Fenton chemistry, to a variety of macromolecules has been recognized *in vitro* for more than two decades. However, such a reaction *in vivo* will be dependent upon not only the availability of a low molecular weight 'catalytically active' iron, to participate in such a reaction, but also to overwhelming the capacity of the cell to combat reactive oxygen species by enzymes and antioxidants.

In normal circumstance iron is carefully sequestered into a variety of proteins (transport proteins or storage proteins) to ensure that there is little free iron either in solution or within the tissues. The nature of the free iron 'pool' remains unknown (Fontecave and Pierre, 1991): a variety of ligands have been suggested, including ATP, ADP, pyrophosphates, citrate and specific amino acids, but this remains to be resolved. It is debatable whether this pool increases in iron overload but it is more than likely that the flux through the pool is enhanced. This small iron pool could initiate and potentiate iron-catalysed lipid peroxidation, while the presence of the cytoprotective enzymes and

antioxidants in the cytosol and membranes of the cell may play a vital role in scavenging and chain-stopping such reactions.

The distended lysosomes containing the bulk of the excessive iron in untreated genetic haemochromatosis patients can be observed by electron microscopy in liver biopsies (Stal *et al.,* 1990), or be measured by the increased activity of the lysosomal enzyme N-acetyl-β-glucosaminidase (Seymour and Peters, 1978) in such specimens. On completion of treatment by phlebotomy, the lysosomes revert to their normal size and the activities of lysosomal enzymes are within the normal range of values.

Damage to the mitochondria could be attributed to the ferrous ions initiating and propagating lipid peroxidation of the mitochondrial membrane, thus increasing permeability and allowing iron to accumulate within the mitochondria. This may alter the steric configuration of the mitochondrial membrane which is essential for the proper functioning of the electron transport system and therefore for aerobic respiration. Alternatively the mitochondrial swelling could result in the shunting of electrons away from the electron transport system. Ferric iron is reduced by an enzyme located on the inner mitochondrial membrane, to provide ferrous iron for insertion into protoporphyrin IX. In the presence of excessive iron this same enzyme may reduce the ferric iron to ferrous, which in turn would be rapidly oxidized back to the ferric state in the presence of intramitochondrial oxygen. Thermodynamic considerations of this system would favour the shunting of electrons away from the electron transport system, thereby diminishing oxidative phosphorylation and reducing ATP. There would be uncoupling of mitochondrial oxidative phosphorylation in which oxygen would be utilized without ATP being formed. The end result of this process would be cellular dysfunction and death due to impaired generation of ATP.

Even though subcellular organelles from experimentally iron-loaded rat livers show increased amounts of lipid peroxidation products, as assessed by thiobarbituric acid-reacting material and diene conjugate species (Britton *et al.,* 1990) or depletion of antioxidants (Ward *et al.,* 1991), to what extent such changes reflect actual damage to the cell is unresolved. Altered mitochondrial function (Bacon *et al.,* 1983, 1985) or increased lysosomal fragility in iron overloaded patients (Seymour and Peters, 1978) and in rats experimentally overloaded with iron (O'Connell *et al.,* 1987; Ward *et al.,* 1991) were considered to be a result of lipid peroxidation of the appropriate organelle membranes. As yet, there is no conclusive evidence that iron-catalysed lipid peroxidation is the cause of the cellular damage (Crichton and Ward,

1992) and it is therefore appropriate to view the problem of iron toxicity in a wider perspective with the possible target for iron or iron catalysed radicals being proteins, DNA, mRNA or Kupffer cell activation.

Highly reactive oxygen species, if generated by the iron species present, could also cause damage to DNA, initiating strand breaks and base modifications. In vitro the hydroxyl radical can induce strand breaks in the phosphodiester backbone of nucleic acids with the formation of 8-hydroxyguanine, 5-hydroxymethyuracil and thymine glycol. The modified base, 8-hydroxyguanine, will induce mutations. It is unlikely that the radicals are able to permeate the several nuclear membranes to access the DNA. Therefore it has been proposed that $H_2O_2$, generated in the cytosol during the dismutation of superoxide, could diffuse across the nuclear membranes to react with DNA-bound iron, generating hydroxyl radicals locally to damage DNA (Halliwell and Aruoma, 1991). In vivo studies of hepatic DNA isolated from iron loaded animals have not shown an increase in hydroxylated DNA bases (Ward et al., 1995c).

Studies of the changes in expression of various mRNAs, from liver specific, growth related and stress induced genes, have identified selective activation of pro-$\alpha_2$-collagen mRNA in addition to that of L-ferritin in the livers of iron loaded rats. This was true even though there was an absence of necrosis and nodular regeneration (Pietrangelo et al., 1990). Other in vivo studies have not confirmed these results (Roberts et al., 1993). Such results indicate that iron can specifically target genes in the liver. Previous studies have also shown that iron-complex induced lipid peroxidation may activate collagen expression in cultured fibroblasts.

More recently studies have been directed towards hepatic cells which undergo activation during experimental iron overload and cause the deposition of extracellular matrix proteins including collagen (Britton et al., 1993). It is unknown what effect this might have in human iron overload but it may be one of the factors controlling the deposition of collagen.

## 8.4  ACUTE IRON POISONING

Accidental iron poisoning occurs predominantly in children who have consumed large numbers (10–50) of iron tablets, usually in the form of ferrous sulphate, over a period of a few hours or less. The pathophysiology of iron intoxication is complex with a variety of processes involved. The ingested iron enters the stomach, where there is a low pH. The ferrous sulphate will be mainly in a soluble form, although a small amount may precipitate, leading to irritation of the gastric mucosa with inflammation and haemorrhage. On leaving the stomach the pH will be drastically altered by the pancreatic bicarbonate in the duodenum. This leads to the formation of insoluble iron complexes, causing further mucosal damage. The severe intestinal lesions in the stomach and upper part of the small intestine cause erythema with areas of sub-epithelial haemorrhage and erosion. In addition, the increased vascular permeability caused by the inflamed gastrointestinal tract leads to the infiltration of many cells including polymorphonuclear leucocytes, macrophages, lymphocytes and plasma cells in addition to the release of cytokines. A variety of highly reactive oxygen species (ROS) including hypochlorous acid, superoxides and peroxides will be generated at the site of inflammation: these exacerbate the inflammatory response, increasing the permeability of the enterocyte and allowing toxins, in addition to the excess iron, to enter the cells and cause cell death. There are losses of intestinal fluid leading to systemic hypotension and circulatory failure. The marked increase in capillary permeability leads to a further plasma loss, a rise in haemocrit level and blood viscosity, while the blood volume, central venous pressure and tissue perfusion decrease. Subsequently decreased cardiac output leads to a state of shock and to circulatory failure and death within hours (Robontham and Lietman, 1980).

## 8.5  FURTHER STUDIES

Clearly it is essential to identify subjects with genetic haemochromatosis who are asymptomatic despite an increasing tissue burden of iron. At the moment such individuals can only be identified by the finding of an altered biochemical parameters, i.e. ferritin and transferrin saturation. Factors which prevent the toxicity of iron in such patients should be identified.

Within the next few years the gene for genetic haemochromatosis should be identified, its function elucidated and the explanation for the defect in the regulation of iron absorption understood. It would then be possible to screen the population, by amplifying the specific mutations within the gene which are associated with the disease.

# 9

# EFFECT OF IRON ON WORK PERFORMANCE AND THERMOGENESIS

## 9.1 EFFECT OF IRON DEFICIENCY ON WORK PERFORMANCE AND FATIGUE

The possible effect of iron deficiency in restricting work performance and a role for iron deficiency in mediating subjective feelings of tiredness have been extensively studied over several decades. Early studies gave conflicting results, and it has been difficult to separate out the effects of anaemia, with a reduced blood haemoglobin concentration, whether due to iron deficiency or not, from potential direct effects of depletion of one or more of the other important functional cellular iron compounds described in Chapter 1 (Dallman, 1989a). Conversely, the possible effects of sustained exercise on iron status and the mechanisms underlying a reduced haemoglobin concentration with endurance sports ('sports anaemia') have been of interest to athletes anxious to reach optimal competitive performance levels (Anonymous, 1985).

### 9.1.1 Work performance

There is no doubt that work performance is limited in the presence of iron deficiency anaemia. In female Sri Lankan tea pickers, and in male Indonesian rubber tappers and weeders who had hookworm infestation, reduced productivity was found to be directly related to the severity of the iron deficiency anaemia (Edgerton et al., 1979; Basta et al., 1979). In these placebo-controlled trials the greatest improvement in work performance with iron therapy was seen in those subjects who were initially most anaemic. Voluntary activity was also reduced in iron deficiency (Edgerton et al., 1979) and iron deficient subjects tired more easily in Harvard Step Tests (Basta et al., 1979).

### 9.1.2 Subjective symptoms of fatigue

In a community survey, Wood and Elwood (1966) found no relationship between subjective feelings of fatigue and haemoglobin concentration, at least in the range which included people who were only mildly anaemic. In addition, Morrow et al. (1968) found placebo as effective as iron therapy in reversing subjective symptoms in a small number of women with depleted iron stores, low serum iron concentration and mild or absent anaemia. However, in another placebo-controlled trial, women who were iron deficient but not anaemic (Hb >120 g/l), had a greater subjective improvement in symptoms of chronic fatigue after iron therapy than after placebo (Beutler et al., 1960). Since this symptomatic improvement was also accompanied by a small rise in haemoglobin concentration in response to iron but not placebo, the study does not provide firm evidence for a role for iron which is independent of increased haemoglobin production. Nevertheless, the studies of work performance described above lend support to the view that iron deficiency, even when associated with only mild anaemia, may be associated with tiring more easily during exercise.

## 9.2 POSSIBLE MECHANISMS

### 9.2.1 Animal studies

Experimental studies in animals have been used in an attempt to determine whether the impaired work performance is due to anaemia and the concomitant reduction in oxygen delivery to the tissues and/or the depletion of functional iron-containing compounds in the tissues, since the oxidative production of energy (ATP) in muscles depends upon the presence of iron-containing electron transport proteins within mitochondria.

Finch et al. (1976, 1979) demonstrated impaired running ability in iron deficient rats, compared with controls, accompanied by a reduction in mitochondrial enzyme function and increased lactate production: the defects were corrected by parenteral iron therapy and unaffected by the haemoglobin concentration adjusted by blood transfusion to be comparable in both groups. Within rat muscle mitochondria, iron deficiency selectively reduces the constituents of the mitochondrial respiratory chain with respect to other matrix enzymes, as well as reducing the number of cristae (Cartier *et al.*, 1986). Others have confirmed that the effects of iron deficiency on exercise performance in animals can be rapidly corrected by iron, independent of haemoglobin concentration (Perkkio *et al.*, 1985; Willis *et al.*, 1990).

### 9.2.2  Human studies

In humans the evidence for an effect of iron deficiency on muscle function which is independent of anaemia is less clear. After production of iron deficiency anaemia in normal subjects (by phlebotomy), no changes in cytochrome oxidase activity were detected in muscle biopsies, and reduced endurance was eliminated on correcting the anaemia with blood transfusion (Celsing *et al.*, 1986). Furthermore, in a small group of patients with polycythaemia, where regular phlebotomy to control the red cell expansion results in iron deficiency without anaemia, there was no clear effect of iron deficiency on muscle function as judged by treadmill test performance (Rector *et al.*, 1982).

Nevertheless, iron deficient subjects have enhanced lactic acid production on exercise (Gardner *et al.*, 1977) and though this is proportional to the severity of the anaemia, it is detectable even when the anaemia is very mild. Iron therapy in iron deficient, minimally anaemic female athletes did not affect maximal exercise performance, but maximal blood lactate levels were reduced after correction of the iron deficiency, an effect not seen in control subjects (Schoene *et al.*, 1983). Detailed exercise studies carried out in a small number of normal subjects, made iron deficient by phlebotomy and subsequently repleted by oral iron therapy, showed a tachycardia which was inversely proportional to haemoglobin concentration. There was a suggestion that the tachycardia may have been more pronounced when the subjects were exercised at a stage in the protocol when, at the same haemoglobin concentrations, there was also depletion of tissue iron (Charlton *et al.*, 1977). Similarly, when severely anaemic iron deficient subjects were given parenteral iron dextran, work performance (maximum work time) improved in parallel with the correction of the anaemia; there was, however, a reduced heart rate response to exercise compared with untreated controls having a comparable haemoglobin concentration (Ohira *et al.*, 1979). These studies suggest the possibility that the iron therapy may be having effects on cellular function/oxidative metabolism which are not simply a result of correction of the anaemia.

### 9.3  SIGNIFICANCE OF EFFECTS OF IRON DEFICIENCY ON WORK PERFORMANCE

Current understanding of the mechanisms by which iron deficiency leads to impaired work performance, with increased exercise-induced lactic acid production and tachycardia, is clearly incomplete. The overall evidence reviewed above suggests that the major effect is through the production of anaemia, with additional effects of tissue iron depletion on muscle function being likely. Iron therapy will correct both the anaemia and tissue iron depletion. There is an additional possibility that iron therapy could enhance work performance in iron deficiency by decreasing susceptibility to infection (Chapter 10) but this is unproven.

Impaired work performance has been calculated to have major economic consequences in areas of the world in which iron deficiency is common, pointing to the need to eradicate underlying causes such as hookworm infestation. In such cases, it has been suggested that iron supplementation or an iron fortification programme could pay for itself many times over by the benefits in improved performance (Dallman, 1989a).

### 9.4  THE EFFECT OF SEVERE EXERTION ON IRON STATUS

#### 9.4.1  'Sports anaemia'

Prolonged severe exertion, including long-distance road running (Dressendorfer *et al.*, 1981), combat training (Lindemann *et al.*, 1978) and hill walking (Milledge *et al.*, 1982), has been associated with a substantial fall in haemoglobin concentration, sometimes up to 20%. This has been described as 'sports anaemia' but its major component is the dilutional effect of an increased plasma volume with a less marked increase in the total red cell mass (Pugh, 1969; Milledge *et al.*, 1982; Weight *et al.*, 1991). It seems reasonable to regard this as a 'physiological response to unphysiological exercise' (Hallberg and Magnusson, 1984). Nevertheless, a low haemoglobin concentration may be a significant limiting factor on exercise capacity at high work intensities, though mild anaemia has little effect on symptoms or cardiac output during normal activities (Woodson, 1984). Furthermore, infusion of stored autologous red cells to increase the haemoglobin concentration produced a significant increase in work capacity (Buick *et al.*, 1980). These findings suggest that the 'sports anaemia' may be a maladaptation for optimum performance, though it may be expected that any attempt to increase the haemoglobin concentration will lead to an associated rise in blood viscosity, eventually carrying its own penalty of impaired blood flow to vital organs, particularly the brain (Thomas *et al.*, 1977).

These considerations have produced a continuing interest in possible pathological contributions to the anaemia seen in endurance athletes, not least because of their desire to ensure optimum oxygen delivery to their muscles in competition. Small long-term reductions in serum iron and ferritin concentrations have been reported in trained runners (Dufaux *et al.*, 1981) and female field hockey players (Diehl *et al.*, 1986) as well as in milder forms of 'fitness' training (Blum *et al.*, 1986). However, detailed studies of iron metabolism have shown no evidence of any associated impairment of iron supply for erythropoiesis (Magnusson *et al.*, 1984). Furthermore, iron deficiency anaemia in both male and female distance runners occurs at a low frequency indistinguishable from that of the general population (Balaban *et al.*, 1989; Weight *et al.*, 1992b). This is in spite of the potential for increased faecal blood loss after marathon races (McMahon *et al.*, 1984; Porter, 1983) and for increased mechanical red cell destruction with loss of iron as haemoglobin in the urine ('march haemoglobinuria'): the latter, at least, is rare even after several days of road running (Dressendorfer *et al.*, 1981). Overall iron losses in distance runners (assessed by radioiron whole body counting) were marginally greater than normal at an average of 2 mg iron/day (Ehn *et al.*, 1980) but in this study appeared to be balanced by a high dietary iron intake. There is little evidence that iron absorption from food is inhibited by exercise but it has been suggested that suboptimal iron intake may contribute to negative iron balance in some athletes (Weight *et al.*, 1992a).

The widely reported finding of reduced haemoglobin and serum ferritin concentrations in association with exercise training (Newhouse and Clement, 1988) apparently conflicts with data showing acute increases in serum ferritin following intense exercise. However, in these reports other measures, including serum C-reactive protein and haptoglobin concentrations, suggest that these changes are part of a transient acute phase response (Chapter 6) to muscle injury (Taylor *et al.*, 1987; Moore *et al.*, 1993) or shortened red cell survival associated with increased degradation of haemoglobin within macrophages (Vidnes and Opstad, 1981). It is clear that a number of factors other than iron status may affect the usual measures of functional, transport and storage iron in different forms of exercise and at different stages of that exercise.

### 9.4.2  Indications for therapy

The changes in iron metabolism in 'sports anaemia' resemble those of pregnancy, in that iron moves from storage sites into an increased total red cell mass, but there is a fall in haemoglobin concentration due to the dilutional effect of an even greater increase in plasma volume. Redistribution to increased myoglobin in trained muscles may also play a role (Magnusson *et al.*, 1984). Unlike pregnancy, there appears to be little increased risk of iron deficient erythropoiesis in 'sports anaemia'. There is no indication for the routine use of iron supplements (Cook, 1994), despite some reports of 'sports anaemia' responding to iron therapy (Hunding *et al.*, 1981). A fall in serum ferritin to less than 10 $\mu$g/l was seen in women athletes undergoing intensive training who were receiving placebo, but far from being accompanied by a fall in haemoglobin concentration, there was simply a smaller rise than there was in a group taking an oral iron supplement (Magazanik *et al.*, 1991). This may reflect the inability of the relatively low initial iron stores in these women to meet training-induced demands for an increased red cell mass. Although the available data are not always consistent, it appears reasonable to restrict the use of iron supplements in endurance athletes to those (probably relatively few) individuals with proven impaired iron supply for erythropoiesis.

### 9.5  EFFECT OF IRON DEFICIENCY ON THERMOGENESIS

Among the metabolic consequences of iron deficiency, impairment of thermoregulation, with related disturbances of thyroid metabolism and catecholamine turnover, has received attention in both animal and human studies.

### 9.5.1  Possible mechanisms – animal studies

Increases in the blood and urine catecholamine concentrations of iron deficient rats were found to be dependent upon exposure to cold, and were associated with an inability to maintain body temperature in the cold (Dillmann *et al.*, 1979). These effects in iron deficient animals are paralleled by a reduction in oxygen consumption and may be related to reduced concentrations of the thyroid hormone triiodothyronine (T3), with impaired peripheral conversion of thyroxine (T4) to T3 (Dillmann *et al.*, 1980), and reduced response to thyrotropin-releasing hormone (Beard *et al.*, 1989). Disturbances in thermogenesis by the sympathetic nervous system may also contribute, with increased fractional turnover of noradrenaline in heart and brown adipose tissue in iron deficient rats (Beard *et al.*, 1988a). Repletion of iron corrected the thermoregulatory disturbance and restored tissue noradrenaline concentrations and T3 production rates: exchange blood transfusion experiments suggested that these changes are, at least in part, independent of correction of associated anaemia, though the data are not clear cut (Dillmann *et al.*, 1979; Beard *et al.*, 1989).

### 9.5.2 Possible mechanisms – human studies

Urinary catecholamine excretion is increased in iron deficient children and returns to normal after treatment with iron (Voorhess *et al.*, 1975). Iron deficient adult subjects also fail to regulate body temperature during cold water immersion, and have increased plasma catecholamine concentrations (Martinez-Torres *et al.*, 1984). However, in a more recent study in carefully matched groups of young women of similar body fatness, lower core temperatures and reduced plasma T3 and T4 concentrations were seen only in iron deficient states associated with anaemia and not in non-anaemic subjects with depleted iron stores only: plasma catecholamines were unaffected by iron status (Beard *et al.*, 1990a,b). The data suggest that, in humans with iron deficiency anaemia, there are disturbances of energy production dependent on thyroid hormone production. The potential role of abnormalities in catecholaminergic energy production in humans remains unclear.

# 10

# IRON IN INFECTION AND IMMUNITY

There are two ways in which iron may influence the course of infection. Firstly, the need for most microorganisms to acquire iron, coupled with a lack of readily available iron in normal body tissues, means that any abnormality (such as iron overload) which makes iron unusually available may predispose to infection. On the other hand, iron is required for various immunological functions and these may be impaired in iron deficiency. The potential role of iron in these activities is summarized in Table 10.1.

## 10.1 IRON AND MICROBIAL GROWTH

### 10.1.1 Microbial iron acquisition *in vitro*

#### (a) Inhibition by transferrin

One of the earliest properties attributed to transferrin was its ability to inhibit bacterial growth *in vitro* due to sequestration of available iron (Schade and Caroline, 1946). This observation has been confirmed in numerous subsequent studies, and an enormous range of microorganisms, including

**Table 10.1** Potential role of iron in immunity (effects shown in parentheses are either controversial or need further study)

| Activity | Involvement of iron | Possible effects of | |
| --- | --- | --- | --- |
| | | Iron deficiency | Iron overload |
| *Non-specific (innate) immunity* | | | |
| Inhibition of microbial growth | Sequestration of iron by host molecules | (Enhanced resistance to malaria) | (Enhances susceptibility to some infections) |
| Phagocytic activity of neutrophils | 1. Involved in oxygen-mediated microbicidal mechanisms | (Depressed microbicidal activity) | – |
| | 2. Membrane lipid peroxidation | – | Phagocytic uptake impaired |
| *Specific immunity* | | | |
| Lymphocyte proliferation | DNA synthesis (ribonucleotide reductase) | Depressed cell-mediated mediated immunity | – |
| T cell subsets | Unknown | – | Decreased CD4:CD8 ratio |
| Macrophage cytotoxicity (immunologically activated) | (Destruction of target iron-proteins by nitric oxide) | – | (Reduced tumouricidal activity) |

many potential pathogens, can be inhibited by transferrin (and lactoferrin) or by human serum *in vitro* (Weinberg, 1984).

### (b) Microbial production of siderophores

Many microorganisms possess mechanisms to obtain iron from host tissues and these may contribute to microbial pathogenicity. The best studied involves the synthesis of low molecular weight, high affinity iron chelating molecules known as siderophores, which are capable, in theory at least, of removing iron from transferrin and other non-haem host iron proteins. The majority of siderophores are derivatives of either catechol or hydroxamic acid (reviewed by Hider, 1984). Microorganisms take up iron–siderophore complexes via specific outer membrane receptors (Neilands, 1982). Production of siderophores and their outer membrane receptors is generally not constitutive but is induced by low levels of available iron (Griffiths, 1987). Some bacteria such as *Campylobacter jejuni* (Baig *et al.*, 1986) and *Yersinia enterocolitica* (Brock and Ng, 1983) can utilize siderophores they themselves do not produce. These may include the *Streptomyces* siderophore desferrioxamine, which is used as an iron chelating drug and is associated with an increased risk of infection with *Yersinia enterocolitica*. In other cases, desferrioxamine may suppress microbial growth, and orally active hydroxypyridone chelators currently being developed for clinical use show less tendency than desferrioxamine to enhance microbial growth (Brock *et al.*, 1988).

### (c) Other mechanisms of microbial iron acquisition

Some bacteria such as *Haemophilus* and *Neisseria* do not synthesize siderophores but instead express receptors for transferrin or lactoferrin, thus enabling the organism to acquire iron direct from the host protein (Otto *et al.*, 1992). Transferrin- or lactoferrin-binding proteins may also be produced by some parasites and it seems likely that these serve to provide iron to parasites. Little is known about iron uptake and metabolism in parasites and this is an area that deserves further study. The extent to which growth and survival of intracellular pathogens depend upon iron availability has also been neglected. In the case of *Shigella*, intracellular haem compounds may provide a source of iron for intracellular growth (Payne, 1989). On the other hand, *Listeria monocytogenes* probably utilizes non-haem iron within macrophages (Alford *et al.*, 1991).

### 10.1.2 Microbial iron acquisition *in vivo*

Evidence that microbial growth *in vivo* under conditions of normal iron status takes place in an iron restricted environment comes from studies showing that siderophores or microbial membrane receptors for siderophores or host iron binding proteins are hyperexpressed during *in vivo* growth (Holland *et al.*, 1992). Convincing evidence for the popular suggestion that increased iron load predisposes to infection due to increased availability of host iron is lacking. Transferrin and ferritin in normal individuals possess considerable reserves of iron binding capacity, and even a substantial increase in total body burden will not saturate either protein. It is only when transferrin in particular becomes saturated that forms of extracellular iron more immediately available to microorganisms are present in plasma. Until this occurs differences in transferrin saturation are not in themselves likely to alter greatly the availability of iron to microorganisms (Gordeuk *et al.*, 1986). The same argument also casts doubt upon the widely held notion that the hypoferraemia of inflammation enhances resistance to infection by reducing the availability of plasma iron. It is perhaps more likely that other effects of the hypoferraemia, such as potentiation of the activity of $\gamma$-interferon (Weiss *et al.*, 1992) are of more importance.

### 10.1.3 The antimicrobial function of lactoferrin

It has long been thought that lactoferrin helps to protect mucosal surfaces against infection, with particular emphasis on the protection of breast-fed newborn infants against gastrointestinal infection.

Lactoferrin is undoubtedly responsible for the *in vitro* bacteriostatic effect of human milk (Griffiths and Humphreys, 1977). However, clinical studies designed to determine whether addition of (bovine) lactoferrin to infant formula can produce a gut flora resembling that of breast-fed infants have failed to show any significant effect (Moreau *et al.*, 1983; Balmer *et al.*, 1989; Roberts *et al.*, 1992). The general conclusion seems to be that manipulation of either iron or lactoferrin levels in formulas based on cows' milk fails to induce a gut flora comparable to that of breast-fed infants.

Lactoferrin in neutrophils probably contributes to the antimicrobial armoury of these cells, as saturation of neutrophil lactoferrin *in vitro* reduced their bactericidal activity (Bullen and Joyce, 1982). Furthermore, patients with lactoferrin-deficient neutrophils suffer from increased incidence of infection (Boxer *et al.*, 1982).

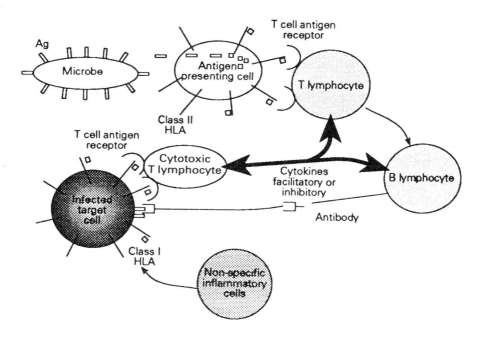

**Figure 10.1** The immune system.

## 10.2 THE EFFECT OF IRON DEFICIENCY ON THE IMMUNE SYSTEM

The possibility that iron deficiency impairs the functioning of the immune system has been the subject of a large number of clinical studies. However, problems of experimental design and adequate controls are particularly evident in this area, and consequently many of the abnormalities reported are controversial, although others appear to be more consistent. The main components of the immune system are shown in Figure 10.1.

### 10.2.1 Lymphocyte function

#### (a) T cell function

T lymphocytes express surface receptors which recognize the unique composition and structure of antigens. They respond to antigens by proliferating, and promoting macrophage activation and antibody synthesis by B lymphocytes. They can also kill cells bearing foreign antigens thereby aiding their removal.

There have been a large number of clinical studies on the effect of iron deficiency and T cells and a more detailed discussion is presented elsewhere (Brock, 1993). Depressed cell-mediated immunity is a fairly frequent finding in iron deficiency, though a few contrary results have also been reported. A reduced proportion of T cells is also commonly found and a likely explanation is an impairment of T lymphocyte proliferation. A failure of peripheral blood mononuclear cells from iron deficient indivi-

duals to respond normally to T cell mitogens *in vitro* is one of the most consistently reported immunological defects associated with iron deficiency, but there are a few contrary reports and in these there were generally complicating factors such as pregnancy (Prema *et al.*, 1982) or hookworm infection (Kulapongs *et al.*, 1974). Studies in iron deficient experimental animals have confirmed the defect in T cell mitogen responses (Mainou-Fowler and Brock, 1985; Kuvibidila and Sarpong, 1990) and suggest that the low transferrin saturation found in such conditions may result in an inadequate supply of the metal to the activated lymphocytes, thus preventing optimal proliferation (Mainou-Fowler and Brock, 1985). Such an explanation is supported by the observation that primed T cells from iron deficient donors function normally in iron replete recipients, whereas cells from normal individuals showed impaired activity in iron deficient recipients (Cummins *et al.*, 1978). Iron deficiency may also lead to intrinsic defects in lymphocytes, as electron microscopy has revealed mitochondrial abnormalities in cells from iron deficient individuals (Jarvis and Jacobs, 1974).

The critical role of iron in lymphocyte proliferation appears to be mainly due to the fact that it is required for the activity of the enzyme ribonucleotide reductase (Hoffbrand *et al.*, 1976), and expression of transferrin receptors on T lymphocytes markedly increases following activation (Larrick and Cresswell, 1979). Changes in iron status could also exert subtle effects on immune function by altering the relative activity of different lymphocyte subpopulations.

### (b)  B cell function

B lymphocytes synthesize antibodies which combine with foreign antigens. B cell function is largely dependent on signals from T lymphocytes which have encountered foreign antigens.

Evidence for impairment of B cell function in iron deficiency is less convincing than that for T cell function. Both normal and reduced (Santos and Falcão, 1990) B cell numbers have been reported in iron deficient individuals but immunoglobulin levels have generally been reported as normal (Bagchi *et al.*, 1980), suggesting that B cell function is not unduly affected. Clinically, the response to vaccination is probably also normal (Chandra, 1975). In contrast, studies in iron deficient experimental animals have generally found an impaired response to vaccination (Kochanowski and Sherman, 1985; Dhur *et al.*, 1990). *In vitro* proliferation of B cells is iron dependent (Brock, 1981; Neckers and Cossman, 1983) but other activities, notably immunoglobulin synthesis, may be less so (Neckers *et al.*, 1984), which may explain the conflicting results from clinical and experimental studies.

### 10.2.2  Macrophage function

Whereas T and B lymphocytes are the main instruments of antigen-specific immune responses, their interaction with antigens recruits non-specific populations of inflammatory white cells such as neutrophils, macrophages and natural killer cells.

### (a)  Effect of iron deficiency on macrophage function

The rather limited studies on macrophage function in iron deficiency suggest that some activities may be impaired. Although one clinical study of iron deficiency reported normal cytotoxic activity and interleukin 1 (IL1) release by monocytes activated *in vitro* (Bhaskaram *et al.*, 1989), studies in animals in which macrophages were activated *in vivo* reported decreased production of IL1 (Helyar and Sherman, 1987) and α-interferon (Hallquist and Sherman, 1989), and reduced tumouricidal activity (Kuvibidila *et al.*, 1983). However, the paucity of studies, particularly in humans, makes it difficult to draw any firm conclusions regarding the effect of iron deficiency on macrophage function.

### (b)  Effect of iron status on the cytotoxic effects of macrophages

Despite the lack of clinical and experimental evidence for the effect of iron deficiency on macrophage function, iron has a critical role in the cytotoxic activity of activated macrophages. Iron is required for the catalytic production of microbicidal hydroxyl radicals via the Fenton reaction (Halliwell, 1989), and reduction of macrophage iron levels below a critical point reduces their activity (Alford *et al.*, 1991).

Many cytotoxic effects of activated macrophages, including some previously ascribed to reactive oxygen compounds, are now thought to be mediated by nitric oxide (NO), which is generated enzymatically from L-arginine (Liew and Cox, 1991) and acts, at least in part, by causing release of iron from iron–sulphur enzymes in the target cell or organism. Nitric oxide is responsible for macrophage-mediated killing of various organisms and excess iron reduces the microbicidal activity.

### 10.2.3  Neutrophil function

A number of studies have shown that, although phagocytic uptake is generally normal, intracellular killing mechanisms are sometimes defective (Chandra, 1975; Walter *et al.*, 1986), a finding that can probably be linked to a defect in the generation of reactive oxygen intermediates by neutrophils (Murakawa *et al.*, 1987; Sullivan *et al.*, 1989). However, two clinical studies failed to detect any abnormality in neutrophil function (Kulapongs *et al.*, 1974; Van Heerden *et al.*, 1983). Other neutrophil activities appear to be unaffected.

### 10.2.4  Other immunological mechanisms

Studies of the effect of iron deficiency on other aspects of the immune system are relatively few. Antibody-dependent lymphocyte cytotoxicity was impaired in iron deficient patients but the defect was not corrected after iron repletion (Santos and Falcão, 1990). Defective natural killer cell activity was found in iron deficient rodents (Hallquist *et al.*, 1992) but this may have been a secondary effect.

### 10.2.5  Immune function in relation to the degree of iron deficiency

Various criteria have been used for diagnosis of iron deficiency in the clinical studies referred to above. In almost all cases the subjects studied were anaemic, with haemoglobin values of 80 g/l or less. Defects have also been reported in subjects with haemoglobin levels up to 100 g/l, though immune responses in a mildly anaemic group with haemoglobin levels between 100 and 120 g/l were normal (Srikantia *et al.*, 1976). Transferrin saturation in subjects with defective immune responses was usually less than 15% and, interestingly, immunological defects are present in association with iron deficient erythropoiesis, where the transferrin saturation is reduced but haemoglobin levels are near normal. Most of the clinical studies were carried out before serum ferritin assays became widely used, but from the more recent studies a serum ferritin of less than 12 μg/l seems to be associated with immunological defects.

## 10.3   THE EFFECT OF IRON OVERLOAD ON THE IMMUNE SYSTEM

The effect of iron overload on the immune system has received less attention than iron deficiency (for a review, see De Sousa, 1989). Most studies have been carried out with patients suffering from hereditary haemochromatosis, or from transfusional iron overload associated with conditions such as β-thalassaemia.

### 10.3.1   Lymphocyte function

#### (a) T cell function

A frequent finding in β-thalassaemia is a reduced ratio of the two T cell subsets expressing the CD4 and CD8 surface markers, i.e. the CD4 : CD8 ratio (Guglielmo et al., 1984; Dwyer et al., 1987). However, chronic alloantigenic stimulation arising from repeated blood transfusions may be responsible for these changes, rather than iron overload (Kaplan et al., 1984). Unfortunately, the situation in hereditary haemochromatosis, in which this complication does not arise, is unclear: both normal and decreased CD4 : CD8 ratios have been reported by the same group (Bryan et al., 1984, 1991).

#### (b) B cell function

The effect of iron overload on B lymphocytes is also unclear: normal numbers were found in haemochromatotic patients (Bryan et al., 1991), and although there is a report of elevated B cell numbers in children with β-thalassaemia (Dwyer et al., 1987) there was no correlation with serum ferritin levels, which again suggests that iron overload per se was not the cause. Increased spontaneous in vitro production of immunoglobulin by B cells has been found in both thalassaemia and haemochromatosis, but serum immunoglobulin levels, though increased in thalassaemia (Dwyer et al., 1987), were normal in haemochromatosis (Bryan et al., 1984).

#### (c) Responses to mitogens

Reduced skin test responses and in vitro T cell responses to mitogens have been reported in β-thalassaemia (Dwyer et al., 1987), though here again the relative importance of iron overload versus chronic alloantigen stimulation remains to be established. In haemochromatosis, there are again conflicting reports, with both normal (Bryan et al., 1991) and depressed (Bryan et al., 1984) responses to mitogens being found.

#### (d) Effect of non-transferrin bound iron

Non-transferrin bound iron may inhibit in vitro lymphocyte proliferation, the CD4 subset of T cells being more sensitive than the CD8 (Djeha and Brock, 1992b), which accords with the reduced CD4 : CD8 ratio noted in iron overload. Ferritin inhibits lymphocyte proliferation in vitro (Matzner et al., 1985) and is immunosuppressive in vivo (Harada et al., 1987), so increased serum ferritin levels in iron overload may also contribute to immunological abnormalities.

### 10.3.2   Phagocytic function

Little work has been carried out to determine whether iron overload impairs the immunological functions of macrophages and neutrophils. There are two reports of defective in vitro phagocytosis of bacteria by monocytes from haemochromatotic patients, which was corrected by phlebotomy (Van Asbeck et al., 1982, 1984). Intracellular killing of microorganisms (but not their ingestion) was defective in thalassaemia and other transfusional iron overload conditions (Van Asbeck et al., 1984; Ballart et al., 1986). Several studies have reported defective neutrophil phagocytosis associated with transfusional iron overload (Flament et al., 1986; Cantinieaux et al., 1987, 1990) but this function appears to be normal in haemochromatosis (Van Asbeck et al., 1982, 1984).

These defects may be caused by binding of iron to the cell membrane, as iron compounds were found to damage neutrophils in vitro due to production of toxic oxygen intermediates at the cell membrane (Hoepelman et al., 1989). Furthermore, a serum factor responsible for the defect in intracellular killing by neutrophils from patients with transfusional iron overload is probably non-transferrin bound iron (Cantinieaux et al., 1990).

## 10.4   IRON AND INFECTION – CLINICAL EVIDENCE

### 10.4.1   Clinical evidence of increased infection associated with iron deficiency

Despite the strong evidence for impaired T cell function in iron deficiency, evidence of increased susceptibility to infection associated with iron deficiency is far from abundant. Two studies reported that the incidence of respiratory infections in iron deficient children decreased after oral administration (Mackay, 1928; Andelman and Sered, 1966) but these were not well controlled, and a third reported no difference (Burman, 1972). However, iron deficiency was found to exacerbate chronic mucocutaneous candidiasis, resistance to which is thought to involve principally cell-mediated immunity (Higgs and Wells, 1972), and a more recent study found increased incidence of upper respiratory tract and gastrointestinal infections in iron deficient infants with depressed levels of T-helper cells (Berger *et al.*, 1992). A link between increased incidence of infection following abdominal surgery and a low serum ferritin level has also been reported (Harju, 1988).

### 10.4.2   Clinical evidence of decreased infection associated with iron deficiency

Suggestions that iron deficiency may actually decrease the incidence of infection, due to reduced availability of iron to the invading microorganisms, are controversial and require further confirmation (Hershko *et al.*, 1988). Iron deficient Somali nomads showed a lower incidence of infection than controls of normal iron status (Murray *et al.*, 1978). An experimental study also reported reduced mortality associated with nutritional iron deficiency in experimental *Salmonella typhimurium* infection in mice (Puschmann and Ganzoni, 1977). Although it has been suggested that iron deficiency may protect against malaria (Murray *et al.*, 1975), the evidence is far from conclusive (Hershko *et al.*, 1988; Snow *et al.*, 1991, also Chapter 6 of this book).

### 10.4.3   Clinical evidence of increased infection associated with iron overload

Although excess iron enhances bacterial growth in serum *in vitro*, there is little evidence that even severe iron overload predisposes to infection. Indeed, infection seems to be a surprisingly infrequent complication of hereditary haemochromatosis (Higginson *et al.*, 1953).

In contrast, increased susceptibility to infection has often been noted in iron overloaded patients suffering from β-thalassaemia, but this correlates more closely with splenectomy than with the degree of iron overload (Eraklis and Filler, 1972). An increased incidence of infection has also been reported in infants with sickle cell anaemia (Powars, 1975), in which free haemoglobin resulting from haemolysis may play a role. Episodes of sepsis have been reported in patients undergoing chemotherapy for non-lymphocytic leukaemia, in whom serum transferrin was fully saturated and non-transferrin bound iron was present (Gordeuk *et al.*, 1986). Nevertheless, on the available evidence it seems likely that infection is not one of the most important complications arising from iron overload disease.

### 10.4.4   Clinical evidence of increased infection associated with iron supplementation

In contrast, injudicious use of iron supplements can markedly increase the risk of infection. Polynesian infants given intramuscular iron dextran prophylactically showed an increased incidence of neonatal sepsis (Barry and Reeve, 1987; Becroft *et al.*, 1977) and this treatment may also exacerbate malaria (Oppenheimer, 1989a). It seems likely that in these cases it is the transient presence of these relatively labile iron compounds, rather than iron overload *per se*, that provides microorganisms with an opportunity for more rapid growth. A similar effect of oral iron supplementation has been reported in Somali nomads (Murray *et al.*, 1978), in whom the incidence of certain infections, particularly malaria, increased after iron supplementation. It is possible that in these severely malnourished individuals a combination of low serum transferrin and a large rise in serum iron following treatment may have enhanced iron availability to microorganisms.

## 10.5   IRON METABOLISM IN INFLAMMATION

### 10.5.1   Possible mechanisms

Inflammation causes alterations in the normal pattern of iron recirculation and is characterized by reduced serum iron and increased storage iron. If the process is prolonged, anaemia may develop. Neither the physiological function nor the mechanisms involved in these changes (known as the hypoferraemia of inflammation) have so far been clearly established (Lee, 1983; Means and Krantz, 1992). One possible factor is a reduced response to erythropoietin during inflammation (Baer *et al.*, 1987), which may account for the limited ability of recombinant erythropoietin therapy to correct the condition (Pincus *et al.*, 1990). Another probable contributory factor is an alteration in iron metabolism in macrophages responsible for processing effete erythrocytes, resulting in reduced iron release. This process is probably due to increased ferritin synthesis (Konijn *et al.*, 1981) and induced by inflammatory cytokines such as tumour necrosis factor-$\alpha$ (Alvarez-Hernandez *et al.*, 1989) or IL1 (Gordeuk *et al.*, 1988). A role for lactoferrin released by degranulating neutrophils has also been proposed (Van Snick *et al.*, 1974), but more recent work casts doubt on such a mechanism (Ismail and Brock, 1993). Whatever the mechanism(s), it is important to note that attempts to correct hypoferraemia and/or anaemia by iron therapy in patients with inflammatory disease is likely to be counter-productive as the majority of such iron will enter the stores and exacerbate the iron loading in these tissues.

### 10.5.2   Hypotheses for the hypoferraemia of inflammation

The most popular hypothesis regarding the physiological role of the hypoferraemia of inflammation is that it reduces the availability of iron to invading microorganisms. However, as mentioned above and discussed in more detail elsewhere (Brock, 1993), there is actually very little direct evidence to support such a theory and strong biochemical arguments against it. Possibly these alterations in iron metabolism could favour diversion of iron to processes involved in immunological activation (Weiss *et al.*, 1992) but further work is required in this area.

# 11

# IRON AND MENTAL AND MOTOR BEHAVIOUR IN CHILDHOOD

## 11.1 INTRODUCTION

There is ample evidence of an association between iron deficiency anaemia and impaired performance in tests of mental and motor function in childhood. A number of reviews dealing with the nature of the association have been published (Parks and Wharton, 1989; Simeon and Grantham-Macgregor, 1990; Pollitt, 1993) and this chapter considers the evidence for a causal association.

Most animal studies have involved rats since their distribution of brain iron is similar to that in the human (Hill and Switzer, 1984). A brief period of severe iron deficiency leads to a deficit of brain iron in the young rat, reinforcing the importance of examining the young child. What is more, the deficit has been shown to be irreversible, due probably to the slow rate of brain iron compound replacement (Dallman and Spirito, 1977). Youdim *et al.* (1977, 1980, 1981) reported that severely iron deficient animals showed alterations in motor activity and sleeping time – findings replicated by some but not by others (Edgerton *et al.*, 1972; Dallman *et al.*, 1984a).

## 11.2 CONCEPTUAL AND METHODOLOGICAL PROBLEMS IN HUMAN STUDIES

### 11.2.1 Conceptual issues

Pollitt and Metallinos-Katsaras (1990) have pointed out that the model invoking a causal relationship between iron deficiency and behavioural effects is questionable because of the non-specificity of the behaviours studied and also because of the lack of definition of the mechanisms involved. Judgement on a single causal relationship is premature and there are serious limitations to the use of developmental or intelligence tests in this field. Long-term effects, given the plasticity of human development, are also open to question.

### 11.2.2 The confounding variable

Confounding variables abound, many of which could have an adverse effect on development; iron deficiency anaemia coexists with and might merely be a marker for a generally poor environment, both nutritionally and psychosocially (Cravioto and Delicardi, 1972). On the other hand, environmental factors can act as protective buffers, mitigating the effects of iron on behaviour. It is, moreover, virtually impossible in human studies to control for genetic factors.

### 11.2.3 The choice of dependent variables

Although both developmental and intelligence tests are widely used clinically, educationally and academically, there is no general agreement on the definition of intelligence and some argue that it cannot be measured (Gould, 1981).

Furthermore, global measures of development and intelligence may be insensitive in the context of iron deficiency. Horowitz (1989) has put forward the view that such instruments are insufficiently sensitive to tease out the precise role of deficits in specific nutrients: 'There are too many general, environmental, ambient forces conspiring to push general development forward. These will often override the differences that are a function of a particular risk factor.' Pollitt and Metallinos-Katsaras (1990) noted also that, although much reported work has relied on the Bayley Scales, they were not designed to tap what these authors see as the very important areas of specific information processing and learning abilities.

Horowitz (1989) agreed that measures of specific cognitive skills (perceiving, remembering, imagining, conceiving, judging, reasoning) are more sensitive and therefore more appropriate when assessing the effects of iron deficiency.

The assessment of social–emotional factors is even harder, yet may be of considerable value; the assessment of school achievement may have the highest face value for policy makers.

A theoretical point which has considerable bearing on the choice of dependent variables, and on the interpretation of changes in them, was proposed by Read (1975) who concluded that the major influence of iron deficiency may be on attentional, arousal or motivational systems rather than basic cognitive abilities. This view has been taken up in research design, for example by Pollitt *et al.* (1983, 1985, 1986). It remains an area of promise at the moment since both the operational definition and the measurement of attention are still to be developed.

### 11.2.4   The establishment of iron status

No single laboratory measure adequately characterizes an individual's iron status (Chapter 5). The greater the number of criteria used, the fewer the number of children likely to be classified as iron deficient. The use of multiple measures, followed in most recent studies, overcomes this difficulty to some extent but it leads to problems when attempting to compare studies.

### 11.2.5   Generalization

Generalizing from one study to another is not easy; generalizing from one country to another may be even harder. Anaemia can be a result of a poor intake of dietary sources of iron but in some parts of the world malaria or chronic blood loss due to hook worm infestation, conditions which could affect arousal and motivation, may play their part.

### 11.2.6   Methodological considerations in intervention studies

The general design of intervention studies involves taking baseline measurements of iron status and behavioural measures, providing iron supplementation for a specified period and then repeating the measurements. If children have shifted significantly both in their iron status and their scores on behavioural measures, and if an adequately matched control group has done neither, it can be reasonably assumed that iron deficiency was causally related to levels of functioning. Ideal intervention trials will include children with and without iron deficiency. Within these two groups, subgroups will be randomly allocated to receive either iron or a placebo on a double blind basis. The inclusion of children without the disorder will indicate the ceiling of achievement which might be reached after correction of the disorder and should determine

whether giving the treatment to children who do not need it is harmless (Wharton, 1992a).

The exception to this design is the prospective cohort study in which supplementation is given to a sample of children from early infancy, with subsequent follow-up assessments. Although ethically problematic, the use of an ABA design, in which subjects return in the third time period to their original iron status and are then assessed behaviourally, gives further weight to findings.

The following test characteristics can affect results of intervention studies and should be noted:

- There can be a ceiling effect which occurs when children have reached more or less their maximum level of achievement and so there would be no expected rise.

- There can be a regression towards the mean, i.e. the further the group is away from the mean on the first test, the more likely they are to approach it on the second test.

- There can be a practice or training effect. If a child is presented with the same test within six months, the materials, demands and general situation will be familiar and the second performance is likely to be better than the first.

### 11.2.7   Confounding variables and statistical analysis in intervention studies

Confounding variables may arise in intervention programmes. For example, iron supplementation may be accompanied by more general health and nutritional advice and other supplementation which may be of greater overall importance. A randomized, double blind, placebo controlled method should overcome this problem.

Negative confounding variables must also be considered. If children have developed a poor attitude to school, to authority and to all tests, this could mask any effects produced by the improved iron status.

Furthermore, very few published intervention studies have used appropriate analytic techniques. It is often assumed that the differences between confounding variables observed in the experimental and control groups can be discounted if they are not statistically significant at the time of baseline measurement. This is not justified; even small differences can affect the outcome measures. An analysis of covariance takes these initial differences into account and should always be used to measure change, rather than the simpler statistical analyses.

### 11.2.8 Conclusions

These methodological problems led Dobbing (1987), when discussing children's diets, to argue that we can say little that is meaningful about the intellectual failure to thrive in those undernourished in babyhood because of the 'sheer impossibility of measuring . . . all the environmental factors . . . and analyzing and expressing their relative power'.

The problems are, indeed, considerable, but if the view expressed above were to have been applied in the face of all difficulties we would have remained in a scientific Stone Age. Much work still needs to be done but recent progress in research design has made it possible to come to some tentative conclusions on the nature of the association between iron deficiency anaemia and performance on mental and motor tests.

### 11.3 OBSERVATIONAL STUDIES ON IRON DEFICIENCY AND MENTAL AND MOTOR BEHAVIOUR

#### 11.3.1 Early studies

The earliest studies suggested that iron deficiency may lead to anorexia, irritability, apathy, listlessness, perceptual restriction and lowered IQ (Werkman et al., 1964; Nelson et al., 1969; Howell, 1971; Sulzer et al., 1973; Read, 1975).

Cantwell (1974) carried out a longitudinal study of 61 full term infants from similar socio-economic groups; during the 6–18 months period, half of them developed iron deficiency anaemia. When examined at 6–7 years of age the anaemic children were less attentive than the non-anaemic group and had more soft neurological signs, although IQs were similar. Controlling only for social class, as in this study, may be too crude to establish that the two groups really were comparable in every factor other than their iron deficiency. The late 1970s and early 1980s saw a marked improvement in experimental design.

#### 11.3.2 Studies on young children indicating no association with development

Several recent studies provide evidence from straightforward comparisons between groups on a one-off basis or from baseline data obtained before an intervention programme. Table 11.1 lists seven studies that have shown iron deficiency not to be associated with development.

There are comments to be made on a number of these studies on several counts:

- Deinhard et al. (1981) used serum ferritin as their only classification of iron status.
- Although Johnson and McGowan (1983) found no significant differences between the anaemic and non-anaemic babies, the anaemic children scored 7.8 and 5.6 points lower on the Motor and Mental Scales respectively.
- Similarly Oski et al. (1983) reported non-significant results but the differences between iron deficient and iron depleted children were around 10 points and the sample was small.
- Pollitt et al. (1985) found no difference in preschool children in Egypt but used a simple, crude vocabulary test, far away from the sensitive measures advocated above.
- Deinhard et al. (1986) also found no significant differences in the mean Mental Development or IQ scores at baseline between the anaemic infants and their controls, but the anaemic infants scored around 7 or 8 points lower. The method of reporting of this study makes it difficult to come to conclusions on the iron deficient non-anaemic children since they were split into two baseline groups.
- The Indian study by Uphadyay et al. (1987) found no significant differences as a function of anaemia but widespread protein energy malnutrition could have masked the more subtle effects of iron.
- The Spanish prospective study on 196 healthy infants found no developmental delay associated with iron deficiency, with or without anaemia, but the Denver Developmental Screening Test is no more than a screening instrument and is regarded as lacking sensitivity.

#### 11.3.3 Studies indicating an association with development

The details of 18 studies are given in Table 11.2. Although together they support the hypothesis that iron deficiency is associated with developmental delay, they do not all point in that direction in every detail. A more or less consistent finding is that the Bayley Mental Scale scores appear to be more reduced in iron deficiency than those of the Motor Scale.

The results from studies using tests of attention and information processing rather than global development (e.g. those reported by Pollitt et al. (1983) and Soewondo et al. (1989)) indicate that this is a fruitful area to follow but since there have been few replications it is hard to come to firm conclusions.

Eleven studies used tests of global development or intelligence, criticized above as possibly lacking

**Table 11.1**  Observational and baseline studies indicating no association between iron and development

| Study | Design | Results and comments |
|---|---|---|
| Deinhard *et al.* (1981) (USA) | 222 infants, 11–13 months Distinguished by serum ferritin values | No significant difference on Bayley Scales, Uzgiris and Hunt Scales or a measure of habituation. |
| Johnson and McGowan (1983) (USA) | 25 infants, aged 1 year<br><br>Controlled for socio-economic factors, mother's education and home environment | No significant difference between anaemic and non-anaemic on Bayley Scales and 8 behavioural measures. |
| Oski *et al.* (1983) (USA) | 38 infants, 9–12 months (18 ID, 10 I deplete, 10 normal)<br><br>All with Hb levels > 110 g/l | No significant difference on Bayley Mental Scale between ID and other groups.<br><br>Motor Scale not given. |
| Pollitt *et al.* (1985) (Egypt) | 157 children, 4–6 years (88 anaemic, 69 normal)<br><br>Controlled for growth retardation, obesity, health history, acute illness, mental and physical handicap, haematologically related disease | No significant difference in discrimination learning or picture vocabulary test results. |
| Deinhard *et al.* (1986) (USA) | 70 children, 16–20 months (25 ID anaemic, 45 ID non-anaemic)<br><br>Matched for mother's education | No significant difference on Bayley Scales. |
| Uphadyay *et al.* (1987) (India) | 469 children, 6–8 years Possible high level of malnutrition in many or all | No significant difference in intelligence test scores using Indian adaptation of Wechsler scale between anaemic and non-anaemic children. |
| Gonzalez de Aledo *et al.* (1990) (Spain) | 196 healthy infants in prospective study | No delay on Bayley Scale associated with ID with or without anaemia. |

ID: Iron Deficient
I: Iron

in sensitivity. However, if sampling is adequate to allow for confounding variables, even a crude test will show up differences.

- The Webb and Oski studies (1973a,b, 1974) were carried out in a deprived area of Philadelphia where the problems of confounding variables are considerable.
- Grindulis *et al.* (1986) found that the anaemic children, when compared with those who were non-anaemic, had less well educated mothers who had been in Britain for fewer years and lived in more crowded homes, although their fathers were of slightly higher social class. However, other potentially confounding vari-

ables such as nutritional status for protein energy and zinc were similar in both groups.
- Lozoff *et al.* (1982a, 1987, 1991), Walter *et al.* (1983), Pollitt *et al.* (1983, 1985, 1986), Soemantri *et al.* (1985) and Idjradinata and Pollitt (1993) all went to some trouble to allow for a number of confounding variables. The Idjradinata and Pollitt (1993) results are stronger for having used an analysis of covariance.
- Prospective studies following children from early in life are rare and Walter *et al.* (1983, 1989a) were able to use children from the same ethnic, socioeconomic and cultural background.

**Table 11.2**   Observational and baseline studies indicating an association between iron and development

| Study | Design | Results |
|-------|--------|---------|
| Webb and Oski (1973a,b,1974) (USA) | 193 children 12–14 years (92 anaemic, 101 non-anaemic) | Non-anaemic significantly better on Iowa Test of Basic Skills (reading, maths concepts, problem solving etc). |
| | | Anaemic children had more deficits in attention and perceptual tasks. |
| Lozoff et al. (1982a) (Guatemala) | 68 children, 6–24 months | 14 and 9 point significant difference between anaemic and non-anaemic children on Bayley Mental and Motor Scales. |
| | Control for birthweight, prematurity neonatal distress, retardation, general malnutrition and illness | Difference also in play |
| Palti et al. (1983) (Israel) | 464 children followed up (184 anaemic, 280 non-anaemic) | Infants anaemic at 9 months had lower developmental and IQ scores at 2, 3 and 5 years (Brunet-Lézine and Wechsler Scales) |
| Pollitt et al. (1983) (USA) | 30 children, 3–6 years (15 I deplete, 15 controls) | Significant difference in discrimination learning (attention) |
| | Controlled for growth retardation, obesity, medical history, acute illness, physical and mental handicap, haematological related disease | No significant difference in Stanford-Binet IQ, conceptual learning or memory. |
| Walter et al. (1983) (Chile) | 37 children, aged 15 months | 12 point difference in Bayley Mental Score between mild anaemic and iron replete non-anaemic infants. |
| | Cohort study, comparable ethnic, socio-economic and cultural backgrounds | ID non-anaemic infants scored between others. |
| | | No significant difference on Motor Scales. |
| Driva et al. (1985) (Greece) | 48 children in Athens orphanage | Scores of anaemic children differed from others by at least one SD on Bayley Mental Scale. |
| | | No significant difference on Motor Scale. |
| Pollitt et al. (1985) (Egypt) | 74 children, 3–6 years (30 anaemic, 44 normal) | Controls faster and more accurate on Kagan Matching Familiar Figures Test. |
| | Control variables as Pollitt et al. (1983) | |
| Soemantri et al. (1985) (Indonesia) | 119 children, 9–11 years (78 anaemic, 41 normal) | Controls significantly better on school achievement test. |
| | Controlled as Pollitt et al. (1983) plus parasite egg count | No significant difference using Raven's Matrices. |
| Grindulis et al. (1986) (Birmingham, UK) | 97 Asian infants, 17–19 months, whose mothers had taken part in a study of nutritional supplementation during pregnancy | Those with Hb levels <110 g/l had lower scores on the Sheridan Developmental Scale with significant differences on social and fine motor scores. |

**Table 11.2**   Observational and baseline studies indicating an association between iron and development (contd)

| Study | Design | Results |
|---|---|---|
| Pollitt *et al.* (1986) (Guatemala) | 50 children aged 30 to 72 months (25 ID anaemic, 25 iron replete) Controls as Pollitt *et al.* (1983) | Significant differences in discrimination and learning tests. No significant differences in memory. |
| Lozoff *et al.* (1987) (Costa Rica) | 191 infants, 12–23 months<br><br>Controlled for birth order, birthweight, Apgar score, weight, maternal age, education and IQ and home environment<br><br>Children grouped according to iron status | Bayley, Mental, Scale: lowest iron group had lower scores than others, none of which differed from each other<br><br>Motor Scores: lowest iron group had lower scores.<br><br>Mildly deficient anaemic group lower than intermediate who were no different from three non-anaemic groups. |
| Pollitt and Kim (1988) (Java) | 119 children, 9–11 years (78 anaemic, 41 non-anaemic) | No significant difference on Raven's Matrices.<br><br>Non-anaemic children significantly better on tests of school achievement. |
| Dommergues *et al.* (1989) (France) | 147 children, 10 months to 4 years<br><br>Followed up in well-baby clinic, account taken of variables known to affect psychomotor development | No significant association between iron status (iron deficiency and ferritin blood levels) and the Brunet-Lézine test at 10 months or 4 years but at 2 years, infants with ID had significantly lower scores on the same test. |
| Pollitt *et al.* (1989) (Thailand) | Children 9–11 years (ID anaemic, I deplete, I replete) | Significant association of iron status with test of Thai language and Raven's Matrices. No difference with maths. |
| Soewondo *et al.* (1989) (Indonesia) | 176 children (mean age 56 months) from ethnically and socioeconomically homogeneous area (ID anaemic, I deplete, I replete) | No significant difference between anaemic and iron replete on picture vocabulary test.<br><br>Anaemic children scored lower on 3 of 9 comparisons on battery of cognitive tests giving 'weak support' for hypothesis of iron effect. |
| Walter *et al.* (1989a) (Chile) | 196 infants in prospective study from birth to 15 months | Bayley Mental and Motor scores of ID anaemic infants lower at 12 months than ID non-anaemic or iron replete controls. Those anaemic at 9 and 12 months had lower scores than those whose anaemia was of shorter duration. |
| Lozoff *et al.* (1991) (Costa Rica) | Follow up of 85% of children seen in 1987 | All had good iron status but those with Hb concentrations of 100 g/l or less had lower scores on intelligence tests. |
| Wasserman *et al.* (1992) (Yugoslavia) | 392 children aged 2 years living near smelter | Difference in Hb concentration at 18 months from 120 to 100 g/l associated with estimated 3 to 4 point decrease in Bayley Mental Scores, independent of lead effect. |
| Idjradinata and Pollitt (1993) (Indonesia) | 126 children aged 12–18 months. 15 exclusion criteria and control for socio-economic factors, mother's education and home environment by analysis of covariance. | ID anaemic children significantly lower scores on Bayley Mental and Motor Scores than ID non-anaemic or iron sufficient. ID non-anaemic not different from I replete. |

**Table 11.3**   Studies with data on dose effects

| Study | Design | Results |
|---|---|---|
| Lozoff *et al.* (1982b) (Guatemala) | Re-analysis of 19–24 month children (see Table 11.2) | Significant correlation between iron status and Bayley Mental Scale but not the Motor Scale. |
| | | Multiple regression showed iron to account for 56% of the variance. |
| Palti *et al.* (1983) (Israel) | See Table 11.2 | Trend towards increments in haemoglobin and IQ at 5 years. |
| Walter *et al.* (1983) (Chile) | See Table 11.2 | Anaemic children had significantly lower Bayley Mental Scale (98) than ID non-anaemic (108) and controls (113). |
| Lozoff *et al.* (1987) (Costa Rica) | See Table 11.2 | Most iron deficient had lowest Bayley Mental score but no linear dose effect. |
| Walter *et al.* (1989a) (Chile) | 196 infants aged 1 year in cohort study of iron supplementation | Bayley Mental and Motor Scores of anaemic infants correlated with iron status. No association of Bayley scores with iron status in non-anaemic babies. |

- Lozoff *et al.* (1987) and Walter *et al.* (1989a) found some indication of difficulty in Bayley Mental tests requiring language comprehension and some gross motor items related to balance.

In a study looking to see whether both lead exposure and iron status affected development at 2 years of age, Wasserman *et al.* (1992) compared infants living near a smelter with a control group. A difference in haemoglobin concentration at 18 months from 120 to 100 g/l was associated with an estimated three to four point decrement in Bayley Mental Development score, an effect present in both groups, suggesting that its effect was independent of lead exposure.

It could be argued that the study by Pollitt *et al.* (1983) gave equivocal results since they found significant differences in some tests but not others. More clearly equivocal is the study by Dommergues *et al.* (1989), who showed that the association is not consistent. A conservative interpretation of this finding is that there may be some factor(s) as well as iron at work.

### 11.3.4   Studies showing cut-off points and dose effects

Five studies have evidence on the issue of the degree of iron deficiency at which behaviour appears to be affected (Table 11.3).

- Lozoff *et al.* (1982b) in their reanalysis of earlier published data of 27 children aged 19–24 months, found a significant positive correlation between iron status and mental test scores and a similar finding from a multiple regression.
- Palti *et al.* (1983) found significant positive trends: for each 10 g/l increment in haemoglobin there was a 1.75 increase in IQ at 5 years, even when the anaemia had been treated at the time of diagnosis.
- Lozoff *et al.* (1987) split their 191 infants into five groups on the basis of their iron status. Eleven possible confounding variables were taken into account. The group with lowest iron status had lower Bayley Mental scores than the other groups, none of which differed from the rest. The pattern was slightly different for the motor score, but again no clear indication of a dose effect was apparent.

Although there is no firm indication of a linear dose effect there is some support from these studies for the notion that the presence of anaemia is a cut-off point at which the adverse effects of iron deficiency are likely to occur.

### 11.3.5   Conclusions on observational data

Although there are methodological and conceptual difficulties surrounding all the studies discussed here, and the results are not always in the same direction, there is considerable support for the

**Table 11.4**  Short-term intervention using standardized tests

| Study | Design | Intervention | Results |
|---|---|---|---|
| Oski and Honig (1978) (USA) | 24 ID anaemic infants, 9–26 months<br><br>Randomly assigned to treatment or placebo | Intramuscular iron for 7–9 days | Bayley Mental Scales significantly higher for treatment group; placebo also higher but not significant. Treated group more alert and responsive with improved motor skills. |
| Lozoff et al. (1982a) (Guatemala) | 68 infants, 6–24 months, with and without ID anaemia | Oral iron for 7 days | In treated group Hb rose to expected levels; no effect on Bayley Scales |
| Oski et al. (1983) (USA) | 38 infants, 9–12 months, matched for age, gender and race (see Table 11.2) | Intramuscular iron (50 mg)<br><br>Tested 7 days later | Increases in Bayley Mental Scale:<br>ID           22<br>I deplete    5.6<br>Normal       6.2 |
| Driva et al. (1985) (Greece) | 48 children, 3–25 months (40 ID anaemic, 8 normal) | Intramuscular iron (50 mg) once | Anaemic infants' iron status corrected and Bayley Mental Scale improved.<br><br>No increase in Mental Scale in normal group. |
| Walter et al. (1983) (Chile) | 37 children, all 15 months (10 ID mild anaemic, 12 ID non-anaemic, 15 normal)<br><br>Controlled for ethnic, socio-economic and cultural backgrounds. | At 15 months, all infants received oral iron 3 to 4 mg for 75 days<br><br>Bayley Scales assessed before and 11–15 days after start of oral treatment | Significant increase in Bayley Mental Scale for anaemic infants, not for others, apart from subgroup of ID non-anaemic with two or more biochemical measures who showed after start of oral treatment. significant increase as well. Improvement in cooperativeness and attention span in anaemic infants only. |

conclusion that iron deficiency sufficient to cause anaemia is associated with impaired performance in children. There is sufficient evidence also to indicate the possibility that iron deficiency without anaemia may impair functioning.

Further evidence from intervention studies involving iron supplementation helps to clarify the picture, to some extent.

## 11.4  INTERVENTION STUDIES RELATING IRON SUPPLEMENTATION TO MENTAL AND MOTOR DEVELOPMENT

### 11.4.1  Short-term intervention studies

The most frequent design, despite all the caveats noted above, uses standardized tests of development or intelligence (see Table 11.4 for intervention studies of less than one month).

The Oski and Honig (1978) results are startling: a 13.6 point rise (almost one standard deviation) in the Bayley Mental Developmental Index in a mean time

of 6.8 days, with the increase in scores inversely related to the initial haemoglobin levels. However, the controls also improved their scores by 6 points indicating the possibility of a training effect explaining development quotient (DQ) changes. The improvement in responsiveness is, however, possibly of greater importance than the rise in DQ, as discussed below.

Commenting on this study, Pollitt et al. (1978) pointed to the wide age range of the children involved and argued that infant development lacks sufficient continuity to allow the same function to be tested across the age range quoted.

Lozoff et al. (1982a) gave one week of oral iron therapy to the children already noted to have differed in their Bayley scores but, once again, both control and treated groups improved in their iron and developmental status.

The later study by Oski et al. (1983) found a surprisingly high mean increase of 22 points (1.5 standard deviations) on the Bayley Mental Scale. They postulated that the biochemical processes underlying the behavioural abnormality are more

**Table 11.5** Longer term intervention studies

| Study | Design | Treatment | Results |
|---|---|---|---|
| Palti *et al.* (1983) (Israel) | 464 infants aged 3 months were followed up over 5 years (184 anaemic, 280 normal) | Not known | Anaemic infants had lower scores on Brunet-Lézine or Wechsler Scales at 2, 3 and 5 years independent of treatment. |
| Aukett *et al.* (1986) (UK) | 97 ID anaemic children, 17–19 months<br><br>Identified via screening programme in community child health clinics. Random allocation to treatment or placebo groups. Hb < 80 g/l excluded | Treated group: Ferrous sulphate and vitamin C for two months<br><br>Controls: Vitamin C only for 2 months | Weak association between rise in Hb and number of new developmental skills achieved on Denver Screening Test.<br><br>Significantly more treated children achieved the expected 6 new skills over 2 months. |
| Deinhard *et al.* (1986) (USA) | Children aged 18–60 months<br>*Treated A*: ID anaemic (25)<br>*Control A*: normal<br>*Treated B*: ID non-anaemic (22)<br>*Control B*: ID non-anaemic (23)<br><br>Matching on gender and at least two of age, race, mother's education<br>Exclusion criteria include: birthweight; height and weight; no disease, mental or developmental retardation<br><br>Treatment and placebo controls | Oral iron 6 mg/kg daily for 6 months to both treated groups<br><br>Placebo to controls | 1. Mental Development (Bayley Beales and Stanford-Binet Form L-M):<br>*Group A*: Treated group showed no significant change. Control group showed significant increase at 3 and 6 months; significantly higher than treated group at 3 months.<br>*Group B*: No significant change in either group.<br><br>2. Behaviour rating:<br>*Group A*: Controls more responsive at 3 and 6 months than in treated group.<br>*Group A*: Mean values for emotional tone (happiness): Control group significantly higher at baseline and 3 months but not at 6 months.<br>*Group B*: No significant difference. |
| Lozoff *et al.* (1987) (Costa Rica) | 182 children, 12–23 months (50 ID anaemic, 42 ID intermediate, 19 ID non-anaemic, 36 I depleted non-anaemic, 35 I replete)<br><br>Double-blind random allocated placebo controlled study for first week<br><br>8 possibly confounding variables controlled (see Table 11.2) | 1. One week treatment: Anaemic and intermediate groups had oral iron, intramuscular iron or placebo<br><br>Non-anaemic had oral iron or placebo<br><br>Placebo injection for those not receiving parenteral iron<br><br>2. After one week: The parenterally treated infants and I replete infants given placebo drops for 3 months; all other infants treated with oral iron | Haemoglobin increased in treated groups.<br><br>No treatment effect at 7 days.<br><br>At 3 months the originally observed lower baseline Bayley Mental and Motor scores were no longer evident in ID anaemic infants whose ID and anaemia were corrected.<br><br>Significant lower scores persisted in those with only partial recovery of ID. |
| Seshadri and Gopaldas (1989) (India) | Study 3: 48 boys, 8–15 years<br><br>Randomly assigned to three groups, 16 in each group | Group A: Daily oral iron 30 mg for 60 days<br>Group B: Daily oral iron 40 mg for 60 days<br>Group C: Placebo | Haemoglobin of all treated children responded to treatment.<br><br>Post-treatment tests of visual recall, clerical skills and 2 subtests of the Wechsler Scale showed significant differences in both treated groups and controls on all but one measure. |
| | Study 4: 65 pairs of girls, 8–15 years<br><br>Matched for age, Hb and cognitive test scores | Oral iron 60 mg for 60 days twice in the school year | Treated group had significant benefit on all measures of iron status and on tests of clerical skills and mazes at 8 months, no significant difference at 4 months. |

**Table 11.5**   Longer term intervention studies (contd.)

| Study | Design | Treatment | Results |
|---|---|---|---|
| | Randomly assigned to treated or placebo groups | | 4 months after supplementation, Hb of treated group almost to baseline with concomitant drop in almost all scores. |
| Walter *et al.* (1989) (Chile) | 196 infants at 3 months | Four dietary groups: breast-fed children assigned to fortified cereal or not; bottle-fed children to fortified or unfortified infant formula | |
| | At 9–12 months | 39 ID anaemic, 127 ID non-anaemic, 30 controls identified from 4 dietary groups above | Lower haemoglobin concentration and longer duration of anaemia associated with lower Bayley Mental and Motor scores |
| | At 12 months, double-blind placebo controlled trial for 10 days | Oral iron or placebo for 10 days | Slight non-significant increase in scores in treatment and placebo group |
| | At 12.5 months, all ID children | Oral iron to all children for 3 months | Slight non-significant increase in scores |
| Idjradinata and Pollitt (1993) (Indonesia) | 126 infants, 12–18 months | Oral iron 3 mg/kg per day for treated children | ID anaemic group who received treatment had significantly higher Bayley Scale Scores than placebo group and the scores were similar to those in iron replete groups. The ID non-anaemic group who received treatment had similar scores to the placebo group at the end of the study. |
| | Selection criteria as Lozoff (1987) Three groups identified: | Placebo for others | |
| | 50 ID anaemic 29 ID non-anaemic 47 I replete, non-anaemic | Tested 1 day before and immediately after 3 months' treatment | |
| | Each group randomly assigned to treatment or placebo | | |

rapidly achieved using parenteral iron but the increase is so great, especially with a test of relatively low sensitivity, that other factors were probably involved.

### 11.4.2   Longer term intervention studies

Table 11.5 shows details of longer term intervention studies. Aukett *et al.* (1986) employed a double blind, randomized, placebo controlled, design over 2 months in iron deficient children, mainly Asians living in Birmingham, aged 17–19 months. Despite the use of a relatively crude screening instrument which would not be expected to pick up fine changes, significantly more of the treated group achieved the expected rate of development, although the difference in scores between the two groups was not significant. A number of children failed to show a satisfactory improvement in development despite a definite rise in haemoglobin concentration indicating that

iron deficiency was only one factor, albeit treatable, in the slower development of underprivileged children.

Perhaps most interesting in the study by Deinhard *et al.* (1986) was not that they found no significant improvement from the baseline mental developmental score at either 3 or 6 months after therapy but that there were significant differences between the experimental and control groups at 3 and 6 months on measures of responsiveness. In this they confirmed the earlier observations of Oski and Honig (1978).

The ceiling effect could have been operating in the Lozoff *et al.* (1987) study in Costa Rica, where all the non-anaemic children had Bayley scores over 100 and the study lacked both a non-anaemic treated group and an anaemic placebo control. Abnormal affective behaviour was associated with anaemia and a poor Bayley Mental Scale score in the anaemic group.

Four sequential supplementation studies were carried out in Baroda, India, by Seshadri and

Gopaldas, 1989. The first two studies had serious methodological problems and are not considered here. The third study compared oral supplementation of 30 and 40 mg iron with a placebo control. The results were far from clear cut: there was no differential effect of haemoglobin correction by dosage and both the supplemented groups improved in their scores on the psychological tests, the effect being largely due to the improvement in scores in the anaemic children. Only on one test did the boys taking the 40 mg iron supplement perform better than those who had received 30 mg iron, but a ceiling effect could explain this finding.

The fourth study was conducted on 83 pairs of girls aged 8–15 years, matched as before. Only 65 pairs remained by the end of the period and there is no information on the drop outs. However, this is the only reported ABA study, in which haemoglobin levels were measured at baseline (A) and then re-examined after intervention had stopped. Haemoglobin levels returned almost to those noted at baseline for months after supplementation ceased and the associated decrease in test scores provides impressive support for the causal nature of iron deficiency.

It is not easy to take clear-cut conclusions away from the prospective study of Walter *et al.* (1989a) but one of their more interesting findings was that the children who had been anaemic for longer showed the greatest deficit in both Mental and Motor scores on the Bayley Scales. Iron supplementation given to all iron deficient children did not lead to an improvement in the scores. This raises the question of whether there is a cumulative effect of iron, or of the environment in which the children are brought up.

Idjradinata and Pollitt (1993) used the Bayley Scales in a randomly allocated, double blind, placebo controlled study of children aged 1–2 years, in Indonesia. An analysis of covariance was used to allow for some between-group differences and the control for possibly confounding variables was thorough. This is the best designed intervention study so far reported and the results confirm the picture of the value of supplementation for iron deficient anaemic children.

Looking at these studies together there is a four to one ratio of iron being causally related to a variety of outcome measures of development, including the results of two randomized, double blind, placebo controlled trials. A previous review (Simeon and Grantham-Macgregor, 1990) concluded that the evidence of a causal link between iron deficiency anaemia and poor developmental performance was stronger in older children than in preschool children but the study from Indonesia is strong evidence of a causal link in young children.

### 11.4.3  Intervention studies using tests of processing and other cognitive skills

Some experts have argued against the use of global aggregate measures in this field as they lack sensitivity. They believe that one should try to tease out the components of functioning likely to be affected by iron deficiency. Table 11.6 summarizes some studies which have tried to do this.

The definition of iron deficiency used by Pollitt *et al.* (1983) in their USA study of young children was that which responds to iron therapy. No differences were found in global IQ (Stanford Binet Form L-M) after the treatment period but there was some evidence that the experimental group performed better on tests of discrimination learning, possibly an effect of iron deficiency on attention. Iron deficient children might have encountered problems in control processes affecting the reception of information rather than in processing or retrieval.

Pollitt *et al.* (1986) achieved somewhat similar results in a study in Guatemala and found that discrimination learning was apparently affected by iron deficiency. However, their study of Egyptian preschool children failed to replicate the findings on discrimination learning (Pollitt *et al.*, 1985).

The study by Soewondo *et al.* (1989) is one of the few to use an analysis of covariance, in this case to take account of differences in Picture Vocabulary Test scores. They also used an anaemic placebo control group who scored badly compared with those who were iron replete. There was some evidence that specific cognitive functions, notably those related to visual attention and concept acquisition, were most affected.

As usual, the conclusions are equivocal. We are presented with a picture of inconsistency which tends to support the notion of noise in the system preventing conclusive findings. Some specific aspects of cognitive functioning are probably influenced by iron deficiency but further research is required to clarify the picture.

### 11.4.4  Intervention studies using miscellaneous outcome measures

Table 11.7 shows details of intervention studies which used miscellaneous outcome measures. Pollitt *et al.* (1989a) looked at the effect of iron treatment on children, already mentioned, who had completed the Matching Familiar Figures Test. After the period of intervention the efficiency of the treated children was significantly greater than that of those given the placebo, the former scoring at a similar level to the controls.

**Table 11.6**   Intervention studies using tests of processing and other cognitve skills

| Study | Design | Treatment | Results |
|---|---|---|---|
| Pollitt *et al.* (1983) (USA) | 30 children aged 3–6 years mildly ID, normal protein energy nutrition status, good health<br><br>Treated group: All those with ID anaemia and half those ID non-anaemic<br><br>No placebo | Oral iron 4 to 6 months | Treated group reached normal iron status.<br><br>Tests based on information processing model plus Stanford-Binet IQ<br><br>No significant difference in IQ.<br><br>Group difference between two treated groups consistent with control process deficit but not learning rate defect. |
| Pollitt *et al.* (1985) (Egypt) | 1. *Preschool children* (88 anaemic, 69 normal)<br><br>Treated and placebo control groups | Treated group 1: oral iron 50 mg daily for 4 months | Substantial rise in iron status in both treated groups (increase also in control groups).<br><br>Picture vocabulary test failed to discriminate between groups.<br><br>No significant effect of treatment on discrimination test or concept learning. |
|  | 2. *6–12 years* (30 anaemic, 40 normal)<br><br>Treated and placebo control groups | Treated group 2: oral iron 50 mg daily for 4 months. | *Baseline*: Matching familiar figures test, anaemic children significantly less efficient than non-anaemic.<br><br>After 4 months, treated groups were significantly more efficient than placebo control group |
| Pollitt *et al.* (1986) (Guatemala) | 3–6 years (25 ID anaemic, 25 I replete) Selection criteria as Table11. 2<br><br>Treated and placebo control groups | Oral iron daily for 11 to 12 weeks<br><br>Or placebo | Substantial rise in iron status in treated groups, some increase in control groups.<br><br>ID children had significant rise in discrimination learning following treatment.<br><br>No effect on conceptual learning. |
| Pollitt and Kim (1988) (Java) | 119 children, 9–11 years (78 ID anaemic, 41 normal)<br><br>Treated and placebo control groups | Oral iron daily for 3 months | Substantial improvement in iron status in treated group only.<br><br>Anaemic treated group did significantly better than controls in school achievement tests after treatment.<br><br>Anaemic placebo group significantly worse after 3 month period. |
| Soewendo *et al.* (1989) (Indonesia) | 176 children, mean age 56 months (49 ID anaemic, 57 I deplete, 70 I replete)<br><br>Half in treated group, half in placebo control groups | Oral iron supplement 50 mg/day for 8 weeks | Significant change in almost all iron indicators in treated groups; none in placebo.<br><br>Analysis of covariance used, with age and initial scores. as covariates.<br><br>No significant effect of treatment in picture vocabulary test.<br><br>3 of 9 cognitive tests showed interaction with iron status.<br><br>Treatment effects related to visual attention and concept acquisition. |

**Table 11.7** Intervention studies using miscellaneous outcome measures

| Study | Design | Treatment | Results |
|---|---|---|---|
| Soemantri *et al.* (1985) (Indonesia) | 119 children, 6–12 years (78 anaemic, 41 normal) | Oral iron 10 mg daily for 3 months<br><br>Worm infestation treatment | Difference between anaemic and non-anaemic children remained after treatment but improvements in iron status of treated anaemic children associated with significant rise in school achievement scores. |
| Heywood *et al.* (1989) (Papua New Guinea) | 96 infants, 2 months<br><br>Cohort study matching for gender, birthweight, mother's education, housing, antenatal care, obstetric history<br><br>Randomly assigned to treatment or placebo group<br><br>All checked for malaria parasites | Intramuscular iron (3 ml Imferon) or placebo | Significant rise in Hb concentrations in treated group.<br><br>Tested at 12 months using visual habituation paradigm<br><br>Treated group had longer fixation time, no effect on habituation.<br><br>Significant interaction with malaria. |
| Pollitt *et al.* (1989a) (Thailand) | 9 to 11 years (ID anaemic, I deplete, I replete)<br><br>Treatment or placebo groups<br><br>Control for gender, school and grade | Oral iron 50 mg/day for first 2 weeks, then 100 mg/day for 14 days<br><br>Or placebo<br><br>All treated with albendazole | Treated group had significant improvement in iron status Improvement also in placebo group.<br><br>No treatment effects on Raven's Matrices, Thai language test or maths |

In Papua New Guinea, Heywood *et al.* (1989) looked at 96 children from a longitudinal cohort with matching for gender and birth weight, with random assignment to treatment or placebo groups, the former receiving intramuscular iron at 2 months. Checks were carried out for malaria parasites. The outcome measure used, based on a visual habituation paradigm, is suggested to be a better predictor of later intelligence than standard developmental tests (Bornstein *et al.*, 1986). Testing at 12 months showed that treated children responded to visual stimuli to a higher degree than those not treated but it is not easy, given our present knowledge, to interpret these findings other than to say that there appear to be some differences as a function of iron. The significant interaction with malaria shown in this study complicates matters further.

High face value is put on tests related to performance in school. Soemantri *et al.* (1985) looked at the effects of iron supplementation on measures of school performance in a poor area of Central Java. Three months of oral iron therapy resulted in substantive increases in haematological indices – changes which, in the anaemic children, were associated with improvement in school achievement scores.

Pollitt *et al.* (1989a) replicated this study in Thailand, without finding similar results. A confounding variable in both studies was the treatment for possible worm infestations which could have brought additional benefits to all children, masking the results of the iron supplementation in the Thai children but not in those from Indonesia. The initial worm burdens and the efficacy of eradication are not clear and may account for the non-replication. Furthermore, deficiencies of thiamin, riboflavin and vitamin A are common in South East Asia and could affect the response to iron therapy.

### 11.4.5 Effect of iron on non-cognitive function

Iron deficient children have often been reported to be irritable, listless, perceptually restricted, distractible and apathetic. Walter *et al.* (1989a) found that iron deficient anaemic infants were more unhappy than controls. Infants with low ferritin values were more fearful, less attentive and more vocal (Deinhard *et al.*, 1986). Lozoff *et al.* (1982a) found that anaemic infants were more withdrawn, hesitant, fearful and tense and less reactive. These behaviours undoubtedly influence test performance and the children's general ability to make use of their environments.

Treated children show increased responsiveness (Oski and Honig, 1978; Deinhard *et al.*, 1986) and it is possible that the improvements in test scores reported in many intervention studies could be a function not so much of cognitive changes *per se* as of responsiveness to the environment, of alertness and other behavioural variables which enable the child to maximize available ability. This goes some way to explaining major changes such as those reported by Oski *et al.* (1983) and lends weight to the notion of correcting deficiency.

## 11.5  AREAS WITH LESS RESEARCH EVIDENCE

There remain a number of questions which have been put but answered only partially in a way that does no more than scratch the surface.

One of these is the topic of the critical period. Is there a time when children are especially vulnerable to a lack of iron, and if so, when is this? Linked with this is the question of the reversibility of iron deficiency: is there an age or a period after which the effects are more or less permanent?

Lozoff *et al.* (1982b) put forward the view that there might be a critical period: their 19–24-month-old anaemic infants had a particular weakness in language, being still at the single word level. The same authors (Lozoff *et al.*, 1982a) argued that the duration of anaemia is of importance, as had Cantwell (1974).

Lozoff *et al.* (1982b) also found an age effect on the Bayley Mental Scales, the older anaemic children doing markedly less well. They concluded that the test is possibly more sensitive in the older age range (i.e. 19+ months) or that the longer the child has been anaemic the greater the effect of iron deficiency. They also mentioned that there might still be a hitherto undiscovered confounder at work.

On the other hand, Simeon and Grantham-McGregor (1990) argued that although there was some evidence to the contrary, there was little clear-cut evidence that children under the age of 2 years benefited from treatment. However, their review was published before the intervention study of Idjradinata and Pollitt (1993). This leaves little doubt concerning the improvement in psychomotor development following treatment in children aged 12–18 months.

If it cannot be demonstrated that at least part of the behavioural deficit is reversible with treatment, then the hypothesis that iron deficiency is a cause and not a mere associate of the deficit is difficult to sustain. But even if it were to be accepted that some deficiency can be reversed in some children, a major question remains: whether deficiency incurred before a certain age or to some degree may be irreversible. Lozoff *et al.* (1987) and Walter *et al.* (1989) speculate on this but there are uncertainties about the efficiency of delivery of supplementation in some cases.

# 12

# IRON AND CORONARY HEART DISEASE

## 12.1 HAEMATOCRIT AND CHD RISK

The suggestion that a poor iron status may be protective against coronary heart disease (CHD) was put forward indirectly by Stokes (1962), who proposed that the lower rates of CHD in women could be partly attributed to their lower haematocrit levels. Subsequent studies confirmed that both the risk and size of cardiac and cerebral infarctions were inversely related to haematocrit and haemoglobin levels (Kannel et al., 1972; Elwood et al., 1974; Bottiger and Carlson, 1980; Harrison et al., 1981; Cullen et al., 1981). These authors, however, ascribed these changes in cardiovascular risks to blood viscosity.

## 12.2 STORED OR CIRCULATING IRON AND CHD RISK

### 12.2.1 Indirect evidence

Sullivan (1981) was the first to suggest that low iron stores per se may explain the reduced risk of CHD in premenopausal women. He supported his argument with evidence that there is a rapid increase in CHD risk in women following the menopause in line with the increase in stored iron as measured by serum ferritin; that the higher risk of CHD in men is associated with a much earlier rise in stored iron; and that iron storage diseases are associated with an increased risk of heart disease which is reduced on treatment with the iron chelator, desferrioxamine. He also postulated that the low rates of CHD in developing countries may be related to the presence of endemic iron deficiencies.

### 12.2.2 International comparisons

International comparisons of liver iron stores and CHD mortality suggested that lower iron stores could protect against CHD (Lauffer, 1991). The correlation between the reported median value of

hepatic storage iron and the CHD mortality rate was moderate ($r = 0.55$, $P < 0.025$).

All of this is weak epidemiological evidence. In spite of the parallels between the gender ratio of CHD mortality and serum ferritin levels by age (Sullivan, 1992b), there are many other factors (e.g. blood pressure, cholesterol levels, sex hormones, smoking, etc.) which may account for differences in CHD risk. The type of tissue damage associated with iron overload is different from that associated with infarction in patients with normal iron stores. International studies are notoriously weak as there are many differences in lifestyle and other risk factors between countries, and these cannot be controlled in such geographical comparisons.

### 12.2.3 Case-control studies

The above epidemiological findings have not been confirmed in case-control studies. Riemersma et al. (1991) found no significant differences in serum iron, iron binding capacity and serum ferritin between 111 Scottish men with angina, aged between 35 and 54 years, and 397 age-matched controls.

A nested case-control study of 238 cases of myocardial infarction (MI), drawn from a 1982 cohort of 14 916 US male physicians aged 40–84 years, showed an insignificant increase in relative risk of MI (Stampfer et al., 1993) with plasma ferritin levels. Mean ferritin levels in cases and controls (250 μg/l vs 222 μg/l) were not significantly different.

### 12.2.4 Prospective studies

(a) Positive associations of iron status and CHD

The first convincing evidence of higher risks of CHD associated with raised levels of stored iron came from a 3-year follow-up of nearly 2000 men aged 42–60 years recruited into the Kuopio Ischaemic Heart Disease Risk Factor Study (Salonen et al., 1992b). Men with serum ferritin greater than or equal to 200 μg/l had a significant

2.2-fold risk of acute MI compared with men with a lower serum ferritin, after adjusting for other risk factors including smoking, $VO_{2max}$, serum high density lipoprotein (HDL) and blood pressure. The risk was highest (5-fold) in those who had both a serum ferritin above 200 μg/l and a serum low density lipoprotein (LDL) cholesterol of greater than 5.0 mmol/l. Even when haemoglobin and haematocrit were included as covariates, the relative risk remained at 2.0 and was the same in relation to serum ferritin in smokers and non-smokers.

The only other prospective study to show any positive association between serum iron levels and CHD has been the 17-year follow-up of a Canadian cohort of nearly 13 000 people aged between 35 and 79 years which included 224 cases of fatal MI (Morrison et al., 1994). An increased relative risk was found among men (RR = 2.18) and women (RR = 5.53) with abnormally high serum iron concentrations (greater than 175 μg/dl), presumed to represent iron overload. Risks for the highest serum iron category were higher for men with elevated serum cholesterol.

(b)   Negative or lack of association between iron status and CHD

In contrast to the results from Finland, several prospective studies have failed to show a positive association between iron status and CHD. A re-analysis of the NHANES (National Health and Nutrition Examination Surveys) data (1971–1987) (Cooper and Liao, 1993) found small but significant reductions in risk of MI and CHD in both men and women with increases in serum iron and transferrin saturation (serum ferritin was not measured).

Another large prospective study following up people from the NHANES did not show a positive association between the amount of circulating iron (measured as transferrin saturation) or dietary iron intake (measured by 24-hour recall) and relative risk of CHD (Sempos et al., 1994). This 13-year follow up of 4518 men and women indicated a possible inverse association between circulating iron stores and total mortality. Circulating iron is poorly correlated with iron stores except when they are fully depleted or overloaded and measurements of serum transferrin saturation are less stable than those of serum ferritin. It would be interesting to know whether levels of transferrin saturation associated with iron deficiency (< 15%) or iron overload (> 60%) were related to CHD.

In another cohort study of 2036 Icelandic men and women aged between 25 and 74 years who were followed for an average of 8.5 years, no association was seen between serum ferritin and risk of MI (Magnussen et al., 1994). Total iron binding capacity (TIBC), however, was found to be a strong

independent negative risk factor in men (RR = 0.95) and the authors have suggested that transferrin might have a protective role in CHD due to its antioxidant role in binding iron. They conclude that the level of circulating iron is more relevant than the level of iron stores.

## 12.3   DIETARY IRON AND CHD RISK

In the Finnish study mentioned in section 12.2.4(a), Salonen et al. (1992a) showed that dietary iron intake was significantly associated with an increased risk of MI (5% increase in risk for each 1 mg increase in dietary iron). Ascherio et al. (1994), however, showed no significant increases in risk of fatal or non-fatal MI or risk of bypass surgery or angioplasty in relation to increases in total dietary iron intake or non-haem intake in a study of nearly 45 000 American health professionals who were aged between 40 and 75 years in 1986. However, haem iron (obtained mainly from red meat) was significantly related to the risk of MI (RR = 1.42) and was also significantly related to the risk of fatal CHD (RR = 2.33).

Can these results be reconciled? There are a few possibilities that must be considered:

- The Finnish study did not calculate haem iron intake but the high serum ferritin values and high meat intakes would suggest a higher than normal proportion of haem iron intake in the Finnish men.
- In the American study, haem iron intake was positively associated with MI but not with risk of bypass surgery and this was interpreted by the authors as showing that high iron stores might worsen the myocardial injury caused by ischaemia rather than promoting atherosclerosis. However, the fact that the relationship between haem iron intake and MI was limited to men who were not taking vitamin E supplements, and was considerably worse in diabetics, also led these authors to suggest that excess dietary haem iron may adversely affect coronary risk only in the presence of oxidative stress from other sources.

## 12.4   HAEMOCHROMATOSIS AND CHD RISK

If iron has a substantial role in the pathogenesis of atherosclerosis and CHD, people homozygous or heterozygous for haemochromatosis might be expected to experience high rates of CHD. Powell et al. (1993) conducted a retrospective analysis of a cohort of 425 subjects with genetic haemochromatosis diagnosed between 1951 and 1986. The mean follow-up was 10 years with 161 deaths and the relative risk for death from CHD was 0.77 (95% confidence limits, 0.46 to 1.20). In contrast, the relative risk for deaths from liver cancer exceeded 200. Furthermore, a study of trace metal content of cardiac muscle confirmed the lack of any relationship between iron content of either serum or tissue and the presence of CHD (Oster et al., 1989). It is therefore unlikely that any association between serum ferritin and CHD is a true reflection of iron-induced disease (Burt et al., 1993).

Sullivan (1995) has challenged this view and believes that the apparent lack of increased CHD in haemochromatosis may simply mean that the modification of low density lipoprotein (LDL) is maximal or nearly maximal at a low level of excess iron. Thus iron depletion could still protect against CHD by making the myocardium less vulnerable to ischaemia, even though there is no evidence of iron overload causing CHD.

## 12.5 POSSIBLE MECHANISM FOR INVOLVEMENT OF IRON IN CHD

Most authors have proposed that iron's putative role in CHD is via free radical production and modification of low density lipoprotein, i.e. it acts as a pro-oxidant. This subject is fully covered in Chapter 6, which also considers the possibility that increased serum ferritin values associated with ischaemic heart disease may be part of an acute phase response rather than simply related to increased iron stores.

Some authors have proposed that iron has the ability to activate the immune system to produce cytotoxic products which cause tissue damage; increased ferritin concentrations might simply result from chronically activated cellular immunity (Weiss et al., 1993a) (Chapter 10).

# 13

# IRON AND CANCER

## 13.1 INTRODUCTION

The idea that iron, and in particular iron overload, plays a role in the development of neoplasia has aroused considerable interest. Nevertheless, despite some persuasive arguments that iron overload favours development of cancer (Weinberg, 1992), much of this area remains controversial. Two separate situations need to be considered:

- Can abnormal forms of iron present in severe iron overload induce carcinogenesis and promote tumour development?
- Can modest iron overload, or even an iron status in which parameters are at the upper end of the normal range, predispose to cancer?

## 13.2 SEVERE IRON OVERLOAD AS A POSSIBLE CAUSE OF CANCER

### 13.2.1 Evidence

It is now widely recognized that in individuals with severe iron overload, such as is found in hereditary haemochromatosis, there is an increased incidence of liver cancer. Indeed this is a significant cause of death in such patients, particularly when cirrhosis is present (Bomford and Williams, 1976; Niederau et al., 1985; Fargion et al., 1992). Sufferers from transfusion dependent thalassaemia have an increased risk of cancer generally (Zurlo et al., 1989). There is also increased risk of lung and laryngeal cancers in miners of iron ore, metal workers and workers in iron foundries (Cole and Goldman, 1975). In some studies, development of sarcoma has been reported at the site of injection in animals receiving parenteral iron dextran (Richmond, 1959; Magnusson et al., 1977) but there has been no more recent confirmation. None of these conditions, however, is likely to reflect the relationship between cancer risk and iron status in the normal range in the general population.

In all the above situations, severe general or local iron overload occurs, and iron is present in plasma and tissue in abnormal forms. Biochemical studies have provided strong evidence that such forms of iron could favour development of cancer, by either initiation or promotion of cancer cell growth.

### 13.2.2 Possible mechanism via initiation

Iron salts can produce strand breaks in DNA when incubated with purified DNA (Imlay et al., 1988) or isolated mitochondria (Hruszkewycz, 1988) and iron overloaded rats show an increased number of strand breaks in hepatic DNA (Edling et al., 1990). Strand breakage in DNA is associated with initiation and promotion of chemically induced carcinogenesis (Hartley et al., 1985). The mechanism of strand breakage probably involves iron-catalysed production of hydroxyl radicals which can abstract hydrogen atoms from both the ribose moiety and purine/pyrimidine bases (Pryor, 1988). This mechanism may be focused on to the DNA strands through the binding of iron to DNA (Halliwell and Gutteridge, 1990).

The oxidative/catalytic mechanisms responsible for these events require the presence of iron atoms with free coordination linkages (Graf et al., 1984) and thus cannot be mediated by transferrin-bound iron, nor by iron resident within the core of ferritin. Consequently these potentially carcinogenic reactions are only likely to occur when iron levels exceed the binding capacity of ferritin or transferrin. Saturation of transferrin occurs only in severe iron overload but potentially reactive iron might be present in hepatic cells under more modest conditions of iron overload.

### 13.2.3 Possible mechanism via promotion of cancer cell growth

In vitro studies suggest that tumour cells show a greater ability than normal cells to grow and/or survive in the presence of high levels of non-transferrin bound extracellular iron (Taetle et al., 1985;

Sturrock *et al.*, 1990), which tends to be toxic for normal cells (Djeha and Brock, 1992a,b). In particular, non-transferrin-bound iron supports growth of human T leukaemia cells but suppresses proliferation of normal T cells (Brock *et al.*, 1993).

Thus excess iron may not only initiate the carcinogenic process but also favour growth of tumour cells over others, such as cells of the immune system, that might otherwise help to limit tumour development.

## 13.3  MODERATE IRON OVERLOAD CONDITIONS AS A POSSIBLE CAUSE OF CANCER

### 13.3.1  Direct evidence from studies measuring iron status

Serum ferritin and transferrin, total iron binding capacity (TIBC) and transferrin saturation have each been used as markers in studies relating iron status with cancer. In general, healthy individuals with high body iron stores show high levels of transferrin saturation and serum ferritin, and low levels of serum transferrin (TIBC).

Four studies have been consistent with the hypothesis that high body iron stores increase the risk of cancer, in men or in women. One study from Finland found no relationship between transferrin saturation and cancer risk but a more detailed analysis of this study does show an increased risk of certain types of cancer in men with high transferrin saturation (Stevens, 1993).

Beasley *et al.* (1981) enrolled more than 20 000 male government workers in Taiwan in a prospective study of the health effects of hepatitis B virus. Two years later, 239 had died of cancer or were alive with a diagnosis of liver cancer. Controls were matched for age and serum hepatitis B surface antigen status. For liver cancer, the serum ferritin was significantly higher among cases; serum transferrin was also higher although not significantly so. In addition, serum albumin was significantly lower among cancer cases and this was true even after adjustment for smoking, which could act as a confounder.

Stevens *et al.* (1988) published the results of an analysis of cancer incidence in the first National Health and Nutrition Examination Survey (NHANES) population. More than 14 000 subjects completed an extensive dietary questionnaire and gave blood samples between 1971 and 1975. There was a significant association of serum iron and transferrin saturation and an inverse association of TIBC with risk of all cancers over an approximately 10-year follow-up period. In particular, cancers of the lung,

colon, bladder and oesophagus appeared most strongly related to transferrin saturation. The relative risk for cancer of the colon was 4.69, comparing people with the highest and lowest transferrin saturation values. Yip and Williamson (1989) have criticised the NHANES study on three points:

- 'Inflammatory processes' could have invalidated the conclusions.
- The top quartile for transferrin saturation (above 36.7%) is not normally considered very high.
- The statistical analyses were not considered valid.

### 13.3.2  Indirect evidence from observational studies

By far the largest body of epidemiological evidence relating cancer risk to iron is from studies which have related meat consumption with cancer risk. The assumption is made that iron status is strongly associated with haem iron consumption from meat (Cook, 1990).

Ecological (correlational) studies have shown consistent positive associations between consumption of meat and cancer mortality (Merlo *et al.*, 1991; Decarli and La Vecchia, 1986). Total meat consumption (1968–69) correlates well with mortality from all cancers (1978–82) in 10 European countries (for males $r = 0.78$, females $r = 0.59$) and with specific cancers: breast (females $r = 0.6$), colon (males $r = 0.79$, females $r = 0.67$), and rectum (males $r = 0.58$, females $r = 0.53$). Similarly in 20 regions of Italy, correlations are strong for red meat consumption and all cancer mortality ($r = 0.83$ in males and females) and for specific sites: breast (females $r = 0.76$), colon (males $r = 0.89$, females $r = 0.81$), rectum (males $r = 0.74$, females $r = 0.65$) and stomach (males $r = 0.64$, females $r = 0.57$).

Ecological comparisons of this type, however, provide only weak evidence for a causal link because of the potential influence of other dietary factors which may affect total iron availability (e.g. non-haem iron from vegetables, grains and pulses) or iron absorption (e.g. vitamin C, dietary fibre, phytate), or which may influence disease susceptibility (e.g. fat intake, antioxidant status).

### 13.3.3  Indirect evidence from cohort and case-control studies

More convincing evidence can be derived from cohort and case-control studies. Tables 13.1 and 13.2 provide a summary of some recent studies of

**Table 13.1**   Summary of some studies relating food consumptiom to breast cancer in women

| Study | Country | Subjects | Cases | Increased risk | Decreased risk | No effect |
|---|---|---|---|---|---|---|
| *Cohort:* | | | | | | |
| Hirayama (1990) | Japan | 26518*  142857 | 243  142 | Meat | Soyabean paste | |
| Kinlen (1982) | Britain (nuns) | 2813 | 62 (deaths) | | | Meat vs non-meat |
| Mills *et al.* (1989) | California (7th Day Adventists) | 16190 | 186 | | | Meat |
| Nomura *et al.* (1978) | Hawaii | 6860 | 86 | Beef | | |
| Vatten *et al.* (1990) | Norway | 14500 | 152 | Meat | | |
| *Case-control:* | | | | | | |
| Ewertz and Gill (1990) | Denmark | 1486 | | Meat | | |
| Goodman *et al.* (1992) | Hawaii | 272 | | Meat, sausages, eggs | | |
| Hislop *et al.* (1986, 1988) | Canada | 436, 846 | | Beef, fatty meat in childhood | | |
| Ingram *et al.* (1991) | Australia | 153 | | Red meats, eggs | | |
| Lee *et al.* (1991), Lee (1992) | Singapore | 216 | | Red meat, animal protein | Soya products, SFA and PUFA, beta-carotene | |
| Levi *et al.* (1993) | Switzerland | 107 | | Pork | Green vegetables | |
| Lubin *et al.* (1981) | Canada | 577 | | Beef, pork, eggs | | |
| Negri *et al.* (1991) | Italy | 2860 | | Fat meat | Green vegetables | |
| Nomura *et al.* (1985) | Hawaii | 344 | | | | Dietary iron |
| Talamini *et al.* (1984) | Italy | 368 | | Meat | | |

SFA = saturated fatty acids; PUFA = polyunsaturated fatty acids.

* Study also includes men

**Table 13.2**  Summary of some studies relating types of food to colorectal cancer

| Study | Country | Subjects | Cases | Risk factors | | |
|---|---|---|---|---|---|---|
| | | | | Increased risk | Decreased risk | No effect |
| *Cohort:* | | | | | | |
| Heilbrun *et al.* (1989) | Hawaii | 7833 M | 951 | | Vitamin C, fibre plus low fat | Dietary iron |
| Hirayama *et al.* (1990) | Japan | 26518 M&W | 563 | Meat with low/no vegetables | Meat with vegetables | |
| Kinlen (1982) | Britain (nuns) | 2813 W | 31 | | | Meat |
| Morgan *et al.* (1988) | California (7th Day Adventists) | 25493 M&W | 182 | Eggs | | |
| Thun *et al.* (1992) | USA | 764343 M | 1150 | | Vegetables, high fibre cereals | |
| Willett *et al.* (1990) | USA | 89538 W | 150 | Red meat, beef, processed meat, liver | Chicken | |
| *Case-control:* | | **Subjects (cases) M & W** | | | | |
| Benito *et al.* (1991) | Majorca | 286 | | Red meat | Cruciferous vegetables, fibre, pulses | |
| Bidoli *et al.* (1992) | Italy | 123 colon, 125 rectal | | Red meat, eggs | Spinach | |
| Hu *et al.* (1991) | China | 336 | | Meat products, low vegetables | | |
| Kune *et al.* (1987) | Australia | 715 | | High beef plus high fat and milk | Beef plus high fibre, vegetables, pork | |
| Manousos *et al.* (1983) | Greece | 100 | | Beef, lamb, rabbit | | |
| Martinez *et al.* (1990) | Puerto Rico | 461 | | Meats, fibre | | |
| Miller *et al.* (1983) | Canada | 348 colon, 194 rectal | | Beef, eggs, lamb | | |
| Negri *et al.* (1991); La Vecchia *et al.* (1988) | Italy | 339 colon, 236 rectal | | Beef, veal | Green vegetables | |
| Riboli *et al.* (1991) | France | 389 | | | Vegetables, fruit, iron | |
| Tuyns *et al.* (1988) | Belgium | 453 colon, 365 rectal | | | Fibre, iron | |
| Vlajinac *et al.* (1987) | Romania | 81 | | Beef, pork, lamb | | |
| Young and Wolf (1988) | USA | 366 | | Luncheon meat | | |

M: men, W: women

diet and cancer which have included an assessment of risk in relation to meat consumption or iron status. In some studies, risks associated with factors which may influence iron or antioxidant status (e.g. intake of vegetables or vitamin C) have also been reported.

### (a)  Breast cancer

For breast cancer, three out of five cohort studies and 10 out of 11 case-control studies have shown an increased risk of mortality or morbidity with increased meat consumption, usually red meat (Table 13.1). In many studies, there is a conflict between the likely enhancement of iron absorption and increase in iron stores which may be associated with the presence of vitamin C in green vegetables, and the protective role that β-carotene, vitamin C and other compounds present in plants may play as antioxidants. In addition, the relationship between breast cancer and iron from meat may be confounded by the actions of saturated fatty acids also present in meat.

### (b)  Colorectal cancer

Similar comments apply to studies of diet, iron and colorectal cancer (Table 13.2). Many cohort studies have reported meat consumption as a risk factor for colorectal cancer. Hirayama (1990) showed an association of cancer risk with meat consumption only when vegetable consumption was low. This is stronger evidence in favour of a meat/iron influence on colorectal cancer risk when it is not countered by the action of antioxidants. Willett *et al.* (1992) showed a positive association between red meat consumption, iron status and cancer but Kinlen (1982) showed no influence of meat consumption. Heilbrun *et al.* (1989) showed no effect of dietary iron intake on cancer risk, but this may be confounded by the antioxidant effects of vitamin C and the potential chelating effects of a high dietary fibre intake, particularly in the presence of a low fat intake.

The case-control studies of diet and colorectal cancer show a less consistent pattern of risk. Two large studies (Riboli *et al.*, 1991; Tuyns *et al.*, 1988) showed increased dietary iron intake assessed by diet histories to be associated with a decrease in cancer risk.

### (c)  Cancer at other sites

Most of the cohort and case-control studies of cancer at a number of sites (stomach, lung, oesophagus, prostate, liver, pancreas) show positive associations between cancer risk and meat consumption or iron intake (Nutrition Society, 1993). There are anomalies, however. Ziegler *et al.* (1981) showed low meat consumption to be associated with increased risk of oesophageal cancer (possibly confounded by low energy intakes in general), and Kaul *et al.* (1987) showed low intakes of dietary iron associated with prostate cancer.

The consistency of findings relating cancer risk to meat consumption is better than that for cancer risk and fat consumption. This points to a factor in meat, which could be iron. The evidence is not wholly consistent, however, either between studies or between sexes (Selby and Friedman, 1988; Takkunen *et al.*, 1989) and the whole of the evidence rests on the assumption that iron stores are directly related to meat consumption.

## 13.4  POSSIBLE MECHANISMS FOR MODERATE IRON OVERLOAD AS A CAUSE OF CANCER

### 13.4.1  Mechanisms via transferrin

While there are clear mechanistic explanations for the high incidence of hepatoma in individuals with severe iron overload, the mechanisms by which moderately increased amounts of iron can increase, or iron deficiency reduce, the incidence of cancer are less obvious. It is well established that transferrin can support the growth of many tumour cells *in vitro* through donation of iron (Sussman, 1992) but the mechanism of uptake is essentially the same as in other normal cells which require transferrin-bound iron for growth and development. Suggestions that transferrin may be a specific tumour cell growth factor (Cavanaugh and Nicolson, 1991) are therefore difficult to sustain and it is unlikely that modest increases in transferrin saturation can significantly alter iron availability to tumour cells, or would favour development of tumour cells over normal cells. The ability of some tumour cells to synthesize transferrin might confer a growth advantage when tumours are starved of circulating transferrin by poor vascularization (Vostrejs *et al.*, 1988) but in such a situation it is transferrin rather than iron that is the limiting factor. It is also possible that tumour cells might be able to use ferritin as a source of iron but, although ferritin receptors have been identified on some tumour cell lines (Fargion *et al.*, 1988), their function is as yet unknown.

Thus it is difficult to see how modest increases in iron status could favour development of tumours; other mechanisms need to be sought to explain possible correlations between iron status and cancer in normal individuals.

### 13.4.2 Possible mechanisms via impairment of immune surveillance

An alternative explanation for a relationship between neoplasia and iron status is through an iron-induced impairment of immune surveillance mechanisms (Chapter 10). Two mechanisms of potential importance in anti-tumour immunity are natural killer (NK) cells and cytotoxic activated macrophages. Although it has been reported that iron overload might impair NK cell activity in β-thalassaemia patients (Akbar *et al.*, 1986), the failure of desferrioxamine therapy to restore NK function (Gascón *et al.*, 1984) and the lack of any defect of NK function in haemochromatosis (Chapman *et al.*, 1988) argue against such an effect.

On the other hand, recent evidence does suggest that the cytotoxic activity of activated macrophages might be impaired in an iron-rich environment. Macrophage cytotoxic activity *in vitro* is impaired by excess iron (Green *et al.*, 1988) and iron reduces the activity of γ-interferon, which is the main macrophage-activating cytokine (Weiss *et al.*, 1992). An early event in the lysis of tumour cells by activated macrophages is a loss of intracellular iron from the tumour cells (Hibbs *et al.*, 1984; Wharton *et al.*, 1988); addition of excess iron interferes with this process.

It is now evident that the factor responsible for iron loss is nitric oxide (NO). This is produced by activated macrophages (Liew and Cox, 1991) and causes loss of iron from iron–sulphur clusters (Lancaster and Hibbs, 1990) and ferritin (Reif and Simmons, 1990). Furthermore, tumour rejection *in vivo* is associated with NO production by intratumour macrophages (Mills *et al.*, 1992) and UV-induced skin cancers progressed more rapidly in mice treated with an inhibitor of NO production (Yim *et al.*, 1993).

Although NO-mediated cytotoxic activity offers an anti-cancer mechanism, which has the potential to be sensitive to iron, further work is necessary before its importance in human cancer can be assessed. In particular, the levels and form(s) of iron capable of reversing cytotoxicity have yet to be established, and to date the role of the inducible NO synthase responsible for NO production by activated macrophages has been much more clearly established in rodents than in humans.

### 13.5 IRON CHELATION AS A POSSIBLE TREATMENT FOR CANCER

Although both normal cells and tumour cells require iron for growth, the possibility exists that subtle differences in iron metabolism between normal and tumour cells might be exploited in cancer therapy.

Desferrioxamine has been used to suppress development of leukaemia cells *in vitro* and *in vivo* (Cazzola *et al.*, 1990), although this chelator also interferes with proliferation of normal cells such as activated lymphocytes (Lederman *et al.*, 1984). An attempt to 'fine tune' this mechanism using a combination of desferrioxamine and an anti-transferrin receptor monoclonal antibody has been reported (Kemp *et al.*, 1992) but its efficacy *in vivo* remains to be evaluated. Desferrioxamine can also enhance the activity of the anti-cancer drug, hydroxyurea, *in vitro* (Cory *et al.*, 1981), and the use of the microbial siderophore, parabactin, to promote cell synchronization and thus enhance the activity of anti-tumour drugs has also been proposed from *in vitro* studies (Bergeron and Ingeno, 1987). The use of gallium, which binds to transferrin and subsequently interferes with cellular iron uptake, has also been suggested (Chitambar and Seligman, 1986). However, the problem of interference with iron uptake by normal cells is again apparent (McGregor *et al.*, 1991).

The use of iron chelators, or interference with tumour cell iron metabolism, requires further research before it can be successfully employed as a therapeutic method.

# 14

# IRON, THE BRAIN
# AND NEURODEGENERATION

## 14.1 INTRODUCTION

The amount, concentration, localization, molecular nature, transport, and cellular function of iron and iron-dependent biochemical interactions in the brain are topics which are of particular interest to understanding the physiological and neurodegenerative processes which occur within the central nervous system (CNS). While neurobiological research into the metabolic and pathogenic role of iron is a developing field of study, the detailed relationship with dietary iron intake remains a largely unexplored area.

## 14.2 ANALYTICAL METHODS

A variety of analytical techniques have been used for the determination of iron and iron metabolism in brain tissue. Histochemical methods have provided valuable semi-quantitative data on the regional and cellular distribution of iron, while sensitive analytical instrumentation, atomic absorption spectrometry (AAS), neutron activation analysis (NAA) and inductively coupled plasma optical emission spectrometry (ICPOES) or mass spectrometry (ICPMS) have permitted the quantitative measurement of many trace elements. In addition, the complementary development of a range of highly sensitive microprobe techniques – namely, energy dispersive analysis of X-rays (EDAX), laser microprobe mass analysis (LAMMA), and proton-induced X-ray emission (PIXE) – has facilitated studies into the distribution of trace elements of nutritional and toxicological interest at a cellular level. Non-invasive *in vivo* analysis of iron is also feasible by magnetic resonance imaging (MRI). Chemical detection of the low molecular mass 'catalytic' or 'free' iron pool within brain tissue has been undertaken using the bleomycin assay. Investigations into iron metabolism have been made possible with the radioisotope $Fe^{59}$; and the location and quantitative assay of transferrin and ferritin has been determined using radioimmune methodologies.

## 14.3 BRAIN IRON CONTENT AND DISTRIBUTION

### 14.3.1 Brain, regional and tissue iron sites

Studies of the regional distribution and concentration of non-haem iron in the brain have demonstrated similar levels in grey and white matter, with preferential localization in the globus pallidus, substantia nigra, hippocampus and choroid plexus. Differences with age and disease states are also evident (Table 14.1).

### 14.3.2 Cellular iron sites

Iron is primarily found in glial cells, with relatively small concentrations in neurones. Although mainly concentrated in oligodendrocytes, significant amounts are found in astrocytes, reactive microglial cells and ventricle tanycytes (Connor *et al.*, 1990).

## 14.4 MOLECULAR SPECIATION

### 14.4.1 Protein-bound iron

(a) Enzymes

Iron is required for a number of important brain biochemical functions, especially those involving the mitochondrial enzymes. Other key iron-dependent functions include synthesis of the key neurotransmitters dopamine and serotonin, and lipid metabolism and associated myelin formation.

(b) Transferrin

Within the brain, transferrin is synthesized by oligodendrocytes and is found in neurones and cerebral spinal fluid (CSF). Experiments in the rat using $Fe^{59}$ and $I^{125}$ transferrin radiolabels have demonstrated that iron uptake into the brain is a two-stage relay transfer process via endothelial and neuronal membrane-bound transferrin receptors, with the

**Table 14.1** Concentrations of iron in the brain

| Analytical technique | Brain | | | | Reference |
| --- | --- | --- | --- | --- | --- |
| | Tissue | Control | Alzheimer's disease | Parkinson's disease | |
| NAA ($\mu$g/g ww) | Grey matter | 59 | 95 | | Ehmann *et al.* (1986) |
| | White matter | 42 | 54 | – | |
| NAA ($\mu$g/g ww) | Grey matter | 247–400 | – | 542–546 | Yasui *et al.* (1993) |
| | White matter | 230–279 | – | 423–548 | |
| NAA ($\mu$g/g dw) | Hippocampus | 456 | 417 | – | Ward *et al.* (1987) |
| | Cerebral cortex | 458 | 467 | – | |
| PIXE ($\mu$g/g dw) | Grey matter | 200–280 | – | – | Duflou *et al.* (1989) |
| | White matter | 90–830 | – | – | |
| SP ($\mu$g/g dw) | Globus pallidus | 550 | – | 580 | Riederer *et al.* (1989) |
| | Substantia nigra | 230–330 | – | 260–420 | |
| | Spinal cord | 20 | – | 30 | |
| ICPOES ($\mu$mol/g dw) | Globus pallidus | 1.4–1.5 | – | 1.0–1.1 | Dexter *et al.* (1991) |
| | Substantia nigra | 1.1–1.4 | – | 1.5–1.7 | |
| | Cerebral cortex | 5.0 | – | 5.5 | |

Key:
dw      =  dry weight
ww      =  wet weight
ICPOES  =  inductively coupled plasma optical emission spectroscopy
NAA     =  neutron activation analysis
PIXE    =  proton induced X-ray emission
SP      =  spectrophotometric

intermediate release of iron at the endothelial abluminal cell surface (Morris *et al.*, 1992a). However, the areas of the brain rich in iron are not coincident with the distribution of transferrin receptors and time sequence studies indicate a subsequent axonal transfer of iron to glial cell deposits. Using blocking antibodies to transferrin receptors, a non-transferrin mediated route of Fe[59] uptake has also been shown to exist (Ueda *et al.*, 1993). Investigations into developmental and age changes in rats (Benkovic and Connor, 1993) and in humans (Hallgren and Sourander, 1958) demonstrate an initial rapid uptake of iron into the brain in the young; a subsequent slow accretion of iron and low turnover in adults; followed in senescence by an accumulation of ferritin iron, particularly in the hippocampal microglia.

(c)  Ferritin

Iron in the brain is predominantly associated with ferritin, and is principally present in oligodendrocytes, astrocytes and microglia. Immunolabelling for the different isoforms of ferritin has revealed localization of the H-chain isoform in neurones and the L-chain isoform in microglia (Connor *et al.*, 1994).

(d)  Haemopexin

The high affinity for haem iron by apo-haemopexin, and its immunochemical detection in neurones, has prompted its proposed role in the axonal transport of iron to glial cells (Morris *et al.*, 1993).

(e)  Haptoglobin

The formation of stable complexes of haemoglobin with plasma haptoglobins provides an important protective mechanism against iron-dependent oxidative stress (Gutteridge, 1987).

**14.4.2  Low molecular mass bound iron**

The proportion of tissue iron not sequestered by proteins is very small. The exact nature and concentration of the low molecular mass 'catalytic' or 'free' iron complex pool is largely unknown. Using the bleomycin technique, samples of human CSF from a variety of conditions have shown normal 'free' iron levels to be 2 $\mu$mol/l, while homogenized gerbil brains contain 20 $\mu$mol/l (Gutteridge et al., 1991).

## 14.5   IRON STATUS AND BRAIN DEVELOPMENT

The functioning of the brain is dependent on several iron-related processes, of which efficient cell mitochondrial oxidative activity and neuronal synaptic transmission are among the most crucial. Behavioural abnormalities in human infants with iron deficiency which may be, but are not always reversed by subsequent iron supplementation, are indicative of iron-related disturbances in neurological function. However, the consequences of low iron intakes on cognitive functioning in adults and senile neurodegenerative processes remain to be determined.

Experimental iron deficiency in rats causes a decrease in brain iron concentration despite adaptive increases in transferrin uptake rate, and demonstrates the special vulnerability of the developing brain (Taylor et al., 1991). Iron deficiency in developing animals is also associated with other changes e.g. in neural lipids and in neurotransmitter metabolism. Reviews include Larkin and Rao 1990, Yehuda 1990, Youdim 1990, and Parks and Wharton 1989.

### 14.5.1   Neural lipids

Iron deficiency in young rats is associated with substantial changes in the fatty acid composition of myelin lipids, e.g. a fall in the proportion of nervonic acid – C24:1 $\omega$9 (Yu et al., 1986).

### 14.5.2   Neurotransmitter metabolism

Iron is involved in the synthesis and degradation of catecholamines and serotonin, which act as neurotransmitters. Monoamine oxidases (MAO) are involved in the catabolism of catecholamines and their activity is reduced in iron deficient rats (Youdim et al., 1980). Symes et al. (1971) found tissue levels of MAO activity were reduced to 60% of control levels in iron deficient rats and were restored to normal with iron supplementation. Reduced levels have also been demonstrated in platelets from iron deficient humans (Youdim et al., 1975). Noradrenaline acts as a neurotransmitter at sympathetic nerve terminals and is metabolized by MAOs. If MAO activity is reduced, an increased

excretion of noradrenaline occurs in the urine. Voorhess et al. demonstrated such increased excretion in iron deficient children which is responsive to iron therapy (Voorhess et al., 1975). Levels of noradrenaline in the urine did not vary directly with haemoglobin, iron or saturation. Children with anaemia due to other causes, e.g. thalassaemia, did not have elevated noradrenaline excretion, which suggests that iron deficiency per se was responsible, rather than anaemia.

Another iron dependent enzyme, aldehyde oxidase, is reduced in brain tissue of iron deficient rats (Mackler et al., 1978). This enzyme is involved in degradation of serotonin, which itself was therefore increased in brain tissue. Levels of both aldehyde oxidase and serotonin returned to normal within 1 week of starting iron. All of the enzyme changes found in iron deficiency are easily and quickly reversed with iron therapy in rats well before there is any change in haemoglobin concentration. This parallels the changes in behaviour noted in the children.

Dopamine is also important as a neurotransmitter in the extrapyramidal system. Dopamine D2 binding sites are reduced in iron deficient rats. These sites are restored to normal with iron therapy in the adult rat, but in newborn rats the binding sites remain depleted despite iron therapy.

## 14.6   POTENTIAL MECHANISMS OF IRON-INDUCED BRAIN TOXICITY

### 14.6.1   Release of protein-bound iron

The involvement of 'free' iron in free radical oxidative reactions includes the catalytic decomposition of lipid peroxides and participation in the Fenton reaction which generates hydroxyl radicals from the superoxide radical and hydrogen peroxide. The formation of the extremely reactive hydroxyl radical means that the level and location of the 'free' iron species needs to be under tight biological control. Although sequestered protein-bound iron is thus rendered incapable of catalysing Fenton oxidative reactions, iron may be released as a result of a number of reductive biochemical processes. Endothelial cells, by a process of superoxide-induced reduction of ferric (Fe(III)) to ferrous (Fe(II)) iron and acidification, are capable of releasing bound iron from transferrin (Brieland et al., 1992). In vitro studies have also demonstrated various reductants to be capable of releasing bound iron from ferritin (Reif, 1992) and haemosiderin (O'Connell et al., 1986), the latter evidently being the more stable. The traumatic release of red cell haemoglobin can also contribute to iron-dependent CNS injury (Sadrzadeh et al., 1987). The diminished capacity of aged brain tissue to bind iron suggests an associated increased susceptibility to 'free' iron induced oxidative stress (Barkai *et al.*, 1991).

### 14.6.2 Generation of free radical reactive oxygen species

Biological reactants which promote the conversion of ferric to ferrous iron and the associated release from bound proteins include superoxide derived from inflammatory polymorphonuclearcytes (PMN), xanthine oxidase activation, and biological thiols. Examples of free radical generating mechanisms more specific to the brain are the monamine oxidase-catalysed metabolism of dopamine, neuromelanin-catalysed reactions and activation of microglial cells (Sonderer *et al.*, 1987). Metabolism of xenobiotic compounds by brain P450 cytochromes may also contribute to oxidative stress (Ghersi-Egea and Livertoux, 1992).

### 14.6.3 Oxidative neuronal damage

The particular anatomical, cell morphology and biochemical characteristics, render brain tissue peculiarly susceptible to the adverse effects of oxidative stress (Evans, 1993). The brain is the most metabolically active organ, consuming more than 20% of the body's oxygen. The high concentrations of potential pro-oxidants (in relation to tissue iron) and of oxidizable substrates (namely polyunsaturated fatty acids (PUFA)), together with the relatively low levels and turnover of antioxidants (particularly catalase and $\alpha$-tocopherol), provide a cellular milieu of enhanced vulnerability. Oxidative injury to the normally tightly controlled blood/brain barrier can compromise its selective permeability; and free radicals may further affect neuronal synaptic and axonal transport function.

### 14.6.4 Iron and trace element interactions

Nutrient and toxic element interactions can play a significant role in the biochemistry of neurodegenerative disease (Evans, 1994). *In vitro* and *in vivo* studies have demonstrated an enhanced rate of aluminium absorption in iron deficiency (Cannata *et al.*, 1991), aluminium gaining access to the brain by binding to transferrin (Roskams and Connor, 1990). Manganese also binds to transferrin and accumulates in the substantia nigra – efferent to regions of high transferrin receptor density (Aschner and Aschner, 1990) – and catalyses the oxidation of dopamine, an action of direct relevance to the manganism/parkinsonian syndrome (Florence and

Stauber, 1989). Oxidation of cobalamin, possibly by iron Fenton mechanism, can adversely affect its enzymic cofactor function in methionine synthetase (Haurani, 1989), which may thus have a deleterious effect on neurological function.

## 14.7 SPECIFIC NEURODEGENERATIVE DISORDERS AND DISEASES

### 14.7.1 Parkinson's disease

That the pathogenesis of Parkinson's disease (PD) may be mediated by enhanced oxidative stress is supported by findings demonstrating a reduction in glutathione, accompanied by increased iron and ferritin levels, particularly in the substantia nigra, an area of the brain severely affected in PD by loss of the neurotransmitter dopamine (Riederer *et al.*, 1989). In contrast, one other study, while confirming the increase in iron levels, has suggested that ferritin levels are decreased – an apparent contradiction possibly due to the use of anti-ferritin antibodies of divergent immunospecificity (Dexter *et al.*, 1991). However, the detection of early falls in glutathione but not iron levels in cases of incidental Lewy body disease (presymptomatic PD) suggests that iron-related pathology is unlikely to be an initial cause of nigral cell death (Jenner, 1993). Microprobe elemental examination of neuromelanin-containing neurones of the substantia nigra using LAMMA have identified accumulation of iron and aluminium, a finding of comparable significance to the proposed toxic role of aluminium in Alzheimer's disease (AD) (Good *et al.*, 1992). It has been proposed that the oxidative interactions of iron, neuromelanin and dopamine are of specific pathogenic relevance in PD (Ben-Shachar and Youdhim, 1993). *In vivo* monitoring of iron-dependent pathology in the substantia nigra of Parkinson's subjects has been studied by MRI (Rutledge *et al.*, 1987).

### 14.7.2 Alzheimer's disease

Early histochemical studies in AD brains indicated an increased level of iron, particularly in the motor and occipital cortex, and specifically located in astrocytic and microglial cells surrounding senile plaques (Hallgren and Sourander, 1960). Immunoreactive detection of ferritin has been reported in AD brains, confirming the strong association with proliferative microglia (Jellinger *et al.*, 1990), and in CSF from AD subjects (Kuiper *et al.*, 1994). Diffuse staining of lactoferrin has also been shown by immunochemical techniques to be associated with pathological extracellular structures like plaques and intracellular neurofibrillary tangles, and with lipofuscin-bearing neurones, especially in the areas of the brain which are primarily affected in AD: the cortex, hippocampus and olfactory bulb (Kawamata *et al.*, 1993). Immunochemical studies of transferrin have found similar extracellular deposition around senile plaques, together with an increased concentration in astrocytes (Connor *et al.*, 1992). Both basal and iron/ascorbate stimulated lipid peroxidative products are increased in the inferior temporal cortex in AD brains (Palmer and Burns, 1994). The increased frequency in AD subjects of the transferrin C2 variant subtype (Van Rensburg *et al.*, 1993), a postulated risk factor for iron-related hydroxyl radical generation (Beckman and Beckman, 1986), implies a possible genetic influence in iron-dependent pathogenic reactions. The *in vitro* demonstration that amyloidogenicity is promoted by iron-catalysed oxidation of $\beta$-amyloid peptides (Dyrks *et al.*, 1992), a determinant of senile plaque formation, is a significant contribution to understanding the potential synergistic interactions which may occur in AD.

Experimental *in vitro* studies have shown that iron-dependent lipid peroxidation is enhanced by aluminium (Gutteridge *et al.*, 1985) and model aluminosilicate particulates – analogous to plaque core components – can promote microglia-derived free radical generation (Evans *et al.*, 1992). However, using tritium-labelled transferrin to examine the detailed distribution of transferrin receptors in the brain, the evident non-coincidence with plaque density suggests that transferrin-mediated co-uptake of aluminium is not the sole determinant of the site of plaque formation (Morris *et al.*, 1994).

### 14.7.3 Other neurodegenerative disorders

The primary pathological feature of Hallervorden Spatz disease, an autosomal recessive genetic disorder, is the gross accumulation of iron in the globus pallidus and substantia nigra (Swaiman, 1991). The reported increased iron concentrations in the brains of multiple sclerosis subjects is paralleled by elevated serum ferritin (Valberg *et al.*,

1989). Increased deposition of iron has also been detected in the brains of subjects with such disparate conditions as spastic paraplegia (Arena *et al.*, 1992), AIDS (Gelman *et al.*, 1992) and Guamanian amyotrophic lateral sclerosis (GALS) (Yasui *et al.*, 1993), while subjects with neuronal-ceroid lipofuscinosis exhibit elevated CSF 'free' iron concentrations (Gutteridge *et al.*, 1982). Toxicological studies with N-methyl-1,2,3,6-tetrahydropyridine (MPTP), which induces parkinsonism in primates, have shown an associated accumulation of iron in the substantia nigra as measured by the PIXE microprobe (Temlett *et al.*, 1994).

### 14.8 IRON OVERLOAD AND CHELATION TREATMENT

The genetic disease haemochromatosis is characterized by increased deposition of iron stores, primarily in the liver and heart, but may very rarely be accompanied by clinical dementia (Jones and Hedley-Whyte, 1983). Experimental studies in iron overloaded rats suggests that increased brain iron may reflect blood plasma iron in conditions where blood–brain barrier permeability is compromised in disease (Morris *et al.*, 1992b).

A number of iron chelating drugs have been used in the treatment of brain disorders in which toxic metal overload has been implicated as a causative factor. Treatment of AD subjects with desferrioxamine, a chelator of both iron and aluminium, has been reported to reduce the progression rate of dementia (McLachlan et al., 1991). Several other comparable antioxidant iron chelators, e.g. the hydroxypyridones, are also under active investigation (Halliwell, 1991) but recent studies have shown that hydroxypyridones alter dopamine turnover in the brain, which could exclude their clinical use (Ward et al., 1995b). The 21-aminosteroid lazeroids have received considerable attention as potentially clinically useful drugs in managing some of the adverse effects of head trauma (Hall and McCall, 1993).

# 15

# IRON IN INFANCY AND CHILDHOOD

## 15.1 EFFECTS OF IRON DEFICIENCY

Iron deficiency in infancy and childhood is associated with a number of non-haematological manifestations, including the delay of mental and motor development. Iron requirements and iron uptake into the brain are highest during periods of rapid growth (Beard et al., 1993) and perinatal iron deficiency alters myelination of nervous tissue (Morris et al., 1992c).

A number of studies of infants and young children have found lower mental scores in children with iron deficiency anaemia compared with non-anaemic children and the balance of investigations suggests a cause-and-effect relationship (Chapter 11).

## 15.2 IRON DEFICIENCY IN THE UK

The prevalence of iron deficiency and iron deficiency anaemia in infants and children in the UK varies depending upon the socio-economic and cultural group studied, the age and the criteria used to establish the diagnosis. Table 15.1 lists recent surveys carried out in the UK. Iron deficiency is seen most commonly in the UK in children aged 6–24 months from ethnic minority groups (particularly Asian) and those from a deprived background; in these populations the incidence of anaemia reaches 25–30% and low plasma ferritin has been found in up to 45%.

A study carried out in 1992/3 on infants aged 8 months from a healthy mixed population found that 27% had haemoglobin values below 110 g/l (Emond et al., 1995). A major survey of young children aged 1.5–4.5 years was completed in 1993 and the results (Table 15.1) show an overall prevalence of 8% in this nationally representative sample (Gregory et al., 1995).

The prevalence of iron deficiency anaemia appears to have decreased over the last two decades in the United States (Dallman, 1990): in children aged 6–60 months from low-income families, the incidence of anaemia was 7.8% in 1975 and 2.9% in 1985 (Yip et al., 1987a,b). Factors contributing to this decline appear to be the increased iron fortification of infant formulas, a decreasing use of cows' milk for children under the age of 1 year, the use of iron-fortified infant cereals and a supplementation programme aimed at low-income families. There are no data to support a declining incidence of iron deficiency in the UK.

The need for a UK national screening programme for iron deficiency has been suggested (Anonymous, 1987; Hall, 1989) and screening at-risk infants at 9–12 months and/or routine supplementation with medicinal iron has been proposed (Addy, 1986). Small-scale screening programmes have been shown to be acceptable in the GP surgery and in community child health clinics (James et al., 1988; Marder et al., 1990). A number of questions about the feasibility of national screening need to be considered: a practical and useful age for screening in this country has not been determined, the most appropriate laboratory tests for diagnosis of iron deficiency remain unclear, and a strategy needs to be developed for children who have been screened. Education programmes coupled with screening are successful in reducing iron deficiency in the short term (James et al., 1989), although the long-term effects appear to be less effective (Walter et al., 1989b; James et al., 1993). The benefits arising from a screening programme have not been evaluated and this has been identified as a priority area for research (DH, 1994).

## 15.3 INFANTS

The placental transfer of iron is discussed in Chapter 17. Neonatal iron status is affected by circumstances around the time of delivery, e.g. early clamping or late clamping of the cord, and fetal blood loss into the placenta or mother. Iron status (comprising both haemoglobin and storage iron) of the infant at birth does not generally reflect maternal iron status (Morton et al., 1988), although infants of severely anaemic mothers (Hb < 60 g/l) may also be anaemic (Singla et al., 1978). Mothers with very high iron stores (ferritin levels > 30 μg/l) have infants with higher cord ferritin values than

**Table 15.1**   Incidence of iron deficiency in infants and children in the UK

| Author(s) | Subjects | Criteria used | Incidence |
|---|---|---|---|
| Grindulus *et al.* (1986) | 22 months, Asian | Hb < 110 g/l | 31% |
| | | ferritin < 10 µg/l | 57% |
| Ehrhardt (1986) | 6 months to 4 years, Asian | Hb < 110 g/l | 28% |
| | | ferritin < 10 µg/l | 45% |
| | 6 months to 4 years, non-Asian | Hb < 110 g/l | 12% |
| | | ferritin < 10 µg/l | 23% |
| Aukett *et al.* (1986) | Mixed inner city, 17–19 months | Hb < 110 g/l | 26% |
| | | ferritin < 7 µg/l | 47% |
| Morton *et al.* (1988) | 6 months, Asian | Hb < 110 g/l | 12% |
| | | ferritin < 10 µg/l | 40% |
| | 6 months, non-Asian | Hb < 110 g/l | 4% |
| | | ferritin < 10 µg/l | 36% |
| | 1 year, Asian | Hb < 110 g/l | 26% |
| | | ferritin < 10 µg/l | 37% |
| | 1 year, non-Asian | Hb < 110 g/l | 12% |
| | | ferritin < 10 µg/l | 21% |
| James *et al.* (1988) | Mixed inner city | Hb < 105 g/l | |
| | All under 5 | | 16% |
| | All aged 1–2 | | 25% |
| | Afro-Caribbean | | 24% |
| Marder *et al.* (1990) | Aged 15 months | Hb < 110 g/l | |
| | Mixed inner city | | 25% |
| | Asian | | 39% |
| | White | | 16% |
| | Afro-Caribbean | | 20% |
| Mills (1990) | 8–24 months, inner city | Hb < 110 g/l | |
| | Asian | | 16% |
| | White | | 9% |
| | Others | | 6% |
| Duggan *et al.* (1991) | 4–40 months, Asian | Hb < 110 g/l | 17% |
| | | ferritin < 10 µg/l | 35% |
| | | FEP > 80 µmol/mol haem | 41% |
| Gregory *et al.* (1995) | Nationally representative | | |
| | 18–29 months | Hb < 110 g/l | 12% |
| | | ferritin < 10 µg/l | 28% |
| | 30–41 months | Hb < 110 g/l | 6% |
| | | ferritin < 10 µg/l | 18% |
| Emond *et al.* (1995) | 8 months, mixed | Hb < 110 g/l | 27% |
| | | ferritin< 12 µg/l | 1.2% |

those born to mothers with ferritin values of less than 10 µg/l (Kelly *et al.*, 1978). However, iron status at 6 months and at 1 year is positively correlated with cord ferritin levels and with maternal ferritin levels at 37 weeks gestation but not with cord haemoglobin values (Morton *et al.*, 1988). It is postulated that cord ferritin levels, although not a good predictor of iron stores in the infant, may indicate those who were able to extract iron more efficiently from the placenta and perhaps from the diet after birth (Morton *et al.*, 1988).

The iron level in the newborn infant is approximately 75 mg/kg body weight (Widdowson and Spray, 1951). There is a significant positive correlation between storage iron (which more than doubles during the last 6 weeks of intrauterine life), birthweight and gestational age (Chang, 1973). Little increase in total body iron occurs in the first 4 months of life, but total amounts increase from approximately 250 mg at birth to about 420 mg at age 1 year (Dallman, 1989b).

### 15.3.1 Iron status after birth

Three stages of postnatal iron metabolism have been described and are illustrated in Figure 15.1 (Dallman, 1986). In the immediate postnatal period, iron is redistributed in the body: haemoglobin levels are particularly high at birth, reflecting the oxygen-poor intrauterine environment; levels decrease during the first 6–8 weeks before stabilizing at 2–3 months (Aggett *et al.*, 1989). There is a sharp decrease in erythropoiesis which is explained by the increased oxygen delivered to neonatal tissues and the shorter lifespan of fetal red cells (about 60% of that of adult cells). Serum ferritin increases, as do reticuloendothelial and liver iron stores. Although total body iron does not change, haemoglobin declines from 170 g/l at birth to 125 g/l by 4 weeks. This period correlates with the period of negative iron balance observed by Cavell and Widdowson (1964). Exogenous iron can, however, be used during early postnatal life. Thirty to 50% of a dose of radioiron was incorporated into erythrocytes of infants aged 25–30 days, 65% at 65 days, 85% at 81–129 days and 70–90% at 223–273 days, compared with around 80% for adults (Garby *et al.*, 1963). The view that iron absorption and erythropoiesis are negligible in the first 2 months of life has been challenged by a recent study examining the incorporation of an iron stable isotope into erythrocytes at 2 months of age. The study found that absorption and utilization did take place; in breast-fed infants iron was stored and utilized at a later stage (Fomon *et al.*, 1993).

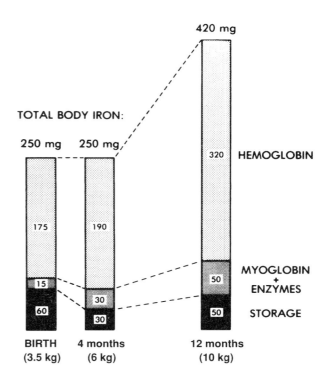

**Figure 15.1** Changes in body iron during infancy. (source: Dallman, 1986.)

### 15.3.2 Iron status at age 2–4 months

The second stage of iron metabolism takes place from 2 to 4 months of age, when erythropoiesis increases. Haemoglobin concentrations change less than in the first 2 months, although there is an increase in total haemoglobin because of the increasing size of the infant. Storage iron deposited in the liver during the earlier stage begins to decrease and by the age of 4 months the liver contains approximately 30 mg iron compared with 60 mg at birth; this is reflected in a decrease in serum ferritin levels. The total body iron shows little change from birth to 4 months of age and is about 250 mg.

### 15.3.3 Iron status after 4 months

After the 4th month of life, increasing body size begins to deplete body iron concentrations and the infant becomes increasingly dependent on exogenous sources of iron. In a normal infant, total body iron increases from 250 mg at 4 months to about 420 mg at 12 months. Liver iron stores are depleted by the age of 6 months unless there is an adequate source of dietary iron.

**Table 15.2** Dietary reference values (mg/day) for iron in infancy and childhood (source DH, 1991)

| Age | Lower reference nutrient intake | Estimated average requirement | Reference nutrient intake |
| --- | --- | --- | --- |
| 0–3 months | 0.9 | 1.3 | 1.7 |
| 4–6 months | 2.3 | 3.3 | 4.3 |
| 7–9 months | 4.2 | 6.0 | 7.8 |
| 10–12 months | 4.2 | 6.0 | 7.8 |
| 1–3 years | 3.7 | 5.3 | 6.9 |
| 4–6 years | 3.3 | 4.7 | 6.1 |
| 7–10 years | 4.7 | 6.7 | 8.7 |
| 11–14 years (males) | 6.1 | 8.7 | 11.3 |
| 11–14 years (females) | 8.0 | 11.4 | 14.8 |

### 15.3.4  Iron status and growth

A number of factors affect iron requirements in infancy. Initial haemoglobin levels and iron stores influence later iron needs since the reduction in haemoglobin in the first 2 months will release iron to be stored. Birthweight and the rate of growth are major influences on iron status during the second half of the first year. While faster growing infants are theoretically at greater risk of early deficiency, the evidence is conflicting: no correlation was found between growth and iron intake or iron status up to 9 months of age in one study (Haschke *et al.*, 1993), whilst Morton *et al.* (1988) found that iron deficiency at 1 year of age was significantly associated with greater weight gain. Iron depletion may impair growth (Owen, 1989) and the effect is likely to be self-limiting. A study in the UK found an improvement in weight gain when anaemic infants were given iron (Aukett *et al.*, 1986). Blood losses and infections will affect iron requirements in early life although these are rare causes of deficiency in the normal infant.

### 15.3.5  Dietary Reference Values

Dietary References Values (DRVs) are shown in Table 15.2. As with most nutrients, they do not apply to exclusively breast-fed infants. The DRVs make no allowance for the differing size of individual children. Advisable intakes have been related to energy intakes and are 0.6–1.2 mg/100 kcal up to the age of 10 years (Wharton and Clark, 1990).

## 15.4  PRETERM INFANTS

The risk of preterm and low birthweight delivery has been found to be higher in iron deficient women (Murphy *et al.*, 1986; Lieberman *et al.*, 1988; Scholl *et al.*, 1992) (Chapter 17). During the third trimester of pregnancy, the fetus rapidly accumulates iron at the rate of 1.6–2.0 mg/kg/day, so the preterm infant is born with a lower total body iron content, although iron concentration in most tissues is similar to that of term infants (Ehrenkranz, 1992). Preterm infants with a size appropriate for gestational age have lower haemoglobin levels than term growth-retarded infants, and higher ferritin levels than preterm growth-retarded infants (Olivares *et al.*, 1992).

Postnatal erythropoiesis undergoes stages similar to those described for term infants, although there are some differences. During early neonatal life haemoglobin concentrations are likely to be 20–30 g/l lower than term infants. The second stage, marked by renewed erythropoiesis, may begin by one month of age. The more rapid growth rate of low birthweight infants means that iron stores become depleted earlier than in term infants, and frequent blood sampling may exacerbate this.

### 15.4.1  Efficiency of iron absorption

The efficiency of iron absorption from the gut is related to:

- postnatal age
- rate of growth and weight gain

- iron intake
- haemoglobin levels.

High absorption rates of 31.5% (Gorton *et al.*, 1963) and 41.6% of exogenous intake (Ehrenkranz *et al.*, 1992) have been reported. Utilization (incorporation into red cells) is more variable with a value of 2.8% found by Oettinger *et al.* (1954), 15.3% (Gorton *et al.*, 1963) and 12% (Ehrenkranz *et al.*, 1992). This suggests that iron can be absorbed and utilized soon after birth (Oettinger *et al.*, 1954). Blood transfusions were found to depress iron absorption (Dauncy *et al.*, 1978) and it has been suggested that infants who have had transfusions of packed red cells do not need supplemental iron as early as others (Austria *et al.*, 1987), although more recent findings suggest that iron absorption and utilization may not be influenced by a history of transfusion (Ehrenkranz *et al.*, 1992).

### 15.4.2   Anaemia in preterm infants

The 'physiological anaemia of prematurity' which may develop soon after birth cannot be prevented by giving iron supplements at this time. This may be due to in part to immaturity of the erythropoietin-producing pathway. There is some controversy over whether recombinant human erythropoietin infusion reduces the need for red cell transfusion: while some studies report success (Ohls and Christensen, 1991) other results are variable and suggest that routine iron supplementation must be given concurrently for optimum results (Shannon *et al.*, 1991). It has been suggested that giving erythropoietin without iron supplements can result in iron deficiency because of increased demands made by rapid red cell synthesis (Stockman, 1988).

Preterm infants have a higher incidence of iron deficiency at 4 months than term infants, and supplements significantly reduce the likelihood of anaemia throughout the first year (Doyle and Zipursky, 1992). Breast-fed preterm infants have lower ferritin levels at 2 months and lower haemoglobin levels at 3 months when compared with infants receiving iron supplements by one month of age (Lundstrom *et al.*, 1977).

### 15.4.3   Iron supplementation

Oral iron supplements interact with a number of nutrients both before and after absorption and may precipitate metabolic disturbance of other nutrients. Preterm infants receiving high levels of iron supplementation (13.8 mg/day) were found to have lower erythrocyte superoxide dismutase activity at age 20 weeks compared with infants on a lower iron

intake, indicating an impairment of copper utilization, although the clinical significance of the differences seen is unknown (Barclay *et al.*, 1991). The anaemia of prematurity may be more severe if iron supplements are given without vitamin E (Zipursky *et al.*, 1987). Infant formulas in the UK contain adequate vitamin E to prevent erythrocyte membranes from peroxidation, so supplements should not be necessary. Infants fed human milk and iron supplements may require additional vitamin E (Melhorn *et al.*, 1979).

A number of recommendations have been made for the timing of iron supplementation in low birthweight infants. In the absence of blood loss, low and very low birthweight infants should not require supplementation until the birthweight has doubled (Oski, 1985), although benefits of iron-fortified formula from birth have been described (Hall *et al.*, 1993) and some centres recommend supplements as early as 2 weeks (Lundstrom *et al.*, 1977). The American Academy of Pediatrics (AAP) (1985) and the European Society for Paediatric Gastroenterology and Nutrition (ESP-GAN) (1987) recommend that iron supplements should be given no later than 2 months of age to achieve an intake of 2.0–2.5 mg/kg/day up to a maximum of 15 mg/day and should be continued until 12 months or until full mixed feeding provides an adequate iron intake.

There is considerable disparity in iron supplementation policies in neonatal units in the UK. One study revealed that about half the units surveyed used gestational age as the sole criterion for supplementation; some used birthweight alone, with thresholds varying from 1500 to 2000 g, and some units used a combination of age and weight (Barclay *et al.*, 1989). Considerable variation was also found in the quantity and duration of supplements given, and current practice should be evaluated.

### 15.4.4   Iron fortified formulas

Iron levels in the milk of mothers delivering preterm do not differ significantly from term breast milk (Lemons *et al.*, 1982). There are no EEC recommendations for the composition of formulas used for preterm infants. ESPGAN recommends that a formula that aims to provide all iron requirements should contain approximately 1.5 mg iron/100 kcal to provide 1.7–2.5 mg/kg body weight per day (ESPGAN, 1987). Brands of preterm formulas for use in hospitals and in the community vary in their iron content (Table 15.3), allowing the paediatrician to adjust the quantity and timing of additional iron. If a sufficient volume of an iron fortified formula is consumed, current recommendations for iron intake may be met by formula, obviating the need for supplements. Breast-milk fortifiers (used to increase the nutrient intake of breast-fed preterm infants) currently marketed in the UK contain no additional iron.

## 15.5   BREAST-FED INFANTS

Maternal iron status and iron supplements taken during breast-feeding do not appear to affect milk levels (Siimes et al., 1984). The iron content of breast milk is within the range 0.06–0.09 mg/100 ml (Table 15.3), giving an average intake in breast fed infants of approximately 0.48–0.72 mg/day, although levels given in milk drop by 50% in the first 6 months and continue to fall. Since the volume of breast milk consumed does not increase significantly after 1–2 months of age, the intake of iron also decreases over this period.

Iron is more efficiently absorbed from human milk compared with formula (Garry et al., 1981). The particular factors aiding iron absorption from breast milk are not clear, although the speciation of the iron and its association with milk components are thought to exert an influence. One third of the iron in breast milk is associated with fat globules, one third with lactoferrin, 10% with casein, and the remainder with low molecular weight compounds (Fairweather-Tait, 1989). This contrasts with formulas based on cows' milk, where a higher proportion of iron is associated with casein. Whole body counting of orally administered radio isotopes of iron in 6–7-month-old breast-fed infants indicated an absorption rate of 38% (Saarinen et al., 1977). The utilization of iron can be more accurately measured by erythrocyte incorporation of stable isotopes of iron; it was found that 20% of an orally administered dose in a single formula feed was incorporated into red cells, indicating an absorption of 31% (Fomon et al., 1993). Absorbed iron from breast milk was more readily incorporated into red cells than was iron from formula milk, although the reasons for this preferential utilization are not clear. These findings have been challenged by studies over several feeds using stable isotopes of iron, which found a mean net iron absorption of 11.8% from externally labelled human milk (Davidson et al., 1994). An investigation using adults showed that inhibitors in some weaning foods, particularly cereals, decrease the bioavailability of iron from breast milk (Oski and Landau, 1980). There may therefore be advantages in giving breast milk feeds separately from the weaning diet.

Despite an apparently efficient absorption and utilization of iron, the absolute amount of iron absorbed from breast milk is less than the absolute amount absorbed from an iron fortified infant formula. Therefore a small proportion of breast-fed infants have lower iron stores at 6 months and a greater proportion appear deficient at 9 months of age or later, when compared with infants receiving iron-fortified formulas (Siimes et al., 1984; Hertrampf et al., 1986; Mills, 1990; Haschke et al., 1993)

Most breast-fed infants should not require additional sources of iron in the diet until 6 months of age. Some infants, particularly if solids are poorly accepted or are low in iron content, may require some fortified foods during the second half of the first year.

## 15.6   INFANT FORMULAS

### 15.6.1   Iron fortification

The optimum fortification level for infant formulas has not been determined; a concentration of 3 mg/l has been shown to prevent iron deficiency in infants up to 9 months of age (Haschke et al., 1993) but fortification of 7 mg/l has been suggested as a minimum (Saarinen and Siimes, 1977b). The ideal age to introduce fortified formulas has also not been defined. Additional iron is needed by 4 months (Aggett et al., 1989) or 5 months of age (Dallman, 1990). Iron can be absorbed and utilised by two months of age (Fomon et al., 1993) and early provision of iron fortified formulas might spare body stores and reduce the risk of later deficiency.

A number of studies have demonstrated that fortification of infant foods with moderate amounts of iron is not associated with an increased incidence of infection (Walter et al., 1986). Iron fortified formulas have been blamed for gastrointestinal upsets such as colic and diarrhoea but there is no evidence to support this belief (Nelson et al., 1988). However, there may be disadvantages in the use of iron fortified foods before they are required; iron fortification of milks may impair the absorption of other minerals and trace elements such as manganese and zinc. Wharton (1989) points out that there is little data on the toxicity of iron in the early months of life.

### 15.6.2   Fortification recommendations

All infant formulas available in the UK are fortified with iron to achieve levels of 4–7 mg/l (Table 15.3). Iron preparations used for fortification are ferrous lactate, ferrous sulphate and ferrous ammonium citrate, all of which have a similar bioavailability. British (DHSS, 1980) and ESPGAN (1977) recommendations are that iron fortification of formulas should ensure levels not less than 0.07 mg/l and not

**Table 15.3** Iron content of milks and formulas available in Britain

| Milk or formula | Iron content (mg/100 ml) |
| --- | --- |
| Human milk | 0.06–0.09 |
| Preterm infant formula (hospital) | 0.04–0.9 |
| Breast-milk fortifier | 0 |
| Preterm infant formula (home) | 0.07–0.65 |
| Whey-dominant infant formula | 0.5–0.7 |
| Casein-dominant infant formula | 0.4–0.65 |
| Soy-based infant formula | 0.65–0.8 |
| Follow-on formula | 0.7–1.3 |
| Cows' milk | 0.06 |

more than 7.0 mg/l. Directives from the European Commission and now confirmed in UK law are that infant formulas used from birth should, if iron is added, contain 0.5–1.5 mg/100 kcal and follow-on milks should contain 1–2 mg/100 kcal (EEC, 1991; Statutory Instrument, 1995).

### 15.6.3 Absorption of iron from formulas

Iron is added to formulas in quantities considerably in excess of those found in breast milk because iron is much less available from cows' milk than from human milk. Seven per cent of the exogenous intake was incorporated into red cells at 8 weeks in formula-fed infants, which indicates a 10% absorption (Fomon et al., 1993). An iron absorption rate of 10% was also found by Saarinen and Siimes (1979). British recommendations for the iron content of formulas (DHSS, 1980) assumes a 10% absorption by infants. A number of factors are involved in the lower availability of iron from formula milks in comparison with breast milk. Bovine casein and whey protein have been shown to exert an inhibitory effect on iron absorption when compared with hydrolysed milk; casein has a greater effect than whey. High concentrations of calcium and phosphorus are also likely to inhibit iron absorption (Hurrel et al., 1989). Formulas are fortified with generous quantities of ascorbic acid and usually contain citrate, both of which increase the bioavailability of non-haem iron. Attempts have been made to increase the bioavailability of added iron by the addition of factors such as lactoferrin: bovine lactoferrin did not improve absorption from formulas in one study (Fairweather-Tait et al., 1987) although higher levels resulted in higher ferritin levels in infants aged 90 days than in those fed formula without lactoferrin (Cherici et al., 1992).

### 15.6.4 Follow-on formulas

Follow-on formulas have an iron concentration that is approximately double that of standard formulas and are intended for children aged 6 months and over. Although they have been shown to be effective in preventing iron deficiency, a study directly comparing the two showed that a standard formula was equally effective (Bradley et al., 1993). Absorption of iron (as a proportion of intake) was found to be lower in formulas fortified to a level of 12.8 mg/l compared with formulas fortified to 6.8 mg/l (7% vs 9%), although the total quantity of iron absorbed per dose was greater for the more highly fortified milk (Saarinen and Siimes, 1977b). Based on this and other work, Dallman (1986) concluded that the absolute amount of iron absorbed from 750 ml (i.e. a day's intake) of breast milk was 0.15 mg; amounts for formula A (containing 0.8 mg iron/l), formula B (6.8 mg/l) and formula C (12.8 mg/l) were 0.05 mg, 0.3 mg and 0.35 mg respectively. Follow-on milks are useful for older infants and young children who drink very little milk, either from choice or because of feeding difficulties, but for most infants they offer little advantage over a standard formula.

### 15.6.5    Soya formulas

Soya formulas that are modified for infants can be used from birth and have been shown to maintain adequate iron status during the first year of life (Hertrampf *et al.*, 1986). It is known that phytate decreases iron absorption in adults; the quantity of phytate remaining in highly modified infant formulas based on soya protein isolates is not known. Apart from phytate, other components in soya products may also decrease iron bioavailability (Hurrel *et al.*, 1992), although ascorbic acid is added to infant soya formulas and increases the bioavailability of iron (Gillooly *et al.*, 1984). There have been few studies of iron availability from soya milk in infants; most studies have been carried out on adults consuming an infant soya formula. In one study comparing iron status in a group of infants, fortified cows' milk and soya milk formulas achieved a superior iron status compared with breast-fed infants, though the same study showed very low iron availability (< 2% absorption) for the formulas when absorption studies were undertaken in adult subjects (Hertrampf *et al.*, 1986). A study using a radio isotope of iron showed a 5.6% iron absorption from soya milk in infancy, although there was a considerable variation, with levels up to 52% (Rios *et al.*, 1975). Using stable isotopes of iron, there was an absorption of 31% from unfortified and 19% from iron fortified soya formula in infants aged 6–10 weeks; the fractional luminal disappearance of iron from the two formulas was 37% and 26% respectively (Mackenzie, 1994). Iron does appear to be readily absorbed from infant soya formulas, and there seems to be no contraindication in terms of iron status to the use of an adapted soya infant formula from birth.

### 15.7    FEEDING COWS' MILK

Cows' milk is low in iron (Table 15.3) and infants of 6–12 months in the USA receiving cows' milk as their main drink failed to meet the Recommended Dietary Intake (RDI) for iron (Tunessen, 1987). Infants achieving recommended intakes for iron from a diet consisting of cows' milk and iron fortified cereal were nevertheless more likely to have a lower plasma ferritin than a similar group fed fortified formula from 6 to 12 months (Fuchs *et al.*, 1993b).

No direct correlation was found between the amount of milk taken and iron status in children under the age of 2 years, but iron deficiency was found to be more common in those consuming over 1 litre of milk and in those children whose diet included cows' milk before the age of 8 months (Mills, 1990). Another study showed that at 1 year iron deficiency was associated with the introduction of cows' milk before 6 months, the consumption of more than 900 ml milk and with only small quantities of solids (Morton *et al.*, 1988). Asian toddlers in Britain drink significantly more milk than white children (Mills, 1990; Duggan *et al.*, 1992) and this may partly explain the greater incidence of iron deficiency in this group.

It is unclear whether drinking large quantities of cows' milk during the second year of life predisposes to iron deficiency because children may consume milk (which is low in iron) to the exclusion of other more iron-rich foods, or whether in addition specific factors in milk are responsible. It has been postulated that milk inhibits the absorption of non-haem iron because of its high calcium and phosphate content. Unmodified pasteurized cows' milk has been reported as causing blood loss from the gastrointestinal tract in infants (Zeigler *et al.*, 1990) but this was not observed in a study of infants aged 6–12 months (Fuchs *et al.*, 1993a). The topic remains controversial.

The American Academy of Pediatrics recommends that unmodified cows' milk should not be introduced as the main drink until the end of the first year of life (AAP 1992). More recently the Department of Health in the UK also recommended that unmodified cows' milk was unsuitable as a main drink for infants under the age of 1 year, although milk can be used as a component of solid foods from the age of 4 months; the use of fortified formulas or breast milk is recommended until the age of 1 year and infant formulas or follow-on milks may be used to advantage after this if there is doubt about the iron content of the young child's diet (DH, 1994).

### 15.8    THE WEANING DIET

### 15.8.1    Iron intakes

Average iron intake in infants in the UK appears to meet the RNI in two out of three recent studies (Table 15.4). In the survey by Mills and Tyler (1992) of infants aged 6–12 months, the mean intake of iron was 104% of the Reference Nutrient Intake (RNI), but the distribution was skewed and the median value was 90% of the RNI. In the study, iron intake was significantly higher in infants who received mainly commercially produced foods compared with those on mainly family foods (10.1 and 10.5 mg/day for 6–9- and 9–12-month-old infants, respectively, compared with 4.6 and 4.9 mg/day).

**Table 15.4**    Iron intakes in British infants and children

| Author(s) | Age (years) | | Intake (mg/day) | % RNI |
|-----------|-------------|--|-----------------|-------|
| Harris *et al.* (1983) | 0.5–1.0 | | 9.6 | 123 |
| | 1.0–2.0 | | 9.4 | 136 |
| | 2.0–3.0 | | 7.0 | 101 |
| | 3.0–5.0 | | 6.0 | 98 |
| Duggan *et al.* (1991) | 0.5–1.0 | iron depleted | 3.17 | 41 |
| | | iron sufficient | 4.15 | 53 |
| | 1.0–1.5 | iron depleted | 4.91 | 71 |
| | | iron sufficient | 3.72 | 54 |
| | 1.5–2.0 | iron depleted | 2.56 | 37 |
| | | iron sufficient | 4.81 | 70 |
| | >2.0 | iron depleted | 3.81 | 55 |
| | | iron sufficient | 3.34 | 48 |
| DH (1989) | Boys 10–11 | | 10.0 | 100 |
| | Girls 10–11 | | 8.6 | 73 |
| Nelson *et al.* (1990) | Boys 7–10 | | 9.1 | 104 |
| | Girls 7–10 | | 9.2 | 104 |
| | Boys 11–12 | | 11.2 | 100 |
| | Girls 11–12 | | 10.0 | 67 |
| Mills and Tyler (1992) | Boys 0.5–0.75 | | 9.6 | 123 |
| | Girls 0.5–0.75 | | 9.0 | 115 |
| | Boys 0.75–1.0 | | 7.2 | 92 |
| | Girls 0.75–1.0 | | 6.4 | 82 |
| Gregory *et al.* (1995) | Mixed 1.5–2.5 | | 5.0 | 73 |
| | Mxed 2.5–3.5 | | 5.6 | 81 |
| | Boys 3.5–4.5 | | 6.2 | 95 |
| | Girls 3.5–4.5 | | 5.9 | 92 |

### 15.8.2    Fortification of weaning foods with iron

Iron fortified weaning foods have been recommended as an effective way of ensuring an adequate iron intake in infancy (Siimes and Salmenpera, 1989), although this view has been challenged (Fomon, 1987) since the iron used in fortifying cereals is often only poorly available. Most commercially produced baby cereals marketed in the UK are fortified with iron, whereas some commercially produced meals or meal constituents are not. There are currently no national or European recommendations for iron fortification of foods designed for the weaning period or for young children.

### 15.8.3    Sources of iron in infancy

In the study by Mills and Tyler (1992), iron intake was significantly higher in infants who received mainly commercially produced foods compared with those on family foods, whose total mean iron intake fell well below the RNI. Table 15.5 taken from this study shows the contribution of various foods to the total intakes of iron in 6–9- and 9–12-month-olds subdivided into two groups: those eating predominantly commercially produced food and those eating predominantly family foods. In the 'commercial food' group, foods in jars/cans, instant dried weaning foods, rusks and infant formula made major contributions to the intake; for the 'family foods' group, breakfast cereals and meat were important sources of iron, although commercial weaning foods made a total mean contribution of 15% to intake in this group also.

### 15.8.4    The Weaning Report

Weaning foods commonly used in the UK are often low in iron and several suggestions have been made in the report Weaning and the Weaning Diet (DH, 1994):

- Meat and fish should be introduced into the diet by the age of 6–8 months in non-vegetarian families.

**Table 15.5**   Sources of dietary iron infants: proportion (%) of total iron intake obtained from food groups (source Mills and Tyler, 1992)

| Foods | Infants 6–9 months | | Infants 9–12 months | |
|---|---|---|---|---|
| | 'commercial'* | 'family'† | 'commercial'* | 'family'† |
| Total cereal | 2 | 39 | 9 | 41 |
| Bread | 0 | 8 | 1 | 10 |
| Breakfast cereals | 1 | 26 | 6 | 24 |
| Cows' milk | 1 | 4 | 2 | 5 |
| Infant formula | 26 | 7 | 21 | 8 |
| Breast milk | 1 | 2 | 0 | 0 |
| Total commercial foods | 70 | 15 | 66 | 5 |
| Infant jars/cans | 19 | 3 | 31 | 1 |
| Instant/dried | 34 | 9 | 18 | 1 |
| Infant rusks | 16 | 3 | 17 | 4 |
| Meat/meat products | 0 | 13 | 0 | 13 |
| Total Intake (mg/day) | 10.1 | 4.6 | 10.5 | 4.9 |

\* Eating mainly commercially produced weaning foods.
† Eating mainly family foods

- Fruit and vegetables, or fruit juice enriched with vitamin C, should be consumed at each meal.
- The consumption of tea and coffee should be limited.

The report recommends that parents and others responsible for feeding young children should receive more education about iron deficiency and its prevention.

### 15.8.5   Bioavailability

The quantity of iron in the diet is not the only factor in determining iron status, since bioavailability from foods varies markedly. In a study of young children in the UK, where intakes were generally less than the RNI, there was little correlation between apparent iron intake and haemoglobin levels, with many cases of iron depleted children consuming more iron than iron replete ones (Duggan et al., 1991). Further research is required to elucidate the role of dietary components in iron nutrition in young children.

### 15.9   PRESCHOOL CHILDREN

A number of studies in the UK have shown a high prevalence of anaemia and iron deficiency in children aged 2 years and under, with poor iron status becoming less common with increasing age. Iron deficiency is more prevalent in children of Asian parentage in the UK (Table 15.1). Healthy Asian children aged 21–24 months in Sheffield had low haemoglobin and ferritin values and high erythrocyte protoporphyrin values; however, the study failed to find an association between iron status and low birthweight, initial feeding method, age at start of weaning, volume of milk intake or iron intake (Duggan et al., 1991). A nutritional survey of Bangladeshi children aged under 5 years living in London revealed high levels of milk intake with late introduction of iron-containing solids. Iron intake in children not receiving fortified formula failed to meet RNI levels (Harris et al., 1983).

Asian children in Birmingham had a low iron intake and other nutrient deficiencies; such children also tended to come from larger families who had lived in England for a shorter time and had a poorer social class and maternal education than the non-

**Table 15.6**    Sources of dietary iron in children: proportion (%) of total iron intake obtained from foods, by age group (years)

| Foods | Mixed 1.5–2.5 | Mixed 2.5–3.5 | Boys 3.5–4.5 | Girls 3.5–4.5 | Boys 10–11 | Girls 10–11 |
|---|---|---|---|---|---|---|
| Total cereal | 44 | 49 | 51 | 48 | na | na |
| Bread | 11 | 11 | 12 | 12 | 13 | 13 |
| Breakfast cereal | 20 | 22 | 22 | 20 | 13 | 10 |
| Biscuits | 5 | 5 | 6 | 5 | 6 | 6 |
| Cows' milk | 5 | 4 | 4 | 4 | 1 | 1 |
| Commercial infant food | 2 | 0 | 0 | 0 | 0 | 0 |
| Meat and products | 14 | 14 | 14 | 15 | 12 | 12 |
| Chips | 3 | 3 | 4 | 4 | 6 | 6 |
| Vegetables/savoury snacks | 11 | 12 | 11 | 11 | | |

Sources: Gregory *et al.* (1995) – children aged 1.5–4.5 years;
DH, 1989–children aged 10–11 years
na, figures not available

anaemic Asian children. As with other studies, there appeared to be no difference in dietary intakes of iron between sufficient and deficient children, with both groups consuming less than recommendations (Grindulus *et al.*, 1986).

A population with a high proportion of Afro-Caribbean children in a deprived inner city area in Bristol was also shown to have a high incidence of iron deficiency (James *et al.*, 1988). Infant feeding practices which predispose to the development of anaemia include late weaning (after 6 months), a prolonged dependence on cows' milk and introduction of low iron weaning foods in Asian children (Duggan *et al.*, 1992) and tea drinking in other cultures (Merhav *et al.*, 1985).

Older preschool age children (2–5 years) in Edinburgh were found to have an intake of iron that was generally lower than the RNI levels, while 7% of 2-year-olds and 3% of 3-year-olds were consuming less than the LRNI; boys tended to take more iron than girls, but iron intakes generally improved with age. The study noted that children with a low iron intake had low intakes of a range of micronutrients. Intakes did not vary greatly over time, and low intakes persisted (Payne and Belton, 1992).

In the recent national UK survey of children aged 1.5–4.5 years (Gregory *et al.*, 1995), mean iron intakes were less than the RNI in children under the age of 4 years; for those aged 4 years and over the average intake matched the RNI, although 57% consumed less than the RNI. The major sources of

dietary iron were cereals and cereal products (Table 15.6). Iron intakes increased with age and boys consumed more iron than girls, although there were no differences in the iron density of diets eaten by different groups.

The incidence of haemoglobin values of < 110 g/l was 12% for children aged 1.5–2.5 years, declining to 2% for children aged 3.5–4.5 years (Table 15.1). There was no correlation between haemoglobin levels and region, social class, employment status, maternal education or number of parents or children in the household, although other indicators of iron status did show a trend of lower iron stores with poorer social circumstance. There was no significant correlation between iron intake and iron status indicators.

## 15.10  SCHOOL-AGE CHILDREN (5–12 YEARS)

Two recent studies have shown that iron intake is generally satisfactory for young boys in the UK, although it decreases to below the RNI in girls as they approach puberty (DH, 1989; Nelson *et al.*, 1990). Table 15.4 outlines iron intakes in children in the UK and Table 15.6 shows the major contributions to iron intake. Studies of iron status in this age group have shown a low incidence of iron deficiency. Inner city children aged 8–12 years in a deprived area in the United States had a 4% incidence of anaemia (Tershakovec and Weller, 1991);

this was concordant with earlier studies in the USA (Cook and Finch, 1976). Iron status was generally satisfactory in an Australian study in 9-year-olds (none were deficient) and 12-year-olds (1.6% deficient) (English and Bennet, 1991).

## 15.11 VEGAN DIETS

The iron intake of vegan children in the UK aged 6–13 years was found to be greater than the RNI, and the diet itself had a higher iron density (mg/kcal) than omnivorous diets for the same age range (Sanders and Manning, 1992). However, in a Dutch study 15% of children aged 4–18 months weaned on to a macrobiotic diet were found to have iron deficiency compared with nil in the control infants (Dagnelie et al., 1989, 1991), so current recommendations may be inadequate for an entirely vegan diet.

## 15.12   OTHER FACTORS AFFECTING IRON STATUS

Children heterozygous for β-thalassaemia are protected against iron deficiency because of increased iron absorption (Galanello et al., 1990) but the two conditions can coexist when iron intakes are very low or where there is blood loss. Heterozygotes for α- and β-thalassaemia both tend to have microcytic red cells so this criterion alone should not be used to diagnose iron deficiency in populations where the thalassaemia trait is common (Earley et al., 1990).

Infections can raise serum ferritin concentrations (Ehrhardt, 1986) and depress haemoglobin levels for a period of about 14 days (Olivares et al., 1989). There is some confusion over whether a low iron status predisposes to infection, or is a result of periods of poor nutrient intake subsequent to illness (Bondestam et al., 1985; Reeves et al., 1984).

In children with lead intoxication, the degree of anaemia produced in combination with an iron deficient diet is greater than that produced by either factor alone. High serum lead levels cause a rise in erythrocyte protoporphyrin. Iron deficiency increases gut permeability to lead (Mahaffey, 1981).

Epidemiological studies have demonstrated a positive relationship between vitamin A levels in serum and iron status, and it is suggested that vitamin A is involved in the release of iron from the liver (Anonymous, 1989). In developing countries where riboflavin deficiency occurs, iron utilization is impaired with riboflavin deficiency (Fairweather-Tait et al., 1992) and microcytic anaemia responds better to iron with riboflavin supplements than with iron alone (Powers et al., 1983). In children on a poor diet who do not respond promptly to iron supplements, other nutrient deficiencies should be considered.

Poor iron status has been described in children with gastrointestinal disease (de Vizia et al., 1992) or who have an abnormal diet because of a medical condition (Bodley et al., 1993); particular attention should be paid to these high risk groups.

## 15.13   IRON SUPPLEMENTS FOR INFANTS AND CHILDREN

It has been proposed that iron supplements cause an increase in infection in childhood, although this has mostly been described in developing countries. Intravenous iron administration to correct deficiency is not recommended because of its association with the exacerbation of infection (Chapter 10). In western countries iron fortification of formulas and foods results in lower infection rates (Oppenheimer, 1989a). High iron levels in the liver have been described in infants dying from Sudden Infant Death Syndrome (SIDS) and there has been speculation that iron supplementation may be a factor in the aetiology (Moore et al., 1994).

The extent of non-prescribed iron supplementation in children has not been well documented. In the national study of preschool children (Gregory et al., 1995), iron supplements provided 2% of the average daily intake, although these tended to be taken by children whose daily intake was already high.

It is common clinical practice to give iron supplements in the UK when haemoglobin values fall below 105 g/l. However, one group of children aged 17–19 months with haemoglobin values of 106–110 g/l were treated for 2 months and increased their haemoglobin levels, indicating a degree of iron deficiency (Parks et al., 1989). Iron supplements have not been used routinely for the prevention of iron deficiency in Britain, although low dose supplements are used extensively in the United States for this purpose. The use of an iron fortified infant formula for the first year of life has been shown to be as effective in preventing iron deficiency as medicinal iron (Irigoyen et al., 1991). Supplementation of children aged 12 to 18 months who were not iron deficient has been associated with slower growth rates compared with a non-supplemented group (Idjradinata et al., 1994).

A number of iron supplements are available on prescription in the UK, containing different amounts of iron: preparations include organic and inorganic ferrous salts and a polysaccharide iron complex. The recommended dosage is 3 mg/kg body weight per day. Supplements are better absorbed if they are taken separately from food in divided dosage, but compliance with this type of regimen may be poor and one dose per day before breakfast is effective (James and Laing, 1994). A once-weekly supplement was found to be effective in iron deficient rats but this regimen has not been evaluated in children (Wright and Southon, 1990).

The duration of administration of medicinal iron supplements should be long enough both to normalize haematological parameters and to replenish body stores, otherwise deficiency will recur unless dietary intakes meet requirements. Supplementation for 3 months was found to have corrected anaemia in 26% of previously anaemic children; the remainder still had abnormal iron stores (Lozoff *et al.*, 1982b). In another study a 2-month supplementation regimen resulted in improvements in iron status but a number of children still had abnormal values despite the improvement (Aukett *et al.*, 1986). There is a recommendation that iron therapy should continue for 6 months in order to replace stores (Brigden, 1993) or for 3 months after haemoglobin values return to normal.

# 16

# IRON IN ADOLESCENCE

Adolescence is defined as the period between childhood and adulthood, i.e. from puberty, the period at which the generative organs become capable of exercising the function of reproduction, to maturity. It is a time of rapid growth and physical and psychological development.

## 16.1 NUTRITIONAL REQUIREMENTS IN ADOLESCENCE

Levels of iron in the body are regulated by the control of iron absorption. There are no specific excretory mechanisms for iron but small daily losses occur through exfoliation of skin and gastrointestinal epithelium (0.14 mg), sweat, bile (0.24 mg) and urine (0.1 mg) (DH, 1991). Basal iron losses in adolescents have been calculated as 0.65–0.79 mg/day and 0.62–0.90 mg/day for females and males, respectively (Commission of the EC, 1993).

From puberty to the menopause, the greatest loss of iron in females is through menstrual blood (median 0.48 mg/day; 95th percentile, 1.9 mg/day) (Chapter 17). The use of oral contraceptives usually results in decreased menstrual blood loss and hence lower losses of iron.

The post-pubertal growth spurt results in an increased requirement for iron. Rapid skeletal muscle development, together with expanding blood volume, create similar extra demands for iron in both adolescent boys (0.55–0.6 mg/day) and girls (0.35–0.55 mg/day). The timing of these extra demands is different for each sex, maturational age being more important than chronological age.

The body tries to meet the increased requirements by increasing the efficiency of absorption of dietary iron and, if necessary, mobilizing body iron stores. However, a long-term deficit may eventually lead to iron deficiency through exhaustion of these stores. It has been estimated that, on average, male adolescents need to absorb 1.05–1.07 mg/day, and females 1.2–1.68 mg/day to meet total iron requirements (Commission of the EC, 1993).

## 16.2 CONSEQUENCES OF IRON DEFICIENCY

Iron deficiency anaemia results in a decrease in the capacity for physical work because of a lowered capacity to transport oxygen and because of the additional impaired oxidative metabolism in iron-depleted muscles (Baynes and Bothwell, 1990). When haemoglobin values fall to just below normal, there may be few problems in sedentary individuals because of compensatory physiological mechanisms that can increase cardiac output and shift the oxygen dissociation curve in favour of improved oxygen delivery to tissues; aerobic work capacity, however, may be affected.

The role of iron in resistance to infection is a controversial area (Dallman, 1987) (Chapter 10). Anaemia has been shown by some investigators to be associated with impaired cell-mediated immune function. A decreased T cell immunity and killing capacity of polymorphonuclear leucocytes, together with some improvement in immune dysfunction with iron supplementation, has been noted by some investigators. Humoral immunity appears to be unaffected by iron deficiency (Heresi, 1986).

Over the last few years there has been increasing evidence that iron deficiency anaemia may affect behaviour (Soemantri *et al*, 1985; Pollitt *et al.*, 1989b) and that deficiencies in early childhood may lead to long-term impairment of psychomotor development and cognitive function (Chapter 11). The question of whether or not the intellectual performance of adolescents is affected by iron deficiency or anaemia remains unresolved.

## 16.3 SPECIAL NEEDS OF SOME ADOLESCENT SUBGROUPS

### 16.3.1 Athletes

Adolescents who are athletes or sportsmen/women must have sufficient iron in the body to ensure maximum oxygen capacity and to avoid the adverse effects of impaired iron status on enzyme systems associated with energy production in muscles.

Failure to maintain an adequate iron status may compromise performance levels and possibly make individuals more prone to injuries. Decreased iron stores have been reported in adolescent runners, particularly females (Nickerson *et al.*, 1985). The early weeks of intense aerobic training are associated with a moderate decrease in body iron reserves, and this is probably the result of mobilization of storage iron for an expanding red blood cell and muscle mass. In the highly trained athlete, iron deficiency may be explained by an accelerated blood loss from the gastrointestinal tract. Adolescent girls who perform intense physical exercise are highly susceptible to iron deficiency (Cook, 1994).

### 16.3.2   Vegetarians

The number of adolescents who are vegetarian is steadily increasing, particularly amongst girls (Gardner Merchant, 1994). There are different types of vegetarian diets, including vegan (which contain no food of animal origin), lacto-ovovegetarian (which include eggs and milk), and 'semi'-vegetarian diets in which some fish and/or white meat is consumed. Avoiding meat may increase the likelihood of several nutritional disorders, principally vitamin $B_{12}$ and iron deficiency. Compared with omnivores, vegetarians often have lower iron stores and are more likely to have iron deficiency anaemia (Chapter 19).

Adolescents who become vegetarian in a household that already contains longstanding, well-informed vegetarians are less likely to be nutritionally deprived since there will be a greater awareness of good vegetarian practices, such as the consumption of an adequate protein mix and of a diet which maximizes iron bioavailabilty. However, 'new' vegetarian adolescents could be more vulnerable to nutritional inadequacies because simply removing meat from a diet is unlikely to result in a well-balanced vegetarian diet. The inclusion of a high proportion of cereal-based products in the diet may appear to provide a sufficient iron intake but may also contain high levels of phytates and other inhibitors of iron absorption; thus the overall bioavailability may be low.

### 16.3.3   Eating disorders and low energy diets

Many adolescents, especially females, experiment with brief periods of food restriction ('dieting') and episodes of binge eating, but only a small percentage of adolescents (1–3%), go on to develop bulimia nervosa (binge eating/purgative cycles) or anorexia nervosa, a self-imposed starvation (Goldbloom and Garfinkel, 1993). Though relatively rare, these eating disorders often appear to originate in adolescence and can lead to a serious impairment in overall health and development. Low energy diets, consumed by adolescents trying to lose weight or by children who are sedentary, are unlikely to provide adequate levels of iron unless the mixture of foods consumed is carefully selected.

### 16.3.4   Pregnancy

There are additional complications from adolescent pregnancies, such as an increased frequency of low birthweight, early delivery and perinatal mortality (Allen, 1993). Infants with low birthweights will have smaller livers, thus total storage iron in the offspring of these mothers will be reduced and this may have future consequences. The assessment of body iron levels in pregnancy is not easy, hence the prevalence of iron deficiency in pregnant adolescents is difficult to quantify (Beard, 1994). However, the estimated high frequency of iron deficiency in female adolescents generally would suggest that pregnant girls are at particular risk and require careful monitoring, and in many cases iron therapy.

### 16.4   PREVALENCE OF IRON DEFICIENCY ANAEMIA AND LOW IRON STORES

During the last decade there have been several studies on the iron status of UK adolescents (Table 16.1). The most useful biochemical indices of body iron levels include blood haemoglobin (Hb), transferrin saturation, serum ferritin and erythrocyte protoporphyrin (EPP; or zinc protoporphyrin, ZPP). The commonly used cut-off points for Hb concentration are 120 g/l for boys aged 11–14/15 years and 115–120 g/l for girls in the same age range (Fairweather-Tait, 1993). There appears to be a high frequency of low (< 20–24 µg/l ferritin) and virtually absent (<10–12 µg/l ferritin) iron stores in both male and female adolescents. The measurement of serum transferrin receptor concentration, a recently introduced test, appears to be of little additional value when assessing iron status in adolescent subjects (Kivivuori *et al.*, 1993).

### 16.5   IRON INTAKES

Mean daily micronutrient intakes in the UK are usually compared with the dietary reference values published by the Department of Health (1991), to allow an assessment of the potential health effects. The measurement of iron intake does not predict

**Table 16.1**    Prevalence of iron deficiency anaemia and low iron stores in UK adolescents

| Study | Sex | Age (years) | n | Anaemia/iron deficiency | Prevalence |
|---|---|---|---|---|---|
| Nelson *et al.* (1993) (London) | M | 12–14 | 202 | Hb < 122/126 g/l | 4% |
| | | | | Serum ferritin < 12 μg/l | 1% |
| | | | | Serum ferritin 12–20 μg/l | 14% |
| | F | 12–14 | 197 | Hb < 120 g/l | 11% |
| | | | | Serum ferritin < 12 μg/l | 4% |
| | | | | Serum ferritin 12–20 μg/l | 16% |
| Doyle *et al.* (1994) (London) | M | 12–13 | 34 | Serum ferritin < 10 μg/l | 8% |
| | | | | Serum ferritin < 20 μg/l | 33% |
| | F | 12–13 | 32 | Serum ferritin < 10 μg/l | 28% |
| | | | | Serum ferritin < 20 μg/l | 64% |
| Nelson *et al.* (1994) (London) | F | 11–14 | 114 | Hb < 120 g/l | 20% |
| Southon *et al.* (1994) (Norwich) | M | 13–14 | 19 | Hb < 120 g/l | 5% |
| | | | | Serum ferritin < 10 μg/l | 11% |
| | F | 13–14 | 35 | Hb < 110 g/l | 0% |
| | | | | Serum ferritin 10 μg/l | 21% |
| Green *et al.* (1991/2) (South Tees) | M | 11–16 | 294 | Hb < 120 g/l | 3% |
| | | | 227 | Serum ferritin < 12 μg/l | 14% |
| | | | 227 | Serum ferritin 12–24 μg/l | 48% |
| | F | 11–16 | 302 | Hb < 110 g/l | < 1% |
| | | | 222 | Serum ferritin < 12 μg/l | 19% |
| | | | 222 | Serum ferritin 12–24 μg/l | 47% |

**Table 16.2(a)**    Studies of iron intakes (mg/day) in UK adolescents

| Age (years) | Sex | n | Mean intake (SE) | Reference |
|---|---|---|---|---|
| *7-day weighed intake* | | | | |
| 10–11 | M | 902 | 10.0 (2.3) C | DH (1989) |
| | F | 821 | 8.6 (1.9) C | |
| 14–15 | M | 513 | 12.2 (3.3) C | |
| | F | 461 | 9.3 (2.5) C | |
| 12–13 | M | 35 | 10.9 (0.5) C | Doyle *et al.* (1994) |
| | F | 30 | 9.7 (0.5) C | |
| 13–14 | M | 19 | 14.9 (2.4) C | Wright *et al.* (1995) |
| | | | 13.2 (1.0) A | |
| 13–14 | F | 35 | 9.7 (0.5) C | |
| | | | 9.8 (0.5) A | |
| *3-day dietary record* | | | | |
| 11–12 | M | 184 | 11.7 (0.2) C | Moynihan *et al.* (1994) |
| | F | 195 | 11.2 (0.3) C | |
| *Food frequency questionnaire* | | | | |
| 12–14 | M | 202 | 12.3 (0.3) C | Nelson *et al.* (1993) |
| | F | 197 | 9.6 (0.3) C | |

C = calculated; A = analysed

**Table 16.2(b)** UK Dietary Reference Values (DRVs) (mg/day)

| Age (years) | Sex | LRNI | EAR | RNI |
|---|---|---|---|---|
| 11–18 | M | 6.1 | 8.7 | 11.3 |
| 11–18 | F | 8.0 | 11.4 | 14.8 |

LRNI = Lower Reference Nutrient Intake
EAR = Estimated Average Requirement
RNI = Reference Nutrient Intake

iron status in groups or in individuals (Southon *et al.*, 1992; Doyle *et al.*, 1994; Wright *et al.*, 1995). However, iron intake data does provide a means of monitoring changes in iron consumption and dietary iron sources over time. There has been a gradual fall in iron intakes in the general population of the UK (Chapter 3), concomitant with changes in life style which have resulted in reduced energy expenditure and hence lower food intake.

The dietary iron intake of British adolescents has been described in several reports (Table 16.2a). These studies show that the most important dietary sources of iron in adolescents are meat and meat products, breakfast cereals, bread and potatoes. Data from Southon *et al.* (1994) indicate that serum ferritin is positively correlated with vitamin C intake ($P < 0.05$) and negatively correlated with milk and milk products ($P < 0.05$).

## 16.6 IRON FORTIFICATION

It has been suggested that one of the most important nutritional modifications which needs to be made during adolescence is an increase in the intake of iron-containing foods (Greenwood and Richardson, 1979). For many years legislation in the UK has required the addition of iron to all flours other than wholemeal. There is also widespread iron fortification of infant formulas and weaning foods, and of breakfast cereals. The latter now provide a significant proportion of the UK Reference Nutrient Intake for several micronutrients, including iron (Table 16.2(b)). Doyle *et al.* (1994) found that adolescents who regularly consumed breakfast cereals (i.e. four or more times each week) had a significantly higher intake of iron (32%) than adolescents who ate breakfast cereals less frequently.

Estimates indicate that 14% of UK adolescents have no breakfast (Bender, 1993). A survey of adolescents living in Hackney (East London) showed that one third of 12–13-year-olds ate nothing before school. Only 20% of adolescents ate cereals for breakfast regularly (many of the others consumed snack foods such as confectionery or crisps) and this was reflected in their significantly higher micronutrient intakes. Apart from the question of whether or not the omission of breakfast *per se* has a detrimental effect on the performance of schoolchildren, it is likely that a significant proportion of adolescents are missing out on a potentially useful source of iron from breakfast foods.

### Acknowledgements

The Task Force would like to express their thanks to Tony Wright (Institute of Food Research, Norwich) for his considerable contribution to the development of this Chapter.

# 17

# IRON AND WOMEN IN THE REPRODUCTIVE YEARS

## 17.1 INTRODUCTION

Menstruation, pregnancy, lactation and, in adolescence, growth influence the iron requirements of women during their reproductive years. Specific systemic metabolic changes associated with these physiological phases affect both the metabolism of iron and the various indices which are used to assess iron status and the risk of iron deficiency. Iron deficiency occurs in developing countries and immigrant or poor communities but there is no clear evidence that it is an extensive problem in developed communities. This was not the case 50 years ago in Europe, and it has been claimed that iron nutriture has improved since 1945 as a consequence of improved nutrition and health, and of fortifying some foods with iron (Scott et al., 1975). However, it has been proposed that as our dietary habits and life style, with lower energy expenditure and dietary energy intakes, depart increasingly from those on which we evolved, we are again increasingly at risk of iron deficiency (Eaton and Konner, 1985; Hallberg and Rossander-Hulten, 1991).

## 17.2 NON-PREGNANT WOMEN

### 17.2.1 Iron deficiency

There are wide geographical and cultural differences in the prevalence of reported iron deficiency in women during their reproductive years. Unfortunately, many studies reporting the prevalence of anaemia in women have used inconsistent or vague criteria to attribute any anaemia to iron deficiency. Using agreed FAO/WHO criteria for iron deficiency anaemia (FAO/WHO, 1988), the prevalence among non-pregnant women varies from 2.3% in the USA, 6.6% in Northern Ireland and 9% in Japan, to 16% in Indian women in Canada, 20% and 21% in Algeria and Benin respectively, 24% in Indian women in South Africa, and 23% in Pakistan (MacPhail and Bothwell, 1992). By focusing on anaemia these data might well underestimate the prevalence of iron deficiency. For example, in Chad

24% of menstruating women were anaemic but 41% had decreased transferrin saturations and increased red cell protoporphyrins compatible with iron deficiency, even though only 10% had serum ferritin concentrations below the threshold equated with iron deficiency (< 12 μg/l) (Prual et al., 1988).

Systematic studies based on other parameters of iron metabolism indicate a high prevalence of iron deficiency as such, and of women at risk of developing iron deficiency. In a cross-sectional study in Denmark, median serum ferritin concentrations were 38 μg/l in women aged 30 and 40 years. As 23% had values between 15 and 30 μg/l and 17% had values less than 15 μg, only 60% of the women had 'replete' iron stores. However, the median haemoglobin values in these women was 137 g/l and only 4% of them had values of less than 121 g/l (Milman and Kirchhoff, 1992). Iron deficiency anaemia (i.e. serum ferritin concentrations < 15 μg/l and haemoglobin levels < 120 g/l) was seen in just 2% of the 30- and 40-year-old women, and in 1% of 50- and 60-year-old women (Milman et al., 1992). In a similar cross-sectional study in England, 25% of women aged 18–44 years had serum ferritin levels between 13 μg/l and 25 μg/l and a further 20% had values less than 13 μg/l; 36% of the same cohort had haemoglobin values below 125 g/l and 5% had values less than 110 g/l (White et al., 1991).

The potential for misattributing anaemia to iron deficiency is illustrated by an intervention study (Garby et al., 1969) in which all women with a haematocrit of < 31% responded to iron supplements whereas only 58% with a haematocrit of < 36%, the customary threshold for anaemia, did so. The fact that about 13% of women with a haematocrit > 36% also responded to iron supplements is a further indication of the dangers of equating iron deficiency with anaemia.

There is cause for some concern about the suitability of customary parameters for the assessment of iron deficiency in women. Many studies, for example, have not allowed for the lower haemoglobin values in black women (mean value 128 g/l) than in white women (134 g/l) (Perry et al., 1992) which is independent of iron intake. It is also important to note that the stage of the menstrual cycle at which indices of iron deficiency are determined can influence estimates of the prevalence of iron deficiency. An analysis of the data from cross-sectional NHANES(II) (Second National Health and Nutrition Examination Survey, USA) demonstrated that haemoglobin values, transferrin saturation and serum ferritin concentrations were lower in samples collected during menstruation than in those taken during the luteal phases of the cycle. These differences probably result from changes in plasma volume since erythropoiesis, as indicated by the percentage level of erythrocyte protoporphyrin, was unaffected. Using assessments based on transferrin saturation, serum ferritin and erythrocyte protoporphyrin, the prevalence of iron deficiency during the menstrual phase was 23±6% compared with 8±2% in the luteal phase; assessments based on transferrin saturation, mean corpuscular volume and erythrocyte protoporphyrin indicated prevalences of 11±3% during menstruation and 4±1% in the luteal phase (Kim et al., 1993). Hormonal influences as well as menstrual losses of iron might explain why women in their reproductive years have lower mean haemoglobin values than men. This difference appears soon after menarche when serum ferritin concentrations are still similar to values in boys (Bergstrom et al., 1995). Such observations support a re-evaluation of the customary reference criteria of iron deficiency in adolescent and, perhaps, in mature women.

As well as problems with reference parameters, there is also concern about the quality of reported data on iron deficiency; few reports make allowance with replicate assays for the considerable day-to-day variation within subjects (Borel et al., 1991; Beaton et al., 1989). Few, if any, of the papers reviewed in this chapter have taken such precautions and for this reason one should be cautious in accepting many data on the prevalence of iron deficiency (Chapter 7).

## 17.2.2  Menstruation

### (a) Blood loss with menstruation

Menstrual blood loss and the associated loss of iron are difficult to measure accurately. The most precise data have been gained using both tampons and sanitary towels to collect losses, with care being taken to avoid inadvertent losses during micturition and defecation (Hallberg et al., 1966). The use of such techniques in co-operative women has shown a skewed distribution of the volume of menstrual loss with mean and median values of 44 ml and 30 ml respectively, and 75th, 90th and 95th centile values of 52.4, 83.9 and 118 ml. There is

considerable variation in blood loss between women but the loss for individual women is very consistent. It is related to body size, is independent of haemoglobin values and is probably genetically influenced (Hallberg et al., 1966; Hallberg and Rossander-Hulten, 1991).

The previous study was done before the general use of oral and intra-uterine contraceptive agents which are known to influence menstrual loss. Intra-uterine devices can double blood loss whereas oral contraceptives reduce it by about 50%; for example, a low dose combined oral contraceptive reduced mean menstrual blood from 60.2 ml to 36.5 ml after 3 months, and 33.7 ml after 6 months. However, serum ferritin concentrations in the women were not altered after 6 months (Larsson et al., 1992). In contrast, a large cross-sectional study from Denmark showed some evidence of iron conservation in women on oral contraceptives, in that they had higher serum ferritin values than non-users (Milman et al., 1992).

### (b) Requirements for iron in menstruating adults

Requirements have been calculated by adding a woman's basal iron requirement to that needed to replace the amounts lost during menstruation. Basal requirements are considered to be about 0.76 mg daily in a 55 kg woman. The conversion of menstrual blood loss to iron loss is done on the basis that 1 g haemoglobin contains 3.34 mg iron (Hallberg and Rossander-Hulten, 1991); thus the median menstrual loss approximates 0.6 mg of iron daily. Adding this to the basal requirement gives an overall requirement of 1.36 mg for a 55 kg woman. However, in 25% of women losses exceed 0.9 mg and in 10% more than 1.4 mg per day, and the requirements of these women are 1.74 and 2.27 mg iron per day. A more detailed derivation of these requirements is given by Hallberg and Rossander-Hulten (1991). Women at the 95th centile of losses would need 2.84 mg iron daily, corresponding to a dietary requirement of 18.9 mg on the basis of a 15% bioavailability of iron from a typical western diet. This is more than twice the median required intake of 9.1 mg iron, and on current diets it would be difficult to achieve such intakes. It might be that intestinal adaptation and consequent improved bioavailability of dietary iron occurs to a greater extent than has been estimated in these calculations because well-nourished women can tolerate menstrual losses of up to 80 ml, i.e. an amount close to the 90th centile of reported losses (Hallberg and Rossander-Hulten, 1991), without developing iron deficiency anaemia (Cohen and Gibor, 1980).

(c)   Requirements for iron in adolescent girls

The iron supply of adolescent girls is of particular concern and has been extensively reviewed (Hallberg et al., 1993b; Beard, 1994; Brabin and Brabin, 1992). Menstruation starts about 1 year after peak growth velocity is reached. At peak growth velocity, systemic iron requirements approximate 1.5 mg/day, and as growth slows, the requirement falls only a little, to 1.3–1.4 mg, because menstruation has started.

If girls have retarded growth because of a general dietary inadequacy they retain the potential for catch-up growth after menarche. Obviously this will increase the need for iron, and menstruating girls on marginally adequate diets are at a particular risk of depleting whatever iron stores they might have; this might compromise the adequacy of their reserves, particularly if they become pregnant during adolescence (Beard, 1994).

On the basis of serum ferritin concentrations, iron deprivation in adolescent girls might be common. In Sweden 40% of girls aged 15–16 years had serum ferritins less than 16 $\mu$g/l, but it was felt that these data underestimated the true risk of iron deficiency, because many of these girls had morphological evidence of defective erythropoiesis (Hallberg et al., 1993b). These findings correspond with the reported incidence of iron deficiency anaemia in 15-year-old girls from Canada, USA and Australia of 40, 42 and 23% respectively (Bergstrom, 1995; Hallberg et al., 1993b). Such age-specific data are not available for the UK but similar prevalences of low ferritin concentrations (i.e. < 24–20 $\mu$g/l) have been reported in groups of girls aged 11–16 years (Chapter 16). It is noteworthy that these values occur against a tendency for the iron intake of adolescent girls and boys to decrease with age (DH, 1989).

Adolescent girls require iron to meet the need for newly synthesized tissue as well as for basal and menstrual losses. Assuming a systemic iron content of 43 mg/kg body weight, about 0.38 mg of iron is required daily between 13 and 16 years of age to meet growth requirements: these vary from 0.36 mg per day at 14–15 to 0.28 mg iron a day at 16 years (Hallberg and Rossander-Hulten, 1991). At 15 years of age menstrual losses are marginally lower than in older women but, overall, adolescent menstrual loss is the same as in adults. On this basis the calculated median iron requirement is 1.73 mg/day with 75th, 90th, and 95th centile requirements of 2.11, 2.65 and 3.21 mg respectively. Assuming a bioavailability of 15%, the latter figure translates to a dietary requirement of 21.4 mg a day (Hallberg and Rossander-Hulten, 1991) with the median dietary requirement at 11.5 mg.

Because of their smaller menstrual losses the iron requirements of women on oral contraceptives are lower (Hallberg and Rossander-Hulten, 1991). However, given current changes in contraceptive practice with possibly more use of barrier techniques, it is probably appropriate to disregard oral contraceptive use when assessing population iron requirements.

### 17.2.3   Effect of exercise

Exercise can alter the parameters of iron status (Chapter 9). Female runners have been shown to have a lower serum ferritin, total iron binding capacity and red blood cell count but significantly higher mean corpuscular haemoglobin levels than inactive women. However, haemoglobin, haematocrit, serum iron, per cent saturation of transferrin and red cell protoporphyrin concentrations were similar, even though the prevalence of serum ferritin concentrations below 20 g/l was higher (Pate et al., 1993). In a study by Lamanca and Haymes (1992) lower haemoglobin concentrations were found in female athletes but this was attributed to the dilutional effect of an expanded plasma volume; anaemia reduced endurance but non-anaemic iron deficiency did not. Iron supplements or increasing meat consumption can prevent the initial decline in haemoglobin and ferritin concentrations which accompany the taking up of exercise (Lyle et al., 1992) but do not necessarily improve endurance (Klingshirn et al., 1992). There would seem to be insufficient evidence that these changes indicate either an increased risk of iron deficiency, or a need for extra iron in female athletes which cannot be met from dietary sources.

### 17.3   PREGNANCY AND LACTATION

The presumed use of iron stores during pregnancy is suggested by the lower serum ferritins in premenopausal multipara than in nullipara and unipara, and by the overall lower serum ferritin values (median 37 $\mu$g/l) in Danish premenopausal mothers compared with postmenopausal women (median 71 $\mu$g/l) (Milman et al., 1992). In English women aged 18–44 years, 45% had serum ferritin concentrations below 25 $\mu$g/l whereas only 15% of women aged 55–64 years had such low values (White et al., 1993).

Iron deficiency anaemia in pregnancy is seen as a significant problem in the developing world (Scott et al., 1975). In Chad, 25% of pregnant women were anaemic and 88% had red cell protoporphyrins and transferrin saturations compatible with iron deficiency, but only 20% had serum ferritin levels < 12 μg/l (Prual *et al.*, 1988). In Johannesburg, 18.9% of pregnant coloured women had haemoglobin concentrations < 110 g/l in the third trimester of pregnancy while 64% had a transferrin saturation of less than 16%, and 68% had a serum ferritin level < 12 μg/l. It was calculated that almost 40% of these women were entering pregnancy with 'seriously depleted' iron stores (Lamparelli *et al.*, 1988a). Similar conclusions were drawn about pregnant Indian women in Johannesburg (Lamparelli *et al.*, 1988b). These data illustrate the possible extent of compromised iron status in economically disadvantaged women in both developed and developing countries.

However, assessing the actual prevalence of iron deficiency in pregnancy and evaluating its effects is difficult for the various reasons outlined earlier, and also because the physiological changes associated with pregnancy alter the customary criteria for diagnosing iron deficiency.

### 17.3.1   The effect of pregnancy on iron metabolism

Adaptations during pregnancy have to supply nutrients to the fetus and to sustain the increased metabolic burden imposed by the products of conception. Although the compositional mass of the latter and systemic changes such as breast development represent just 6–8% of the mother's normal nutrient intake, the actual metabolic demands are larger, reflected by an overall 15–16% increase in oxygen consumption. The plasma volume and the red cell mass adapt to support this increased metabolic activity with changes during early pregnancy, thereby anticipating the greater changes which occur during the latter half of gestation for the metabolic needs and rapid growth of the fetus (Letsky, 1991). These changes have both a direct and an indirect effect on the metabolism of iron, and on parameters of 'iron status' and anaemia.

#### (a)   Plasma volume

Plasma volume increases steadily until 32–34 weeks of pregnancy and plateaus in the last 8 weeks. The increase in plasma volume is related to the size and health of the conceptus. In a 55–60 kg non-pregnant woman, the plasma volume is about 2.6 litre; in a singleton pregnancy this increases by 1.27 l; and with twins, or triplets, the increase approximates 1.96 l and 2.4 l respectively. Conversely, intra-uterine growth retardation is associated with smaller increases in plasma volume. Plasma volume expansion might be larger (1.5 l) with subsequent pregnancies but the evidence for this is inconclusive. There is really no

correlation between the non-pregnant plasma volume and the absolute increase which occurs during pregnancy (Letsky, 1991). Thus, the smaller the initial plasma volume, for example in small women, the larger will be the subsequent relative increase during pregnancy, and the corresponding dilutional effect on circulating blood components and hence indicators of iron metabolism will also be greater. Similarly women with the heaviest infants have the greatest falls of haemoglobin during pregnancy (Higgins *et al.*, 1982), and the customary fall in haemoglobin concentrations during pregnancy does not occur in mothers whose fetuses are not growing normally (Bissenden *et al.*, 1981; Steer *et al.*, 1995). Without consideration of such factors it would be misleading to interpret the lower concentrations of particular constituents as evidence of anaemia or deficiency.

#### (b)   Red cell mass

The red cell mass increases linearly from the end of the first trimester of pregnancy and is possibly related to the size of the fetus. In women the mean red cell mass is about 1.4 l with an 18% increase to 1.64 l during pregnancy; in women given iron supplements during pregnancy a 30% increase to 1.8 l has been reported. The effects of age and parity on increases in red cell mass are not certain (Letsky, 1991).

Despite the increase in the red cell mass, the concomitant increase in plasma volume reduces the blood viscosity; this along with decreased peripheral resistance to blood flow improves the oxygenation of peripheral tissues. The haemoglobin concentration, haematocrit and red cell count also fall during pregnancy because of the predominant effect of plasma volume expansion over the increase in red cell and haemoglobin mass. Maternal HbF production is also increased, perhaps specifically or as reflection of increased erythropoietic activity. Since the relative balance between plasma volume and red cell mass changes constantly during pregnancy, the US Centre for Disease Control (CDC) has set aside the standard cut-off for diagnosing anaemia of 110 g/l and has produced a series of reference values or cut-off points for the 'physiological anaemia of pregnancy' based on data from European surveys on women taking iron supplements. These set 5th centile haemoglobin values for 4-weekly intervals from 12 weeks gestation with, for example, 12, 20, 24 and 40 week values of 110, 110, 105 and 119 g/l respectively (Morbidity and Mortality Weekly Report, 1989).

### (c)    Erythropoietin (EPO)

During pregnancy, maternal circulating EPO values increase 2–4-fold. This change probably occurs in the first 16 weeks of gestation. In non-pregnant women there is a negative correlation between the haematocrit or haemoglobin and serum EPO, but this relationship is lost during the first and second trimester of pregnancy, when EPO concentrations are lowest. The inverse relationship is apparent again in the third trimester, but only at haemoglobin concentrations below 90 g/l, and is re-established along with normal erythropoiesis about 3 weeks after delivery, as the circulation and blood volume return to non-pregnant conditions. The early loss of the customary relationship is possibly secondary to loss of a hypoxaemic drive for renal EPO production, perhaps a consequence of improved arterial oxygen saturation and tissue perfusion (Huch and Huch, 1993; Beguin et al., 1991). Maternal EPO does not cross the placenta and amniotic fluid EPO is probably of fetal hepatic origin. Fetal circulating levels of EPO react autonomously to intra-uterine hypoxia (Huch and Huch, 1993; Beguin et al., 1991).

Changes in erythropoietic activity are thought to underlie increases in the mean corpuscular volume during pregnancy (Chanarin et al., 1977). Some of the circulatory and erythroid changes are also anticipatory of the loss of blood at delivery. During an uncomplicated vaginal delivery some 500 ml of blood is lost, and the estimated loss following delivery of twins, or after a caesarean section, is 1 litre. In contrast to non-pregnant women, this loss does not induce hypovolaemic shock; this only seems to happen if more than 25% of plasma volume is lost. Most blood loss occurs within 3 days of delivery, after which only about 80 ml of blood is lost in the lochia (Letsky, 1991).

After delivery there is a fall in blood volume due to diuresis. The red cell mass falls with the natural senescence of the red cells and the iron released from degraded red cells is stored and reused. Thus pregnant women can tolerate a loss at delivery of up to 1 litre of blood with very little change in their haemoglobin in the first five days post-partum. During the puerperium there is a temporary erythroid hypoplasia until the non-pregnant plasma volume and EPO response are regained, approximately 3 weeks post-partum (Letsky, 1991).

### (d)    Transferrin and transferrin receptors

Transferrin saturation falls during pregnancy independent of other parameters of iron status as do serum transferrin concentrations. However, the latter is a dilutional phenomenon and both the total transferrin pool and the turnover of transferrin bound iron are increased during pregnancy (Morgan, 1961). Serum transferrin receptor levels are thought to be a sensitive index of iron deficiency or, perhaps more accurately, tissue iron requirements during pregnancy. Serum transferrin receptor concentrations are low in the first two trimesters, but in the third trimester and post-partum transferrin receptor levels match the degree of anaemia and might indeed be a sensitive indicator of increased iron requirement and deficiency (Carriaga et al., 1991). There is no information as yet about the size of the transferrin receptor pool during pregnancy.

The increased synthesis of transferrin during pregnancy is accompanied by a fall in the proportion of tetra-sialo transferrin and an increase in the amount of penta- and hexa-sialo forms. Hexa-sialo transferrin has a higher affinity for placental transferrin receptors than does the tetra-sialo form. Interestingly, similar changes are seen in women on oral contraceptives but not in those with iron deficiency. It has been surmised therefore that these alterations represent a specific adaptation of the systemic metabolism of iron to ensure its effective delivery to, and uptake by, the placenta (de Jong et al., 1992).

### 17.3.2    Placental transfer of iron to the fetus

The maternal surface of the placental syncytiotrophoblast is rich in transferrin receptors which take up iron from transferrin by receptor-mediated endocytosis. This transfer is regulated by the expression of the placental transferrin receptors and by the level of placental ferritin, as well as by the changes in transferrin and the characteristic affinity of placental transferrin receptors. Collectively these changes achieve a unidirectional transport of iron to the placenta and the fetus. The capacity for this uptake and transfer of iron increases with placental growth but the regulatory mechanisms are unclear. Certainly the relatively moderate reductions in the fetal iron stores associated with marked iron deficient pregnancies can be seen as an argument that transfer of iron to the fetus is under feto-placental control (van Dijk, 1988; Harris, 1992).

### 17.3.3    Intestinal absorption of iron

The intestinal absorption of iron, as for other nutrients, increases during pregnancy. For example, one study showed that the absorption of radio-iron from a standard meal was 3.1% in non-pregnant women, 0.8% in early pregnancy, and then 4.5% and 13.5% at 24 and 36 weeks gestation respectively (Svanberg et al., 1975). This impressive adaptation has been confirmed using stable isotopic markers which demonstrated absorptions from a solution containing 5.23 mg inorganic iron of 7.6% (range 1–22%) at 12 weeks gestation, 21.1% (9–58%) at 24 weeks, 37.4% (18–56%) at 36 weeks and 26.3% (range 8–54%) 12 weeks after delivery (Whittaker *et al.*, 1991). The practical relevance of these findings has subsequently been demonstrated in a longitudinal study of 12 normal pregnant women in whom the geometric mean absorption of a stable isotopic marker of iron added to a meal was found to increase from 7% at 12 weeks gestation to 36% and 66% at 24 and 36 weeks respectively, and to fall to 11% at 16–24 weeks post-partum (Barrett *et al.*, 1994).

### 17.3.4    Iron intakes in pregnancy

There is generally no change in dietary intakes of iron during pregnancy. For example, a dietary assessment of women in north-west London found mean daily intakes of iron to be 11.3 mg in Europeans, 10.9 mg in Hindu non-vegetarians, 10.2 mg in Hindu vegetarians and 11.3 mg in Muslims (Abraham *et al.*, 1987). None altered their intake of iron during pregnancy and the intakes of the groups were similar, albeit for possible differences in foodstuffs and associated bioavailability of the iron. In Birmingham, England, a study of women attending a maternity unit using combined 7-day weighed and 24-hour recall methods found mean daily iron intakes of 11 mg (range 3–32) in Pakistani Moslems, 12 mg (5–28), in Sikh women, 10 mg (6–17) and 9 mg (4–15) in Bangladeshi women (Wharton *et al.*, 1984). A longitudinal study of these Asian women found that intakes did not change during pregnancy and confirmed the particularly low iron intakes of some women (Eaton *et al.*, 1984). The intakes of some women were, in fact, disconcertingly incongruous with the calculated requirements, but the intakes of most women were between the British Lower Reference Nutrient Intake value (8.0 mg/day) and Estimated Average Requirement value (11.4 mg/day). A similar range of intakes (9–12 g/day) was found in a comparative study of English and Scottish women; there were, however, significant regional and social class differences in intakes with English women in 'non-manual' social groups having the highest intakes and Scottish 'manual' groups having the lowest intakes (Schofield *et al.*, 1989).

### 17.3.5    Iron requirements of pregnancy

The net additional iron requirements during pregnancy are estimated to be 1040 mg (Table 17.1).

This factorial approach relates the requirements to the amounts needed by the fetus, the placenta, alterations in maternal red cell mass and changes in basal iron losses. Losses at delivery are thought to be greater in primipara than in multipara. Overall, taking into consideration the iron conserved by the cessation of menstruation (280 x median daily iron loss), about 1000 mg of iron is needed to meet the demands of pregnancy. In the first trimester requirements are less than those of menstruating women, that is around the basal requirement of 0.8 mg daily, but subsequently the daily needs increase to about 4.4 mg in the second trimester and to 6.6–8.4 mg in the last trimester. These requirements have to be met from iron stores and from dietary or supplementary sources, or both. The ideal amount of stores, as represented by serum ferritin concentrations, needed to meet these requirements is unknown: probably the amount required will vary with the type of diet eaten and the degree of intestinal adaptation to absorb dietary iron. Women who enter pregnancy with a serum ferritin above 80 μg/l appear to have a low risk of becoming iron deficient whereas those with a value < 50 μg/l may develop iron deficient erythropoiesis (Romslo *et al.*, 1983).

### 17.3.6    The effects of iron deficiency on reproductive efficiency

Reproductive efficiency, from fertility through to the subsequent health and development of the child, might be vulnerable to iron deficiency during pregnancy. Women with low serum ferritin concentrations might be at risk of infertility and of having early abortions. This could be related to the postulated role of placental isoferritin as a down-regulator of cellular immunity (Sirota *et al.*, 1989). In animal models severe iron deficiency is associated with congenital defects but evidence for this in women is not strong, possibly because such severe degrees of iron deficiency are not usually encountered. Additionally a single nutrient deprivation is uncommon and it is possible that effects attributed to iron deficiency are caused by concomitant, but overlooked, deficiencies of other essential nutrients. Some of these deficiencies interact; for example, iron deficiency alters the utilization of folate and, in animal models, the teratological effects of iron and folate deficiencies are synergistic (O'Connor, 1991). It is difficult to interpret most studies of the effect of iron deficiency on the fetus or child because many do not consider socio-economic and possible nutritional confounders, and also because the actual criteria used to characterize the iron deficiency are variable and often fail to allow for the changes in haemoglobin and indices of iron

**Table 17.1**   Iron requirements in pregnancy (based on Letsky, 1991)

|  | Loss of iron (mg) | Range |
| --- | --- | --- |
| Fetus | 300 | (200–370) |
| Placenta | 50 | (35–100) |
| Expansion of maternal red cell mass | 450 | (400–570) |
| Basal iron losses | 240 | (200–270) |
| **Total:** | **1040** | **(835–1310)** |
| Post-delivery: |  |  |
| Contraction of maternal red cell mass | 450 (gain) | (+400–570) |
| Maternal blood loss | 175 | (−100−250) |
| **Net iron cost of pregnancy** | **765 mg** |  |

metabolism which occur during pregnancy. Thus one cannot observe effects of iron deficiency on pregnancy outcome by comparing values in samples collected from mothers delivering preterm with those from women delivering at term. Studies should compare indices derived from women at similar durations of gestation; many studies have not done so (US Preventive Services Task Force, 1993). Maternal ethnic origin, body size and parity should also be controlled for in such studies but, again, they seldom are (Allen, 1993).

The possible effect of iron deficiency on reproductive efficiency and the difficulty of isolating its effects from other associated confounding factors are illustrated by a recent prospective study involving 826 inner-city ethnic minority women in New Jersey (Scholl et al., 1992). This found that 3.5% of them had both anaemia, as defined by the CDC, and serum ferritin concentrations less than 12 µg/l. This was associated with lower energy and iron intakes early in pregnancy and a tripled risk of a low birthweight infant and a doubled risk of preterm delivery compared with anaemia from other causes. An interplay with other nutritional and socio-economic factors was suggested by the increased prevalence of inadequate pregnancy weight gain among those with anaemia of other causes as well as in those with iron deficiency anaemia. Furthermore, when vaginal bleeding had occurred at or before entry to the study, the risks of a preterm delivery were increased 5-fold for iron deficiency anaemia and doubled for other anaemias (Scholl et al., 1992).

In general the efficient feto-placental acquisition of iron protects the fetus and neonate from the effects of maternal iron deficiency. Neonates have higher serum ferritin concentrations than their mothers (Wong and Saha, 1990; Milman et al., 1987) and mothers with low serum ferritin concentrations can still transfer iron to their fetus. For example, no difference was found in the serum ferritin concentrations of babies delivered to mothers with low (< 10 µg/l) or high (> 20 µg/l) levels of serum ferritin (Wong and Saha, 1990). A correlation between cord and neonatal serum ferritin concentrations and maternal values can sometimes be found but often this is not the case. Other studies have suggested that fetal iron reserves are dependent on maternal iron stores (Milman et al., 1987), and that the neonatal iron stores of babies born to mothers with iron deficiency are at risk of being depleted prematurely (Singla et al., 1985). However, there is no evidence that such babies actually become iron deficient.

Information linking gestational iron deficiency with later problems in infancy, childhood and adulthood is of variable quality. A retrospective case-control study of maternal diet and the risk of primitive neuroectodermal tumours of the brain in children under the age of 6 years identified several possible maternal dietary factors, including iron supplements, as protective (Bunin et al., 1993) but this is not evidence for a causal link. Similarly an association, in a retrospective analysis of 8684 pregnant women, between anaemia and iron deficiency with increased placental weight and ratios of placental weight to infant birthweight may merely reflect that iron deprivation is a surrogate for other nutritional and socio-economic deprivations, even though the highest ratio of placental weight to birthweight occurred in the most anaemic women. It is noteworthy that both of these features were also independently associated with a high maternal body mass index (Godfrey *et al.*, 1991). Nonetheless, such observations cannot be ignored, especially since increased placental weight, and an increased ratio of placental weight to infant birthweight have been associated with an increased risk of hypertension in later life (Barker *et al.*, 1990).

Also of concern is the possibility that infantile iron deficiency might be associated with disturbances in cognitive development and affective behaviour (Chapter 11). However, given the efficient supply of iron to the fetus, it is not clear if maternal iron deficiency during pregnancy would contribute to this.

There appear to be no investigations into the effect of iron deficiency during pregnancy on subsequent maternal health and reproductive success.

### 17.3.7  Intervention studies

There are extensive changes in maternal iron metabolism during pregnancy to ensure effective uptake of the element from food and delivery to the fetus. Pregnancy is not, in itself, an adequate reason for supplementing the diet with extra iron but there have been many attempts to investigate the potential benefits of giving women extra iron during pregnancy.

Trials of iron supplementation during pregnancy have been reviewed recently by the US Preventive Task Force (1993). Most were criticized for not making allowances for other factors, such as those mentioned earlier, which either are associated with poor pregnancy outcomes or should be controlled when assessing haematological outcomes. The statistical power of many studies was also criticized. Iron supplements can certainly improve haematological outcomes such as haemoglobin and serum ferritin concentrations and transferrin saturation, and they can prevent an increase of serum EPO in the last trimester, but evidence of beneficial outcomes for the infant and the mother is less clear (Milman *et al.*, 1991; Thomsen *et al.*, 1993).

If birthweight is an intended outcome, allowance must be made for birth order, maternal size, intakes of other essential nutrients, adverse factors in appropriate controls and confounders arising from co-existent nutritional and socio-economic factors. Often iron (sometimes with folic acid) is the only supplement given, and there appear to be no studies using a broad nutritional supplement with and without iron to investigate the specific effect of iron whilst avoiding other potential nutritional confounders.

### 17.3.8  Lactation

The iron content of human breast milk falls during lactation: at 2 weeks post-partum milk contains 0.56 mg/l; 0.4 mg/l at 6–8 weeks; and about 0.3 mg/l at 3–5 months (Siimes *et al.*, 1979). The volume of breast milk increases during the first month of lactation and it is difficult to calculate maternal iron losses in this period with confidence, but assuming a median daily volume of 400 ml, this would average an iron loss of about 0.23 mg per day in the first month, after which milk production approximates 800 ml which would entail a loss of 0.32 mg per day. These are smaller quantities of iron than the mother is conserving by the typical lactational amenorrhoea; in addition, the intestinal absorption of iron is increased during lactation (Svanberg *et al.*, 1975). The iron 'status' of the mother has little effect on the iron content of the breast milk, and the provision of iron supplements during lactation does not increase the volume or iron content of breast milk (Vuori *et al.*, 1980).

In a dietary survey Schofield *et al.* (1989) found that, by and large, lactating women had similar iron intakes to women choosing to bottle-feed their infants and that these intakes did not differ from those during pregnancy. In contrast with these women, and with American lactating women who had iron intakes 12.1 ± 1.5 mg daily, Nepalese lactating women had daily iron intakes of 32.7 ± 3.0 mg. However, their plasma ferritin concentrations were only 21 ± 3 μg/l, compared with 41 ± 4 μg/l in the American women. It was suggested that this difference arose from differences in the bioavailability of the dietary iron and the coexistence of infection and intestinal parasites in the Nepalese (Moser *et al.*, 1988).

### 17.3.9  Iron supplementation

#### (a) Routine supplementation

Despite evidence of maternal adaptation during pregnancy to enable more efficient use of dietary iron (i.e. the increase in bioavailability), routine supplementation is practised widely. The efficacy of such supplements in relation to beneficial outcomes, if any, can be compromised by the frequency and severity of side effects, which reduce the compliance of the mothers (US Preventive Services Task Force, 1993), as is exemplified by recent studies in Burma and Thailand (Charoenlarp *et al.*, 1988) and in Indonesia (Schultink *et al.*, 1993).

The routine use of iron prophylaxis during pregnancy is the subject of considerable debate. Some favour selective supplementation only when clinically indicated (Hibbard, 1988) whereas others advocate iron as a matter of routine (Horn, 1988).

It has been proposed that 30 mg iron daily would avoid perceived side effects (Pippard and Chanarin, 1988) and keep supplements at a reasonable level, avoiding the profligate and uncritical use of high doses. This policy is advocated by the USA Institute of Medicine (Institute of Medicine, 1990). Even so, the US Preventive Services Task Force, after an extensive and critical appraisal of the literature, concluded that in the USA 'there is currently little evidence from published clinical research to suggest that routine iron supplementation during pregnancy is beneficial in improving clinical outcomes for the mother, fetus, or newborn. The evidence is insufficient to recommend for or against routine iron supplementation during pregnancy.' (US Preventive Services Task Force, 1993).

(b)   Adverse effects of iron supplementation

Oral iron–folate supplements have been associated with reduced plasma appearances of zinc after an oral load of the element (Simmer et al., 1987) but consequent concerns that iron supplements could compromise zinc nutriture during pregnancy can be allayed. Although iron supplements do reduce plasma zinc concentrations, they have no significant effects on tissue zinc contents as manifest in neutrophils and mononuclear leucocytes, or on serum alkaline phosphatase activity (Hambidge et al., 1987; Bloxam et al., 1989).

The risk of haemochromatosis from oral supplementation is not well proven (Chapter 8) but there are some reports of acute iron toxicity arising during pregnancy (Lacoste et al., 1992).

# 18

# IRON STATUS IN OLDER PEOPLE

## 18.1 THE DIAGNOSIS OF ANAEMIA

The World Health Organization (WHO) criteria for anaemia (haemoglobin concentrations < 130 g/l blood for men and < 120 g/l in women) are used by most researchers to assess prevalence in all adults, including older persons (WHO, 1968), although a cut-off of < 115 g/l is sometimes used because of the high prevalence of low haemoglobin concentrations in the elderly (Joosten et al., 1992). Whatever cut-off is chosen, it should be recognized that the distribution of haemoglobin in the elderly is not bimodal, and the diagnosis of anaemia is essentially an arbitrary attempt to categorize populations into groups (anaemic and non-anaemic) (Hodkinson, 1985). The cause of low haemoglobin (Hb) values can be obtained from the microscopic examination of a blood smear and further haematological investigations as indicated. Hypochromic red cells together with a low reticulocyte count, serum iron, transferrin saturation (high total iron binding capacity – TIBC) and ferritin levels suggest iron deficiency anaemia. In contrast hypo- or normochromia and low reticulocyte count with normal or high ferritin, low iron and normal transferrin saturation (low TIBC) frequently indicate anaemia of chronic disease (Joosten et al., 1992; Fairley and Foland, 1990; Scott, 1993). These are the two most frequent causes of anaemia in elderly persons (Joosten et al., 1992) (Table 18.1).

## 18.2 SURVEYS IN GREAT BRITAIN

National surveys to assess nutritional status in elderly people in Great Britain were done between 1967 and 1974 (DHSS, 1972, 1979) and in 1991 (White et al., 1993). The earlier studies obtained information on elderly men and women, above and below 75 years old, living at home in England and Scotland. The sample was biased to ensure approximately equal numbers of the four groups. Two rural areas (Cambridgeshire and Angus) and four urban areas (two in the north and two in the south) were surveyed. In contrast, the study in 1991 surveyed approximately 2000 adults aged from 18 to 74 years in England only, of whom 210 men and 240 women were older than 65 years.

### 18.2.1 1967 survey results

There were 879 persons who participated in the 1967 survey. Haematology was done centrally except for haemoglobin estimations which were done locally. Anaemia was defined as a haemoglobin < 130 g/l in men and < 120 g/l in women and was observed in 7.3% (approximately the same in men and women). Mean values, the ranges and the differences between the sexes were approximately the same as in persons aged less than 65 years. However, there was a significant relationship between Hb concentrations, mode of living and serum iron. Of the persons living alone, 9% were

Table 18.1  Anaemias in elderly persons

| Type of anaemia | Red cells | Reticulocyte count | Serum iron | Transferrin saturation | TIBC | Ferritin |
|---|---|---|---|---|---|---|
| Iron deficiency anaemia | Hypochromic | Low | Low | Low | High | Low |
| Anaemia of chronic disease | Hypo- or normochromic | Low | Low | Normal | Low | Normal or high |

**Table 18.2**   Percentage distribution of low serum ferritin and low haemoglobin (from White *et al.*, 1993)

| Factor | Men | | Women | |
|---|---|---|---|---|
| | 65–74 years | 75+ | 65–74 years | 75+ |
| **Serum ferritin *(µg/l)*** | **127** | **80** | **138** | **99** |
| <13 | 6% | 6% | 8% | 12% |
| <25 | 11% | 15% | 17% | 23% |
| **Haemoglobin *(g/l)*** | **127** | **81** | **137** | **99** |
| <110 | 2% | 10% | 4% | 9% |
| <125 | 8% | 17% | 29% | 39% |

anaemic compared with 6% in those living with spouses or relatives. There was no relationship between Hb and social class but the numbers may not have been large enough (DHSS, 1972).

Serum iron was low (< 60 µg/dl) in 17% of the men and 23% of the women. There was a significant relationship between Hb and serum iron, but of those with low serum iron, only one fifth were anaemic. Analysis of all the haematological data suggested that iron deficiency was the sole cause in only 12.9% of those with anaemia. Subnormal folate or vitamin $B_6$ status was present in 30% of those with anaemia, and 20% showed evidence of poor folate and vitamin $B_6$ status as well as low serum iron. A high proportion of those with anaemia had normal serum iron (36%) and some of those (15%) had normal folate, vitamin $B_6$ and vitamin $B_{12}$ levels.

Mean values for TIBC in both men and women were essentially the same as in the population aged below 65 years. The distribution was wider than found in younger persons, with approximately 40% outside the normal range – men mainly below and women above. Iron deficiency anaemia (serum iron < 60 µg/dl and TIBC > 400 µg/dl) was present in 4.7% men and 13.2% women. The extent to which chronic disease was lowering TIBC and masking iron deficiency could not be estimated.

The mean values of serum folate, red cell folate, vitamin $B_{12}$ and vitamin $B_6$ were lower than those found in populations below 65 years but very few of the subjects with subnormal blood levels were anaemic (DHSS, 1972).

### 18.2.2   1972 survey results

The survey in 1972 was a follow-up on the survivors of those surveyed in 1967 who were still living in their own homes; 365 subjects participated fully. Between 1967 and 1972, the mean values for haemoglobin fell from 149 and 139 to 145 and 135 g/l for men and women respectively. Likewise the number of cases of anaemia increased from 7.3% to 12.5% (men 16.9% and women 8.8%) in all areas. TIBC values were unchanged and serum iron, folate, vitamin $B_{12}$ and vitamin $B_6$ were all low but almost the same as found in the earlier survey. However, the number of serum vitamin $B_{12}$ values in the range associated with pernicious anaemia (< 100 ng/l) had increased from 1% to 2.5%. The higher prevalence of anaemia in men than women was believed to be due to a higher prevalence of chronic disease in men since there was a higher proportion of lower TIBC values in men (20%) than women (6.4%).

### 18.2.3   Health survey for England, 1991

The health survey for England included data on haemoglobin and ferritin measurements taken for 444 subjects over the age of 65 (Table 18.2). The number of women with low haemoglobin values was markedly higher than that reported in the earlier surveys. Haemoglobin in both sexes was significantly associated with age (negative), Body Mass Index (BMI) and smoking (positive). In contrast, ferritin in both sexes increased with age, alcohol consumption and BMI but was negatively associated with physical activity.

**Table 18.3**    Dietary reference values for iron (mg/day) (from DH, 1991)

| Group | Lower reference nutrient intake | Estimated average requirement | Reference nutrient intake |
|---|---|---|---|
| 19–50, males | 4.7 | 6.7 | 8.7 |
| 19–50, females | 8.0 | 11.4 | 14.8 |
| 50+ years | 4.7 | 6.7 | 8.7 |

#### 18.2.4   Other survey results

In the 1970s, the prevalence of anaemia in the elderly was approximately 10% (DHSS, 1972, 1979). Similar prevalences of anaemia in the elderly were reported in community studies in Coventry (8%, n = 221), Wales (10.5%, n = 533) and among Asian immigrants (8%, n = 87) (Elwood *et al.*, 1972). In all cases, the prevalence of anaemia in men was greater than that in women as had been reported in the government surveys. In the 1991 survey (White *et al.*, 1993), while the haemoglobin data suggested a slightly greater prevalence of anaemia than previously observed, the number at risk of anaemia on the basis of ferritin results appeared unchanged.

Some other recent results have reported lower prevalences of anaemia. A study of 416 persons over 65 years recruited from a 'retired persons' club in Newcastle found only 4% with anaemia (Rodger *et al.*, 1987). A study of healthy, free-living persons with a mean age of 70 years in southern Ontario, Canada, found only one woman with anaemia (2%) (Martinez, 1988) and there were only 4.4% with anaemia in those aged 65–75 years in the NHANES II study in the United States (Dallman *et al.*, 1984b). Likewise, only 5.2% of men and 5.7% of women (70–75 years) were found to be anaemic in the European SENECA studies. However, although the sample size was in excess of 2000 and included subjects from centres all over Europe, the participation rate was only 50% and the results are probably an underestimate of the true prevalence of anaemia in Europe (Dirren *et al.*, 1991).

In contrast, the prevalence of low haemoglobin values in institutionalized patients or hospitalized admissions tends to be higher. For example, Morgan *et al.* (1973) reported 37% (men) and 21% (women) with anaemia from 93 acute geriatric admissions in northern England. Others found anaemia in 19% of elderly women in a nursing home in Colorado, USA (Jansen and Harrill, 1977)

and 24% of patients consecutively admitted to an acute geriatric ward in Belgium (Joosten *et al.*, 1992).

### 18.3   IRON INTAKES AND REQUIREMENTS OF OLDER PEOPLE

Dietary requirements for iron in women after the menopause match requirements in men (DH, 1991) (Table 18.3). Iron consumption was approximately 11 mg/person/day in the National Household Survey in 1988 (MAFF, 1991). Intakes of 14.0 (men) and 12.3 mg (women) were reported in a survey of 2000 British adults; iron intakes in older people differ very little from those seen in younger adults (Gregory *et al.*, 1990). The low prevalence of iron deficiency in healthy elderly persons suggests that such iron intakes are also adequate for that group. While the efficiency of iron absorption can adjust to meet most requirements, the COMA Panel issuing the dietary reference values cautioned that iron in diets containing little or no meat or accompanied by habitual tea consumption might be less well absorbed, and such people might need a higher iron intake (DH, 1991).

### 18.4   ANAEMIA IN OLDER PEOPLE

Iron, folic acid and vitamin $B_{12}$ are the main nutrients linked with haemopoiesis. Anaemia in the elderly tends to be mainly associated with chronic disease (Joosten *et al.*, 1992). In healthy free-living elderly persons, iron deficiency anaemia was found in only 13% of those with anaemia and was three times more common in women than men (DHSS, 1972). Similar results were reported from the NHANES surveys in the USA who reported iron deficiency anaemia to be rare in elderly men (Yip and Dallman, 1988).

### 18.4.1   Folate status

Many of the subjects in the British elderly surveys had low blood folate levels and about 15% of subjects were regarded as folate deficient (low levels of either plasma or red cell folate). However, there was no relationship between folate status and haemoglobin, and only two subjects were mildly anaemic (Hb 110–120 g/l) in those whose red cell folate levels were below 100 μg/l (DHSS, 1972). Other workers (Elwood *et al.*, 1972; Kahn *et al.*, 1990) have found no evidence of anaemia linked to low folate levels in elderly Caucasian subjects but did find some association in Asian subjects who had the lowest folate levels and the highest proportion of raised mean cell volumes (MCVs) (Elwood *et al.*, 1972).

### 18.4.2   Vitamin B$_{12}$ status

Pernicious anaemia has been reported to occur in about 0.1% of the general population (Scott, 1960) but is predominantly a disease of advancing age associated with an increase in atrophic gastritis (Denham and Chanarin, 1985). Megaloblastic changes are associated with serum vitamin B$_{12}$ levels < 100 ng/l (Elsborg *et al.*, 1976) seen in 1% of elderly subjects in 1967 and 2.5% in 1972 surveys (DHSS, 1972, 1979). Two men (1%) showed evidence of anaemia linked to low vitamin B$_{12}$ levels in 1972 (DHSS, 1979).

### 18.4.3   Other causes of anaemia

Nutrition *per se* appears to play a relatively minor role in the aetiology of anaemia in the elderly. There is some evidence of iron deficiency in elderly women (DHSS, 1972) and of folate deficiency in elderly Asian women (Elwood *et al.*, 1972) but these may be a feature of chronic deficiencies of these nutrients in earlier years extending through into later life. In contrast, stronger evidence exists for the view that inflammation is the predominant cause of anaemia in the elderly (Yip and Dallman, 1988). Anaemia frequently accompanies a variety of chronic disorders including inflammation, infection, malignancy and connective tissue disease such as rheumatoid arthritis (Lipschitz, 1990). Poverty (Yip and Dallman, 1988) and deprivation (DHSS, 1972) are associated with more anaemia in the elderly and although it is suggested that malnutrition is frequently associated with poverty and isolation in the elderly (Ahmed, 1992), there was no association between social class and markers of iron status in either of the two British surveys (White *et al.*, 1993; DHSS, 1972). Thus, evidence suggests that inflammatory or chronic (malignant) disease is the main cause of anaemia in the poor and deprived (Yip and Dallman, 1988).

## 18.5   IRON AND VITAMIN C STATUS

The mobilization of iron from ferritin entails a one-electron reduction of iron from the ferric to the ferrous state and may involve ascorbate (vitamin C). The concept that iron mobilization requires ascorbate has been used to explain the inverse association between vitamin C and iron status, since higher levels of leucocyte ascorbate have been associated with iron deficiency (Jacobs *et al.*, 1971; Bingol *et al.*, 1975) and conversely, in South African Bantu with haemosiderosis and scurvy there is very rapid decarboxylation of administered ascorbate (Hankes *et al.*, 1974).

Plasma ascorbate concentrations are frequently reported to be lower in elderly than in young people (Newton *et al.*, 1985; Burr *et al.*, 1974; Thurnham, 1992b). There is evidence of constitutive disease in the elderly and more in men than in women at the same age although elevated ferritins are relatively modest. At all ages, ascorbate is frequently lower in persons with disease (Hume and Weyers, 1973; Cunningham *et al.*, 1991) or exposed to trauma (Crandon *et al.*, 1961; Irvin *et al.*, 1978) and this may be a protective mechanism to reduce the conversion rate of Fe(III) to Fe(II) (Stadtman, 1991; Thurnham, 1994). In the presence of trauma, acute phase responses may promote iron uptake and avoidance of Fe(II) formation (Stadtman, 1991). Ascorbate has been shown to be potentially lethal if administered to persons who are sensitive to intravascular haemolytic conditions such as glucose-6-dehydrogenase deficiency or paroxysmal haemoglobinuria (Campbell *et al.*, 1975; Iwamoto *et al.*, 1994). Thus low plasma ascorbate will tend to accompany low TIBC or haemoglobin and elevated ferritins, which in turn tend to accompany disease.

## 18.6   AGE-RELATED CHANGES IN HAEMATOLOGICAL VARIABLES

It is well documented that serum iron levels fall with advancing age in both sexes (Rodger *et al.*, 1987; Elwood *et al.*, 1971). Low serum iron together with a high TIBC (> 400 μg/dl) is an indication of iron deficiency anaemia. However, in the elderly, low TIBC (< 280 μg/dl) is more commonly found (DHSS, 1972, 1979; Morgan *et al.*, 1973) which again suggests that these changes are more likely to be associated with chronic or occult disease.

Folate and vitamin $B_{12}$ levels are also frequently low in the elderly. Lower folate levels may indicate reduced intakes of folate-containing foods, but neither low folate nor low vitamin $B_{12}$ levels are usually associated with anaemia in community surveys (DHSS, 1972, 1979). In contrast, in hospital studies there is more evidence of low folate in those with anaemia (Morgan et al., 1973) but as there is usually more anaemia, the association may be more of a consequence than a cause.

# 19

# IRON STATUS OF VEGETARIANS

## 19.1  VEGETARIAN DIETS

A vegetarian is usually defined as someone who does not eat animal flesh (meat, poultry, fish) but who does include eggs and dairy products in their diet. Usually the first step is to give up red meat and then poultry and fish. Some go a step further and become vegans who exclude all food of animal origin. There has been a strong growth in the popularity of vegetarianism: estimates suggest there are more than 2 million vegetarians in the UK and another 3.5 million who avoid red meat. The trend towards vegetarianism has always been strongest in the younger age groups and particularly in women. Some 'new' vegetarians in the USA and Europe follow macrobiotic diets, which in practice are vegetarian and contain only small amounts, if any, of milk. Table 19.1 summarizes the main types of vegetarian diets.

### 19.1.1  Religious beliefs

The term 'Asian' is generally used in the UK to describe people originating from India, Pakistan, Bangladesh or Sri Lanka. People from the Indian subcontinent are the largest subgroup of the non-white UK population, and originate mainly from Gujarat and Punjab. The majority of the Gujarati community are Hindus. Orthodox Hindus believe in the doctrine of 'ahimsa' (not killing animals) and this, together with the sanctity of the cow, forms the basis of their vegetarianism. There are also other religious groups where some form of meat avoidance is prescribed (Table 19.2).

### 19.1.2  Sources and bioavailability of iron

Although a significant portion of iron in the diet of omnivores is derived from the consumption of meat and liver, there are plenty of vegetable and cereal sources of iron (Table 19.3). Bread, wholegrain and fortified cereals, nuts and dark green leafy vegetables are good sources of iron but fruit, dairy products and most starchy foods, such as white rice and potatoes, are poor sources. Iron from plant foods is poorly absorbed compared with haem iron found in some foods of animal origin, especially meat; the presence of dietary fibre, carbonates, oxalates, phosphates, phytates and/or tannins which form insoluble complexes with non-haem iron

**Table 19.1**  Main types of vegetarian diet

| 'Semi' or 'demi' vegetarian (meat-avoiders) | Vegetarian (Lacto-ovo) | Vegan | Macrobiotic |
|---|---|---|---|
| Exclusion of red meat. Occasional consumption of fish/poultry. | Exclusion of all meat, fish and poultry and ingredients derived from these sources e.g. gelatine, rennet. Consumption of dairy products and eggs. | Exclusion of all animal products and derivatives. | Exclusion of all meat and dairy products and eggs. Occasional use of (mainly lean) fish Progressive levels of the diet become increasingly restrictive with gradual elimination of animal-origin foods, fruits and vegetables. |

**Table 19.2**    Food exclusion in religious groups

| Religion | Pork | Beef | Lamb | Chicken | Fish |
|---|---|---|---|---|---|
| Hindu | x | x | s | s | s |
| Muslim | x | Halal only | Halal only | Halal only | s |
| Sikh | x | x | s | s | s |
| Jewish | x | Kosher only | Kosher only | Kosher only | s |
| Buddish (strict) | x | x | x | x | x |
| Seventh-Day Adventist | x | x | x | s | s |
| Rastafarian | x | x | x | x | x |

x = food avoided; s = specific criteria may apply.

may further limit its absorption. However, absorption may be improved by the presence of vitamin C, which occurs in many fruits and vegetables.

## 19.2    DIETARY IRON INTAKE OF VEGETARIANS

Dietary surveys of vegetarians living in the UK have generally found total iron intakes to be as high, or higher, than intakes among omnivore groups (Draper *et al.*, 1993; Reddy and Sanders, 1990; Nathan *et al.*, 1994)). Total iron intakes to match intakes in omnivores have also been reported in vegetarians in New Zealand (Alexander *et al.*, 1994), the USA (Calkins *et al.*, 1984) and Canada (Janelle and Barr, 1995). Vegans further exclude eggs, milk and other dairy products from their diets, but these foods are not useful sources of iron and so vegans who follow balanced diets need not be considered at any additional risk of low dietary intakes of iron.

## 19.3    HAEMATOLOGICAL DATA IN VEGETARIANS

### 19.3.1    Haemoglobin levels

Haemoglobin concentrations are generally within the normal range in vegetarians (Anderson *et al.*, 1981; Gear *et al.*, 1980; Reddy and Sanders, 1990; Fordy and Benton, 1994) but this may be the case only in those who have been vegetarian for some time. Nelson *et al.* (1994) showed that being a vegetarian increased the likelihood of having a low Hb level amongst white 11–14-year-old girls, whereas

this risk was reversed in girls of Indian origin. It was suggested that many of the white vegetarians in the study had only recently adopted the practice and may not have found suitable alternative dietary sources of iron, previously supplied by meat.

### Effect of vitamin C

Nelson *et al.* (1993) found a significant association between low vitamin C intakes and the presence of anaemia in British children aged 12–14 years. The beneficial effect of high vitamin C intake could be seen more clearly when the iron intake was low: the percentage of those with anaemia in the low iron/low vitamin C intake group was approximately twice that in the low iron/high vitamin C group. Reddy and Sanders (1990) studied UK adults and found the vitamin C intake of Caucasian vegetarians, but not Indian vegetarians, to be significantly greater than that of Caucasian omnivores. The Hb concentrations were significantly lower in Indian vegetarians than in both Caucasian vegetarians and omnivores; no significant differences were found in total iron intakes among the groups.

### 19.3.2    Iron deficiency

Although many studies have shown Hb concentrations to be normal in vegetarians, serum ferritin concentrations are often reported to be lower than in omnivores. Reddy and Sanders (1990) found serum ferritin concentrations were significantly lower in vegetarians (both Indian and white) than in omnivores. Geometric mean values for ferritin were below 12 µg/l and 8 µg/l for white and Indian vegetarians, respectively; the mean value for white

**Table 19.3**  Iron content of plant foods (based on Holland *et al.*, 1991; Crawley, 1988)

| Foods | Iron (mg/100 g) | Serving size (g) | Iron (mg per serving) |
|---|---|---|---|
| Bran flakes | 20 | 30 | 6.0 |
| Wholemeal bread | 2.7 | 70 (2 slices) | 1.9 |
| White bread | 1.6 | 70 (2 slices) | 1.1 |
| Baked beans in tomato sauce | 1.4 | 135 | 1.9 |
| Red kidney beans, cooked | 2.0 | 90 (3 tblsp) | 1.8 |
| Peas, boiled | 1.6 | 70 | 1.1 |
| Brown rice, cooked | 0.5 | 200 | 1.0 |
| Broccoli, boiled | 1.0 | 100 | 1.0 |
| Peanuts | 2.5 | 50 | 1.3 |

omnivores (excluding those taking supplements) was 18 μg/l. Intake of haem iron, which supplied approximately 25% of total iron intakes of the omnivores, was found to be positively correlated with serum ferritin concentrations.

In a study where 50 vegetarians were 'matched' with omnivore controls, serum ferritin concentrations were found to be significantly lower in vegetarians (Alexander *et al.*, 1994). While more male vegetarians than omnivores had low ferritin levels (< 12 μg/l), no such differences were seen in women. There was no correlation between serum ferritin and dietary factors (iron or vitamin C intakes) although heavier menstrual losses in women were associated with lower ferritin levels in vegetarians only.

The predominantly vegetarian Indian population in the UK and North America has been reported to have a higher prevalence of iron deficiency, particularly among women (Bindra and Gibson, 1986; Reddy and Sanders, 1990) and infants (Ehrhardt, 1986), compared with the predominantly omnivore general population. There are suggestions that the high prevalence of iron deficiency in Indian vegetarian females may be due to diets low in available iron concomitant with high intakes of dietary fibre, phytate and tannins (Bindra and Gibson, 1986).

An increased prevalence of iron deficiency anaemia has also been reported in 'new' vegetarians and those following macrobiotic diets (Helman and Darton-Hill, 1987; Nelson *et al.*, 1994).

Iron deficiency anaemia, defined as a combination of low Hb, low ferritin, and high free erythrocyte protoporphyrin (FEP), was observed in 15% of infants being fed macrobiotic diets but was not evident in the control groups ($P = 0.003$) (Dagnelie *et al.*, 1989).

# 20

# PUBLIC HEALTH CONSIDERATIONS

## 20.1  INTRODUCTION

Public health involves the maintenance, protection and promotion of health of the population as a whole, rather than on an individual basis. These aspects are not mutually exclusive, as some policies may seek to reduce the number of those who require individual attention. There are three broad areas which are fundamental to public health in respect of diet and nutrition, including iron:

- Scientific understanding of the links between diet and health.
- Monitoring of dietary intake and nutrient status.
- Public health policies to maintain/improve nutrient status and health.

As for other nutrients, the population requires access to enough appropriate food to provide adequate amounts of bioavailable iron for the avoidance of deficiency without the possible risks of excess, both of which are potential problems (Chapters 7 and 8). Monitoring iron status in the population requires criteria relevant to functional end points, including health. Haemoglobin and haematocrit values are important measures of health which can be altered by changes in iron status. However, because these values are also other determinants, their measurement should not be the sole criteria for assessing iron status in the population. Furthermore the various physiological or biochemical functions of iron have different relationships with haematological measures. This means that it is currently difficult to define an optimal iron intake, or level of haemoglobin, for any individual.

Clinically, the most objective goal is probably the avoidance of iron deficiency anaemia. Intakes of iron which maintain low body stores, but without anaemia, might also be associated with adverse effects.

The adverse effects of regular and prolonged intakes of iron in excess of the requirement to avoid anaemia may be subtle, and become manifest only as interference with the bioavailability of other elements, such as copper and calcium.

There are some problems which are peculiar to the assessment of iron and health. Although some men have high requirements, a larger proportion of women have particularly high iron requirements due to heavy menstrual losses. Many women's requirements are often not met by diet alone. Iron is unusual in that those with the higher requirements – menstruating women – tend to be those with the lower intakes, simply because women eat less food overall than adult men.

## 20.2  SCIENTIFIC UNDERSTANDING

Inadequate dietary iron can lead to anaemia, with evident clinical consequences. It is not clear precisely what haemoglobin concentration, or haematocrit, is 'optimal', because the apparently optimal levels for one function (e.g. work performance) might be high and those for another (e.g. risk of cardiovascular disease) might be low (Chapters 9 and 12). While it is probably desirable to have some stored iron, in case of unexpected excessive losses or an interruption of dietary supply, it is difficult to define the minimum required level of stores – or how to measure them. A variety of possible measures of iron status is available, each of which has some value, but there is no single measure, or combination of measures, which defines overall iron status. Nevertheless, because of the undisputed importance of iron deficiency, it is necessary to make pragmatic decisions to allow monitoring and policy development. There is a need for research on the best measures of iron status in different population groups.

## 20.3  MONITORING

Monitoring involves measurements on representative samples of the population, or of population subgroups, and should be distinguished from screening, which involves measurements on all members of a population (see later). It is necessary to monitor the intakes of iron and of the foods containing it. It is also vital to monitor the iron status of the population as a whole, and especially of vulnerable population subgroups. Because of the clinical importance of anaemia, measures of haemoglobin concentration and haematocrit are essential. It is also valuable to estimate iron stores, and serum ferritin is most commonly used for this.

### 20.3.1   Monitoring iron intake

In the UK there are monitoring programmes both for diet and for health. The National Food Survey, carried out by the Ministry of Agriculture, Fisheries and Food (MAFF), collects data on purchases of food in a representative sample of households in Great Britain. The data are analysed to provide information on average intakes, and trends, of foods and nutrients by the British population but cannot describe the distribution of individual intakes within the household (MAFF, 1993). More precise data on the distribution of individual intakes in the population is obtained by the National Diet and Nutrition Survey Programme jointly sponsored by MAFF and the Department of Health. This programme is based on representative samples of 1500–2000 individuals from four different age groups (children aged 1.5–4.5 years, children aged 5–15 years, adults aged 16–64 years and people aged 65 years and over) in Great Britain. Each group is surveyed in turn about every 2 years, so that all ages will be covered in a decade. A survey of adults aged 16–64 years was conducted in 1986/7 (Gregory et al., 1990), and fieldwork on a sample of children aged 1.5–4.5 years was completed in 1993. Fieldwork on a sample of adults over 65 years was started in 1994.

### 20.3.2   Monitoring iron status

Since 1991 the Department of Health has conducted an annual Health Survey for England on a representative sample of adults aged 16 years and over, which gives limited information on dietary habits but also measures haemoglobin and ferritin levels. The results of the first survey of about 3000 individuals were published in 1993 (White et al., 1993). Subsequent surveys will be based on a sample of some 17 000 individuals.

Interpretation of dietary monitoring data is complex. Apart from the inherent difficulties of measuring food consumption and nutrient intake, a fundamental problem is the identification of reference values as cut-off points for judgements on the adequacy or inadequacy of intakes, because of the complex relationship between iron intakes and iron status. In the UK the Recommended Daily Amounts (RDAs) have now been superseded by the Reference Nutrient Intake values (RNIs) (DH, 1991). Conventionally these are set at the estimated average requirement of a nutrient plus 2 standard deviations, and where there is a 'normal' distribution, this is assumed to meet the needs of about 98% of the population.

Because of the skewed distribution of iron requirements, particularly in women of childbearing age, the method of setting cut-off points may not be appropriate. Some argue that the cut-off should incorporate the needs of nearly all women – and therefore be set at the 98th centile, which is around 18 mg/day (Hallberg and Rossander-Hulten, 1991). Others argue that because such intakes are rarely achieved by diet, the cut-off for dietary monitoring in women should be set at 2 standard deviations above the mean, which would exclude those with high menstrual losses. This results in a figure which is around the 75th centile of iron requirements of all women; those whose requirements exceed this should take iron supplements (DH, 1991). These two positions are not based on different information on iron intakes or on the link between intake and status, but represent different philosophical approaches. For public health purposes, the lower value is appropriate. The upper value is useful in the diagnostic setting where individuals may need personal counselling and possibly medication or clinical investigation.

A further problem in deriving cut-off values for dietary iron intakes is the variable absorption of iron, depending not only on the habitual diet but also on an individual's iron status. It is not possible to predict iron absorption in an individual and research is needed to identify features of individuals and their diets to enable better prediction of the adequacy of iron supply in the short and long term.

### 20.4. POLICY OPTIONS

Public health activities address primary prevention, secondary prevention or both. Primary prevention activities target high risk groups or the whole population. Foods peculiar to some high risk groups (e.g. weaning foods for infants) can be fortified. Alternatively routine supplementation could be advised. Fortification of staple foods can increase intakes of a nutrient by the population as a whole. Fiscal measures or social benefits could be specifically devised. Secondary prevention (i.e. the detection and treatment of iron deficiency at an early stage to prevent possible sequelae) depends on screening. Screening the whole population would not be feasible but can be useful in high risk groups. Criteria for intervention, and the most effective and cost-effective interventions, remain to be determined.

## 20.5  GENERAL CONSIDERATIONS

The broad objective of policies is to ensure an adequate but not excessive intake in the population at large and in subgroups who might be vulnerable to inadequate iron intakes. Education – of the population about how to select diets conducive to health, and of professionals in order to give appropriate and more detailed advice – is a foundation of public health policy. In addition to education, mechanisms are needed to ensure adequate access to appropriate foods by all sectors of the population.

Anaemia is uncommon in the adult population in England. In 1991, 4% of women and 1% of men over 18 years of age had haemoglobin concentrations below 110 g/l. A larger proportion of those aged 75 years or more were anaemic. Nevertheless levels of serum ferritin indicative of low iron stores were found in a substantial number of women of childbearing age (White *et al.*, 1993). About 20% of women aged between 18 and 54 years had serum ferritin concentrations less than 13 μg/l. This is particularly important for those women who become pregnant and therefore enter pregnancy with insufficient stores of iron to meet the demand from fetal and placental growth (Chapter 17). It has been suggested that maternal iron deficiency may lead to relative fetal growth retardation, which may have implications for the health of the fetus in adult life (Godfrey *et al.*, 1991).

There are other population subgroups who are at risk of iron deficiency. These include infants and preschool children, adolescents and some vegetarians. It is important for monitoring programmes to cover vulnerable subgroups as well as the population at large, and for education to be appropriately targeted through the various channels for information – e.g. directly from government, or via schools, special interest groups or general media.

For the first few months of life, before weaning, infants rely on the iron stores with which they are born since breast milk contains only limited iron, although it is very efficiently absorbed. It is important that the weaning diet contains adequate bioavailable iron from foods such as meat or from fortified weaning foods or milk formulas. Cows' milk is not a good source of iron. Infants in the second 6 months of life are at risk of iron deficiency (Oski, 1993), which may have consequences for later development (Barker *et al.*, 1993; Pollitt, 1993) (Chapter 11).

Adolescence is a period when girls are particularly vulnerable due to increased requirements for growth and to the start of menstrual losses, sometimes combined with dietary restriction for slimming. Iron deficiency anaemia has been reported in 11% of females and 4% of males aged 12–14 living in London (Nelson *et al.*, 1993).

Vegetarians should ensure that their diet is varied and supplies sufficient essential nutrients. Some vegetarians may simply avoid meat and so may be at risk of iron deficiency (Chapter 19).

## 20.6  SPECIFIC POLICIES

### 20.6.1  Food fortification

Fortification of staple foods can be valuable in ensuring adequate intakes of specific nutrients for the population at large. However, except where a particular food is eaten exclusively by a particular subgroup, fortification cannot be used to target vulnerable groups. By its very nature fortification targets all consumers of a food, and therefore increases overall intakes of a nutrient but is less effective in achieving high intakes in a subgroup with very high requirements. Furthermore, such an approach has the potential for those with lower requirements to be exposed to unnecessarily high intakes of a nutrient.

Several foods in the UK are fortified with iron. Since 1953 all **flour,** except wholemeal, has had to contain not less than 1.65 mg iron/100 g (Statutory Instrument, 1984), which means that iron has to be added. This level, which is that naturally found in flour of 80% extraction, was originally required because 50 years ago white bread could provide up to one-third of the food energy eaten by poorer families. In 1981 the Committee on Medical Aspects of Food Policy (DHSS, 1981) recommended that such fortification was no longer necessary because:

- there was evidence that little if any of this iron is absorbed;
- bread had become less important in the diet;
- the nutritional quality of the rest of the diet had improved.

Nevertheless, iron must still be added to all flours, except wholemeal, usually as powdered iron of a specified particle size and solubility. The addition of iron to flour is also required in some other countries, while in others it is not allowed or is required only in flours labelled as 'enriched'.

**Infant formulas** and follow-on formulas must meet compositional requirements of the European Commission Directive on these products (Statutory Instrument, 1995). Infant formulas where iron has been added must contain specified amounts and forms of iron (Table 20.1). Where iron is not added, the label must carry a statement to the effect that infants over 4 months of age must obtain iron from

**Table 20.1**   Compositional standards for the iron content of formulas with added iron (mg)
(based on Statutory Instrument, 1995)

| Formula | per 100 kJ | | per 100 kcal | |
| --- | --- | --- | --- | --- |
| | Minimum | Maximum | Minimum | Maximum |
| Infant formula | 0.12 | 0.36 | 0.5 | 1.5 |
| Follow-on formula | 0.25 | 0.5 | 1 | 2 |

other sources. In practice, all formulas used in the UK currently have added iron.

Iron is often added voluntarily to other foods in the UK, especially **breakfast cereals,** which typically contain between one sixth and one quarter of the reference amount per serving. Iron is also added to several low-energy foods for slimming, some vegetarian dishes and some confectionery.

Claims for nutrients are allowed on food labelling provided that the food contains a 'significant amount' of that nutrient (Statutory Instrument, 1994). Iron from fortification contributes on average 1 mg per day to dietary intake but can provide much more. For example, six slices of bread (1.8 mg) and one serving of breakfast cereal (2.4 mg) would provide about 4.2 mg iron.

### 20.6.2   Screening

Since only a small proportion of anaemic individuals seek medical treatment, it has been suggested that there should be screening programmes to detect and treat anaemic people. The feasibility and cost effectiveness of such an approach has not been tested, even when confined to high risk groups. An effective screening programme would require a cheap, simple, sensitive and specific test to identify individuals who require particular attention, and the infrastructure to obtain and analyse samples, process the data and report results to the responsible health care workers. Research is needed to identify in which situations such a scheme would be cost effective, and the best treatment for identified cases.

### 20.6.3   Dietary supplements and prescribed medicines

In 1986/7 about 1% of adults aged 16–64 in Great Britain were taking medication prescribed for anaemia – three times as many women as men. In addition about 2% of men and 4% of women took non-prescribed dietary supplements containing iron (Gregory et al., 1990). For men, there was little difference in mean or median iron intakes, or in intakes from food alone or from all sources, including supplements. For women, average intakes of iron from food were 10.5 mg/day, and from all sources 12.3 mg/day. Median intakes from food were 9.8 mg/day, and from all sources 10.0 mg/day. These data suggest that the supplement takers were those whose intakes of iron from food were higher.

This is supported by the observation that the lower 2.5th centile of intake was 4.7 mg/day, (of which 0.1 mg/day was from supplements), compared with the upper 97.5th centile of 30.7 mg/day (with 9.6 mg/day from supplements). Voluntary supplement-taking seems therefore to be ineffective in reaching those with the lowest dietary intakes.

There is no formal guidance on the prescription of iron supplements during pregnancy. Usually, pregnant women are screened for anaemia, and iron supplements are recommended if haemoglobin concentrations are low. Occasionally serum ferritins are measured. Some argue that such screening is unnecessary, as many women have inadequate iron stores prior to pregnancy and so would benefit from supplementation in any case. About a quarter of obstetricians in the UK routinely prescribe iron supplements, either alone or in combination with other micronutrients, from the 20th week of pregnancy.

# 21

# CONCLUSIONS AND SUGGESTIONS FOR FUTURE PROGRESS

## CHAPTER 1  IRON BIOCHEMISTRY

### Conclusions

- Iron is a transition metal and exists predominantly in two valencies, Fe(II) and Fe(III). It reacts readily with oxygen and can promote the production of free radicals. Iron takes part in redox processes and plays a central role in energy metabolism in cells.
- Iron is present in the body as functional or transport and storage forms. The functional compounds include haemoglobin in the red blood cells (oxygen transport), myoglobin in muscle (oxygen storage), cytochromes in mitochondria (oxidative production of energy), iron–sulphur proteins in mitochondria (electron transport), and haem and non-haem iron enzymes involved in a variety of metabolic activities.
- There are interactions between iron and other micronutrients, e.g. copper, vitamin A and riboflavin, during absorption and also at all stages of metabolism thereafter.

### Suggestions/questions

- More research is needed on the interaction of iron with other micronutrients, both pre- and post-absorption into the body.

## CHAPTER 2  IRON ABSORPTION

### Conclusions

- Iron homeostasis is achieved by modifying the efficiency of iron absorption. Body iron stores have a marked influence on this. There is also a short-term control which appears to depend on the previous exposure to iron of the intestinal mucosal cells.
- Haem and non-haem iron are absorbed by two different pathways. Approximately 25% of haem iron is absorbed and this is fairly constant. In contrast, only 10–15% of non-haem iron is normally absorbed from the diet. The absorption of non-haem iron from different foods and meals is very variable, and depends on a number of diet- and host-related factors.
- The main site of iron absorption is the duodenum. Non-haem iron is taken up into the enterocytes by means of an unidentified carrier. It is then either transferred from the cell to the circulation, where it is carried

by plasma transferrin, or remains in the mucosal cell bound to ferritin or other proteins.

● Although many dietary factors/constituents affect iron uptake by the gut, the relative effect of each one (alone and in combination with others) has not been fully described.

● Methods used to assess the availability of iron include *in vitro* measures of 'absorbable' iron and *in vivo* isotope techniques, including [59]Fe whole body retention, faecal monitoring to measure isotopic balance, plasma appearance (following iv and oral administration of isotopes), and incorporation into haemoglobin.

● Estimates of dietary iron bioavailability from single test meals might not accurately reflect how much iron can be taken up from the diet as a whole or over a 'long' period, because single meal studies do not show the effects of intestinal adaptation.

### Suggestions/questions

● A simple *in vitro* technique to measure the proportion of iron in foods and meals that is available for absorption needs to be developed and validated. These results together with a list of biochemical constituents of meals could then be used to predict the quantity of 'absorbable' rather than total iron when assessing the quality of diets.

● The nature and regulation of iron uptake by the body should be further assessed as a means of better defining individual needs for the element and to improve recommendations on iron intakes.

● Stable isotopes for measuring iron bioavailability from foods avoid the need to administer radio isotopes. Thus emphasis should be placed on the further development of stable isotope methodology.

● Further research is required to elucidate the mechanism and regulation of iron absorption. The protein(s) and other factors in the brush border membrane that reduce Fe(III) and subsequently carry iron into the mucosal cells need to be identified, and the role of iron-binding proteins and ligands in the movement of iron within the mucosal cell and subsequent transfer into the circulation requires clarification.

● Iron absorption is inversely related to the size of body stores and the level and form of iron in the gut to which the mucosal cells have been exposed. How does the information reach the absorptive cells? What is the mechanism whereby mucosal cells change their efficiency of iron uptake and/or transfer to the circulation in order to maintain body homeostasis?

## CHAPTER 3   DIETARY SOURCES AND INTAKES OF IRON

### Conclusions

● The most important sources of iron are those foods which are rich in iron, eaten in reasonable quantities, and from which the iron is reasonably well absorbed. This means that the major sources for most people are meat and meat products, fortified cereals and cereal products and vegetables.

● Since the 1950s, UK law has required iron to be added to most wheat flour. Iron fortification of breakfast cereals is widespread but voluntary. Infant formulas and some weaning foods are also fortified with iron and are a major source of iron in infancy. Current fortification practices contribute about

10% of the normal adult daily intake of iron and more than double this for preschool children.
- Iron intakes in children, adolescents and women of child-bearing age are most likely to be below estimated requirements. Average iron intakes in preschool children are below reference nutrient intake levels.
- Iron intakes have been declining in the UK during the last 30 years.

### Suggestions/questions

- More data are needed on the haem iron and non-haem iron content of foods and this should be included in food composition tables. These tables should also include more details on fortified sources of iron, i.e. the amount and form of added iron and that present naturally.
- So that a full assessment of current fortification and supplementation practices can be made, more data are needed on the absorption, safety and efficacy of the different chemical forms of iron used in fortification and supplement manufacture.

# CHAPTER 4   MECHANISMS OF CELLULAR IRON HOMEOSTASIS

### Conclusions

- Transferrin is the major protein used for iron transport. The single polypeptide chain of the glycoprotein has N-terminal and C-terminal domains, each of which bind a single molecule of Fe(III). The delivery of iron to cells involves the interaction of transferrin with its specific membrane-bound receptor.
- Most of the storage iron in the body is present in cells as ferritin, a hollow soluble spherical protein, the centre of which contains iron. With increasing iron accumulation, the iron cores aggregate and the protein shells partly disintegrate forming a (modified) ferritin called haemosiderin.
- The major pathway of iron metabolism involves haem synthesis and breakdown. In mammals, haem is largely present in haemoglobin in the red blood cells. Biosynthesis is linked to erythroblast development with erythroid ALA synthase (eALAS) being the rate limiting enzyme. Haem catabolism is controlled by the enzyme haem oxygenase.
- There is a synchronized regulation of the synthesis of the transferrin receptor, ferritin and eALAS, the key proteins involved in cellular iron uptake, storage and utilization respectively. This takes place at the post-transcriptional level by a cytoplasmic iron regulatory protein (IRP) which binds to iron responsive elements on the messenger RNA (mRNA) of these proteins. In the presence of iron, IRP stimulates translation of ferritin and eALAS mRNAs. In iron deficiency, IRP stabilizes transferrin receptor mRNA. Thus homeostasis is maintained.

### Suggestions/questions

- The control of cellular iron homeostasis *in vivo* within different tissues (e.g. intestinal mucosa, brain, kidney) requires further clarification.
- The speciation of non-transferrin bound iron (low molecular weight iron) within the cell remains unclear. Do ADP and ATP act as possible physio-

logical ligands? The transport of iron across mitochondrial membranes requires elucidation, although it may be via iron–ATP or iron–ADP receptors.
- The mobilization of iron from ferritin and haemosiderin, and its possible regulation, is poorly understood. Are such mechanisms altered in iron loading anaemias?
- The physiological role, if any, of plasma ferritin needs elucidation.

# CHAPTER 5  MEASUREMENT OF IRON STATUS

## Conclusions

- Normal iron status implies the presence of erythropoiesis which is not limited by iron and an adequate reserve of 'storage iron' to cope with normal physiological functions. Other definitions for iron deficiency anaemia, iron deficient erythropoiesis and iron overload are given in the text and the glossary.
- Assessments of iron status do not always make it clear that arbitrary cut-off points are used to indicate risk of deficiency. Cut-off points which are appropriate for epidemiological studies may be inappropriate in a clinical setting.
- Convenient methods for assessing the adequacy of the functional iron compartment are the measurement of haemoglobin levels and red cell measurements such as mean cell volume and mean cell haemoglobin. Transport iron is best measured by serum iron concentration and total iron binding capacity (TIBC) from which transferrin saturation of iron can be calculated. Alternative ways of looking at adequacy of iron supply to the bone marrow are erythrocyte protoporphyrin and serum transferrin receptor concentrations. Storage iron is routinely assessed by measurement of serum ferritin levels but there is concern about the variability of the assays.
- Representative reference ranges for all these indices of iron status are available for adults; they show the importance of considering men and women separately. Reference ranges for children, adolescents and elderly people are also available.
- Indices of iron metabolism are affected by disease processes and do not necessarily indicate either recent iron balance or iron stores; this needs to be considered when interpreting the data.

## Suggestions/questions

- The assessment of the levels of storage iron in the presence of infection and inflammation need improvement. Simple tests which are not influenced by infection and which accurately reflect the 'nutritional factors' modifying iron status are needed.
- Further studies on serum transferrin receptor levels are needed to determine their value in identifying people with impairment of iron supply even in the presence of infection or inflammation.
- What are the best parameters for defining iron deficiency and iron deficiency anaemia? Are multiple analyses better than single measures?
- Standardization of the assay of serum ferritin is essential if the proportion of people with low or absent iron stores is to be compared in different population groups.

# CHAPTER 6    IRON AS A PRO-OXIDANT

## Conclusions

- Iron is capable of acting as a pro-oxidant and catalysing the formation of reactive oxygen species.
- The pro-oxidant effects of iron are prevented by the binding of iron to various proteins both in the circulation and in the tissues to restrict the accessibility of free iron ions.
- In disease, there is evidence that inflammation may disturb the structural integrity of tissues, potentially exposing tissue components to the oxidative effects of iron.
- Changes possibly involved in the movement of iron in disease include reductions in serum transferrin iron and vitamin C and increases in intracellular ferritin and serum caeruloplasmin.
- A shift of iron from the circulation to storage compounds in the macrophages of the spleen, bone marrow and liver is a common feature of many diseases and may assist in reducing the risk of pro-oxidant damage. In the long term this redistribution of iron in chronic disease is likely to contribute to the development of the 'anaemia of chronic disease'.
- The anaemia of cancer, the high serum ferritin in heart disease and the elevated iron stores in kwashiorkor may be different facets of long-term alterations in iron mobilization to reduce the immediate effects of disease, but may eventually have their own pathological consequences.

## Suggestions/questions

- More needs to be known about:
    - the influence of disease and inflammation on the distribution of iron;
    - the interaction of tissue antioxidants and iron in disease;
    - the importance of vitamin C as a pro-oxidant and its interaction with iron both in the gut and in tissues; and
    - the influence of dietary antioxidants on iron absorption.
- Better methods are needed to detect 'free' iron in inflamed tissues.
- Is the iron loading of macrophages associated with infections always non-toxic or can tissue breakdown promote a pro-oxidant effect?

# CHAPTER 7    IRON AND ANAEMIA

## Conclusions

- Anaemia is usually defined on the basis of arbitrary threshold criteria for haemoglobin concentration, based on population studies, but is more correctly defined as a reduction in haemoglobin concentration to below that which is optimal for that individual.
- The commonest cause of anaemia is impaired haem synthesis resulting from iron deficiency. The number of people thought to be affected worldwide by iron deficiency anaemia (IDA) is about 500 million, but many more have absent iron stores and/or iron deficient erythropoiesis. Prevalence is particularly high in developing countries. In developed

countries, it remains high in infants, preschool children, adolescents and in women of child-bearing age.

- Treatment of iron deficiency anaemia is usually through oral iron supplementation.
- Iron loading anaemias are much less common than iron deficiency anaemias. In these cases, the iron overload results from either excessive iron absorption from the diet or the need for regular blood transfusion.
- Treatment of iron loading anaemias is with iron chelating agents.
- In the anaemia of chronic disorders, there is malutilization of iron, with redistribution from functional haemoglobin to iron stores.

## Suggestions/questions

- What is the significance of iron deficiency anaemia in 'at risk' groups? What are the deleterious effects, and are they outweighed by any benefits?
- What are the underlying mechanisms for the impaired proliferation of red cell precursors which result in the anaemia of iron deficiency being relatively hypoplastic?
- Further investigations are needed of novel forms of oral iron supplements and methods of delivery to reduce adverse effects and enhance patient acceptability.
- There is a need for the development of a safe alternative parenteral iron preparation, for use in the relatively infrequent cases where oral iron therapy is inadequate.
- There is a need to develop techniques to predict which patients, with anaemias known to be potentially associated with iron overload, will actually develop life-threatening iron overload. This applies particularly where the anaemia is relatively mild and the iron overload results from excess iron absorption rather than blood transfusion. Is it possible to use combined measures of the concentration of serum transferrin receptors and blood reticulocytes to define the degree of expansion of the erythroid marrow and its effectiveness in delivering red cells to the circulation, respectively? Is excessive iron absorption more common where the expanded red marrow is the result of ineffective erythropoiesis?
- What is the mechanism for the enhanced iron absorption seen in many iron loading anaemias? Further studies assessing the availability of the existing iron stores as well as the relationships between iron absorption and erythropoiesis are required.
- Studies are needed relating tissue iron distribution to the nature of the underlying anaemia (e.g. in sickle cell disease, where iron accumulation in macrophages may sometimes predominate over the more toxic parenchymal iron accumulation).
- An oral iron chelating agent is needed to replace the existing cumbersome use of subcutaneous infusions of desferrioxamine to remove toxic iron from tissues. A non-absorbable oral iron chelator might also be of value in reducing iron uptake where excessive iron absorption is the primary mechanism of iron overload.

# CHAPTER 8   IRON OVERLOAD AND TOXICITY

## Conclusions

- Chronic iron overload of the tissues, especially the liver, can occur as a result of the primary genetic defect, genetic haemochromatosis. Secondary iron overload is observed in a variety of clinical conditions, including tha-lassaemia.

- The prevalence of genetic haemochromatosis in a number of European populations is about 0.3% with a heterozygote frequency of approximately 10%. In the homozygous disease state, there is an abnormally high intestinal absorption of iron which is inappropriate for the systemic iron stores. The gene responsible, which has not yet been identified, is located on chromosome 6.
- Acute iron poisoning leads to profound metabolic disturbance and can be fatal.
- As there is no mechanism for the excretion of excess iron, progressive iron overload develops when there is a sustained increase in iron absorption or there is regular blood transfusion causing iron accumulation in the tissues. This results in increased amounts of catalytically active iron which can be fatal due to extensive tissue damage (particularly heart, liver and endocrine glands).

## Suggestions/questions

- Once the gene for genetic haemochromatosis has been cloned, it will be necessary to elucidate its function, in order to understand its role in iron absorption. The risk of iron overload in heterozygotes can then be assessed, together with the possible risk of the gene in other diseases associated with iron overload.
- Genetic screening will make it easier to identify subjects with genetic haemochromatosis before they accumulate excess iron.
- More work is needed to define the role of iron in exacerbating cellular damage, either directly by causing mitochondrial or lysosomal dysfunction, or indirectly, by altering RNA or DNA expression.
- More work is required to define the possible factors which augment or reduce the toxicity of iron in tissues overloaded with iron in humans.

# CHAPTER 9   EFFECT OF IRON ON WORK PERFORMANCE AND THERMOGENESIS

## Conclusions

- Iron deficiency anaemia, even when mild, limits work performance and is associated with tiring more easily.
- People with iron deficiency anaemia can show disturbances of thermogenesis (heat production) dependent on the thyroid hormones.
- Prolonged severe exertion can lead to a substantial fall in haemoglobin levels which is often referred to as 'sports anaemia'. This is usually due to the dilutional effect of an increased plasma volume. Impairment of iron supply for red blood cell formation is only rarely a contributory factor.

## Suggestions/questions

- Iron supplementation or iron fortification programmes can improve work performance in some groups but is this due to a direct effect of iron on muscular performance or to the correction of anaemia? The possibility that a concomitant reduction in infections contributes to the improvement needs to be explored.

- Iron deficiency in animals may impair cellular function and/or oxidative metabolism independently of anaemia. Does this occur in humans?
- Do abnormalities in the sympathetic nervous system catecholamine metabolism contribute to impaired heat production in people with iron deficiency anaemia?
- What are the mechanisms for the reduction in storage iron levels in the body typically observed with prolonged/severe exercise?

# CHAPTER 10 IRON IN INFECTION AND IMMUNITY

## Conclusions

- Iron is required for the growth of many pathogenic microorganisms, but under normal conditions iron binding proteins in the host prevent the metal becoming available to the organisms. This probably prevents growth of many pathogens *in vivo*.
- In severe iron overload, when serum transferrin is fully saturated, other forms of iron may be present in plasma which could be more readily available to microorganisms.
- Metabolic activity associated with activation of the immune system is a process requiring iron.
- Iron deficiency, and perhaps also iron overload, may impair immune function. Severe iron deficiency is associated with reduced T lymphocyte and neutrophil function.

## Suggestions/questions

- What is the range of 'normal' iron status which will not upset the balance between adequate immune function and minimal bacterial growth? This is the critical factor in understanding whether changes in iron status predispose to infectious disease.
- Both the mechanisms and the physiological function of the changes in iron metabolism associated with inflammation remain enigmatic. Which inflammatory mediators regulate the genes involved in iron metabolism and how?
- A large well-controlled study of the effect of iron deficiency on immune function is needed. A set of internationally agreed guidelines for the selection of patients and controls should be produced for such a study.
- The role of lactoferrin in protecting breast-fed infants against gastrointestinal infections requires further study, and a clinical trial using human (rather than bovine) lactoferrin would be of value.
- The role of nitric oxide as an antimicrobial system in humans and the degree to which it is affected by iron needs elucidation. This may indicate the extent to which abnormal iron status may compromise macrophage microbicidal activity.
- Does iron play an important role in the development of parasitic diseases which are of world-wide importance, such as malaria? How do intracellular blood-borne organisms as opposed to extracellular organisms take up iron, and to what extent is this important in regulating killing by phagocytic cells?

## CHAPTER 11   IRON AND MENTAL AND MOTOR BEHAVIOUR IN CHILDREN

### Conclusions

- Studies using varying criteria for iron deficiency and statistical analyses (of varying quality) have shown that iron deficiency anaemia in young children is associated with impaired performance in a number of developmental and psychological tests.
- The balance of evidence from intervention studies supports a causal relationship between iron deficiency anaemia and impaired mental and psychomotor performance.
- Even though iron deficiency anaemia is only one factor affecting mental and motor development in children, it should be treated and if possible prevented.
- Most of the available evidence suggests that iron deficiency without anaemia does not affect mental and motor behaviour.

### Suggestions/questions

- Research is needed to discover if there is a critical period of development when children are especially vulnerable to a lack of iron. If so, when is this, and is the duration of the anaemia important?
- How permanent is the impairment in mental and psychomotor function resulting from iron deficiency in infancy?
- What is the most efficient way to supplement children with iron deficiency anaemia? To what degree does successful treatment depend on the severity of the iron deficiency anaemia and on the mode of delivery of the supplement?
- The mechanisms whereby iron deficiency anaemia results in psychomotor delay, possibly via alterations in neurotransmitters and neural lipids, require further study.
- The effects of iron deficiency anaemia on mental function in adults, particularly the elderly, need to be determined.

## CHAPTER 12   IRON AND CORONARY HEART DISEASE

### Conclusions

- Early evidence on the association between low haematocrit values and lower risk of coronary heart disease (CHD) suggested that the low blood viscosity in premenopausal women could explain their low risk of CHD compared with men.
- Some recent epidemiological evidence has suggested that increased iron stores (as reflected by serum ferritin levels) are associated with the increased risk of CHD, but this view is not universally accepted. The link between levels of dietary iron and CHD risk is even weaker and no link has ever been proposed in the UK.
- The increased serum ferritin levels found in some CHD patients do not necessarily reflect iron stores and are much more likely to reflect the role of ferritin as an acute phase marker.

### Suggestions/questions

- Further prospective studies are needed to explore the relationship between serum ferritin and tissue iron levels and the risk of CHD.

## CHAPTER 13   IRON AND CANCER

### Conclusions

- There are plausible scientific explanations for the increased incidence of hepatoma (liver cancer) in severe iron overload.
- The direct evidence associating modest iron overload conditions (i.e. high iron status) with cancer risk is very limited. There is a positive association between meat consumption and some sites of cancer, particularly breast cancer and colorectal cancer. The evidence is not consistent, however, and rests on the assumption that iron stores are directly related to meat consumption.

### Suggestions/questions

- Is there a causal role for dietary iron rather than intake of meat in some cancers?
- The relative importance of iron overload *per se* versus cirrhosis in the aetiology of hepatoma requires further study.
- The differences in intracellular iron metabolism between tumour cells and normal cells requires elucidation.
- Further studies are needed on chelation and interference with iron delivery as a possible therapeutic regimen for cancer, with emphasis on protocols that will preferentially inhibit growth of tumour cells rather than normal cells.

## CHAPTER 14   IRON, THE BRAIN AND NEURODEGENERATION

### Conclusions

- Iron performs a crucial role in the functioning of the brain involving key metabolic enzymic reactions associated with mitochondrial oxidation, neurotransmitter synthesis and myelin formation.
- There is accumulating evidence that localized perturbations in iron-mediated free radical oxidative injury contribute to the pathogenic progression of a number of neurodegenerative disorders including Parkinson's and Alzheimer's diseases.
- Direct evidence of the injurious role of 'free' iron in the pathogenesis of neurodegenerative disease is still lacking.

### Suggestions/questions

- There is a need to identify the non-transferrin mediated iron transport mechanisms and their quantitative contribution to iron uptake into the brain.
- Additional studies are necessary to determine the levels of iron distributed among the various iron-binding proteins, and in particular the location, concentration and identity of 'free' iron present in normal and diseased brains.

- A description of the mechanisms involving the release, relocation and deposition of iron within the brain is required.
- An investigation into the role of dietary intakes of iron, antioxidants and oxidizable substrates in modulating free-radical induced neurodegenerative injury should be carried out.
- A study of the interactions between iron and other nutritional and toxic trace elements in the brain is needed.
- Clarification should be sought of the extent to which observed pathological alterations in brain iron are incidental or are causally related to the initiation and exacerbation of neuronal dysfunction and damage.

# CHAPTER 15   IRON IN INFANCY AND CHILDHOOD

## Conclusions

- Infants are born with sufficient iron stores to last 4–6 months, unless the mother is severely anaemic or unless there are abnormal blood losses during delivery. The age at which infants require extraneous iron depends upon their birthweight and growth rate.
- Preterm infants have reduced iron stores and then, if well, they grow rapidly so requirements are high. They become iron deficient by around 3 months chronological age if no additional iron is given.
- Generally, term infants require iron from the diet from the age of 4–6 months onwards. This can be obtained from breast milk or from fortified infant formula/follow-on milks and from solid foods. It is difficult to achieve an adequate iron intake without the use of a fortified infant formula or fortified solid foods in children under the age of 2 years.
- In some surveys of children aged 1–2 years living in inner city areas, the prevalence of iron deficiency anaemia was 10–30%. A nationally representative government survey of children aged $1\frac{1}{2}$–$4\frac{1}{2}$ years has shown that a significant proportion have iron intakes well below the lower reference nutrient intake (LRNI) level. A disturbingly high proportion (12%) had haemoglobin levels below 110 g/l and 28% had low ferritin values (below 10 $\mu$g/l) suggesting absent iron stores.
- Iron supplements should be considered for infants and children with a haemoglobin value of less than 110 g/l. Supplements should be continued for 3–6 months in order to build up iron stores.

## Suggestions/questions

- The prevalence of iron deficiency in the UK in particular areas and groups should be regularly monitored. The usefulness and effectiveness of screening for iron deficiency in children needs to be assessed.
- The best indicators for assessing iron status for young children in the clinical situation need to be identified. Reference values and commonly used 'cut-off points of normality' are not the same as for adults. Are there other identifiable risk factors for iron deficiency apart from low birthweight, early introduction of unmodified cows' milk and prolonged exclusive breast feeding?
- The benefits of giving iron supplements to all preterm babies and to exclusively breast-fed babies aged more than 6 months needs evaluating. What are the long-term effects of iron supplements on

gastrointestinal function and body content of iron? What is the optimal dose of iron given to young children? What is the utilization of medicinal iron supplements using different iron compounds?

- The appropriate levels of iron added to infant formulas and follow-on milks need consideration. What is the value of adding iron to weaning foods?
- The role of lactoferrin in breast milk needs further elucidation.

# CHAPTER 16    IRON IN ADOLESCENCE

## Conclusions

- Information from surveys suggests that low iron stores are common in adolescents and that iron deficiency anaemia may also be present.
- The most important sources of iron in adolescents are meat and meat products, fortified breakfast cereals, bread and potatoes. Total iron intake cannot predict iron status in an individual. In adolescent boys there is a positive correlation between vitamin C and serum ferritin concentration, and a negative correlation with milk and milk products.
- Adolescent subgroups most at risk of iron deficiency are boys or girls who consume low energy diets, and girls who become pregnant.

## Suggestions/questions

- There is a need for more information on the iron intake, requirements and status of adolescents in the UK.
- The relationship between diet and iron status requires more detailed study in order to make dietary recommendations for improving iron status.
- The effect of iron deficiency on the cognitive performance of adolescents needs to be established.

# CHAPTER 17    IRON IN WOMEN IN THE REPRODUCTIVE YEARS

## Conclusions

- Women who lose iron regularly via menstrual blood loss typically have lower serum ferritin values, compared with postmenopausal women. Intra-uterine devices tend to increase blood loss and oral contraceptives tend to decrease blood loss.
- During pregnancy, accompanying physiological changes reduce the concentration of haemoglobin, leading to apparent anaemia in spite of an increase in red blood cell mass. Because of these events, there is not a universally acknowledged cut-off point for defining anaemia in pregnancy.
- About 1000 mg of iron is needed to meet the additional demands of pregnancy which must come from iron stores and from increased absorption from dietary sources or from supplements.
- The changes in iron metabolism that occur in pregnancy support an efficient delivery of iron to the feto-placental unit. For this reason, except for the grossest maternal iron deficiency, there is a poor correlation between the maternal iron status and the iron status of the newborn. When anaemia does occur in the newborn, it is rarely due to iron deficiency in the mother.
- There is little evidence to show that iron supplementation on a routine basis

during pregnancy is warranted, but identification of those who may benefit from iron supplements is difficult.

- Few studies focus on the long-term health of the mother as an outcome measure of iron status in pregnancy, e.g. maternal morbidity or subsequent health and reproductive efficiency.

### Suggestions/questions

- Are there any stages in pregnancy when the fetus is particularly susceptible to an abnormal iron supply and what factors influence this?
- How does maternal iron metabolism adapt to ensure that the fetus gets sufficient iron? What is the molecular basis of the adaptive phenomena and how is it regulated? What are the effects of this adaptation on the long-term health of the mother?
- What is the significance of the high prevalence of low plasma ferritin levels in pregnancy? Do we need new markers to monitor systemic iron metabolism?
- How can we best identify women, before and during pregnancy, in whom iron supplementation would prove beneficial?

# CHAPTER 18    IRON STATUS IN OLDER PEOPLE

## Conclusions

- Low haemoglobin is more common in persons over 65 years than those who are younger. In older people living at home, the prevalence of anaemia varies from 1 to about 10%: it increases with age and is more common among poor or socially deprived persons. In institutions, the prevalence of low haemoglobin values can be much higher. Anaemia in the elderly tends to be that associated with disease.
- Folate, vitamin $B_{12}$, vitamin C and other nutritional factors linked with erythropoiesis are frequently abnormal in elderly persons but vitamin status does not appear to be associated with the level of anaemia.
- Pernicious anaemia due to malabsorption of vitamin $B_{12}$ is present in about 1% of older people.

## Suggestions/questions

- More research is needed to enable a better interpretation of the indices of anaemia in older people, in view of the increasing prevalence of disease with age.
- Research is needed to determine whether there are advantages to be gained from low levels of TIBC and elevated ferritins associated with disease.
- Research is needed to determine what levels of anaemia are functionally significant in the elderly.
- More research is needed to determine the nature of the interaction between vitamin C and iron in the body, particularly in disease.
- What types of diet containing significant sources of bioavailable iron are suitable for elderly people with poor dentition and appetite?

## CHAPTER 19   IRON STATUS OF VEGETARIANS

### Conclusions

- There are many non-meat sources of iron available to vegetarians although it is less well absorbed than the haem iron found in meat. There is enhanced absorption of non-haem iron if large amounts of vitamin C are consumed at the same meal.
- Amongst vegetarians, there is a greater likelihood of low serum ferritin levels than in omnivores and they may be at increased risk of developing anaemia when iron requirements increase (e.g. in pregnancy). This is probably a reflection of lower iron bioavailability from diets containing no meat.
- Some vegetarians in the UK, particularly in the Indian population, are at increased risk of iron deficiency anaemia.

### Suggestions/questions

- What is the best dietary advice for vegetarians to ensure that they have adequate iron stores?
- The reasons for the higher risk of iron deficiency anaemia in Indian vegetarians should be explored.

## CHAPTER 20   PUBLIC HEALTH CONSIDERATIONS

### Conclusions

- One percent of men and 4% of women in the UK have haemoglobin levels below 110 g/dl. These low haemoglobin levels are more common in women of reproductive age (5%) and in men and women of 75 years or over (6% and 12% respectively). Although not all anaemia is due to iron deficiency, low iron stores (serum ferritin below 13 $\mu$g/l) are also common in women of reproductive age (16–28%). Low iron stores may be a particular problem in these women if they become pregnant. Iron deficiency anaemia is more common in vegetarians, particularly in Indians living in the UK.
- 'At risk' groups for iron deficiency anaemia include: infants who are exclusively breast-fed for more than 6 months; bottle-fed infants given cows' milk as their main drink rather than an iron fortified formula; toddlers; adolescents; and women in their reproductive years, particularly when pregnant.
- The assessment of the functional significance of iron deficiency requires further information on the relationship between haemoglobin or serum ferritin, or other markers of iron status, and measures of functional impairment. National survey data are available on population levels of haemoglobin and serum ferritin in adults.
- The principal objective of public health policy is to prevent the occurrence of iron deficiency or excess (primary prevention). Public education and the wide availability of sufficient foods with adequately bioavailable iron are the mainstays of this. Some people choose to take iron supplements or multivitamin preparations, although those with lowest iron intakes from the diet tend not to take them. The advantages and disadvantages of this practice are not

clear. In addition, fortification of staple foods increases the iron intakes of most people, but the bioavailability of the iron is not known.

- In some groups, there is a higher likelihood of iron deficiency. For these 'high risk' groups, specific approaches can be developed – either educational (e.g. dietary advice for vegetarians) or influencing the diet by manipulating food composition (e.g. fortification of certain foods).

- The detection of iron deficiency at an early stage by screening to prevent clinical problems (secondary prevention) would be difficult for the general population but might be possible for 'high risk' groups. However, the appropriate criteria for intervention and the most efficient and cost-effective strategies remain to be determined.

- All infant formulas currently available in Britain are fortified with iron, but legislation allows low iron formulas; follow-on milks must be fortified. The weaning diet should contain foods rich in bioavailable iron: many weaning foods are fortified. A recent COMA report recommended that the introduction of cows' milk as a main drink should be delayed until after the first year of age.

- The fortification of staple foodstuffs such as bread and cereals makes a reasonable contribution to total dietary iron intakes but there is little evidence as to what extent this is bioavailable. There is therefore insufficient data to recommend either the removal or the extension of current iron fortification of foods in the UK.

## Suggestions/questions

- Standard criteria should be used for the interpretation of measurements of iron status in epidemiological surveys.

- The efficacy and cost effectiveness of various options for screening and intervention for iron deficiency and/or anaemia in 'high risk' groups should be determined.

- A critical evaluation of the nutritional significance of current UK iron fortification policy is needed. How bioavailable are the forms of iron being used? Are groups at 'high risk' of iron deficiency anaemia benefiting?

- The relative risks of iron deficiency and iron overload need evaluating. What is the risk in the population from increasing iron intakes? Do heterozygotes for genetic haemochromatosis absorb significantly increased amounts of iron? If so, caution would have to be exercised in increasing levels of fortification.

- The advantages and disadvantages of the consumption of iron supplements within the general population need to be considered.

# 22
# GENERAL CONCLUSIONS AND RECOMMENDATIONS

- Iron deficiency anaemia (IDA) is a common nutrient deficiency in the UK and worldwide. Apart from the adverse manifestation of anaemia itself, there are particular concerns about the effects of iron deficiency anaemia on mental and psychomotor development in children, and in work performance in adults. The degree to which these concerns apply to less severe iron deficient states such as iron deficient erythropoiesis without anaemia or depletion of iron stores is unclear. Although there is a need to know more about the relationship between these iron deficient states and IDA, it seems prudent to reduce all states of iron depletion in society.
- 'At risk' groups for IDA include: infants who are exclusively breast-fed for more than 6 months; bottle-fed infants fed cows' milk as their main drink rather than an iron fortified formula; toddlers; adolescents; and women in their reproductive years, particularly when pregnant. The Task Force considers there is some urgency in introducing programmes to reduce the prevalence of IDA in toddlers.
- Most of the body's iron is present in haemoglobin and iron stores. Assessment of both these compartments is essential in defining an individual's iron status. Assessment of iron supply to the tissues, particularly to the bone marrow (transport iron), may give additional information.
- The quality of the methodology used for the assessment of iron status is far from perfect and some of the inadequacies have been highlighted by this Task Force. All methods are affected by factors other than iron nutrition. Researchers use a variety of reference ranges and a variety of cut-off points to define poor iron status in different groups. This Task Force has suggested guidelines for cut-off points to indicate risk of deficiency for future comparisons of prevalence data from epidemiological studies.
- Studies of the effects of abnormal iron status are often difficult to compare because of the populations chosen for investigation, the methods used and the cut-off points adopted. This makes it difficult to evaluate data and to standardize multi-centre prospective studies. Guidelines should be established for characterization of subjects and their controls for the quality asssessment of laboratory measurements.
- There is no correlation between total dietary iron intake and iron status in individuals. One reason is that iron absorption is inversely related to iron stores in normal healthy people. Further reasons are the different efficiencies of absorption for haem iron and non-haem iron and the wide range of dietary enhancers (e.g. vitamin C, animal protein) and inhibitors (e.g. phytate, calcium) of non-haem iron absorption. There is a need to know how dietary practices in the long term, affect iron absorption and status. To what degree can adaptation occur?
- Although iron can act as a pro-oxidant, tight control of systemic and cellular iron homeostasis usually prevents any deleterious effects. Disease can alter the systemic distribution of iron between the functional and storage compartments and tends to be accompanied by a reduction in the amount of transport iron which may reduce the potential for pro-oxidant activity. There

is a need for more information about the availability of 'free' iron to participate in the catalytic formation of reactive oxygen species during normal or pathological states.

- Some epidemiological evidence suggests that higher serum ferritin levels are associated with increased risk of coronary heart disease (CHD). Other studies do not support this view, as high ferritin levels might indicate some other underlying disease. The link between increased levels of dietary iron and CHD risk and/or some cancers is tenuous.

- Genetic haemochromatosis is an autosomal recessive disorder in which intestinal iron uptake is inappropriately high in relation to body stores, resulting in iron overload. About 1 in 300 people in the UK may be homozygous for the abnormal gene and be at risk of developing iron overload. Detection of the gene responsible should allow identification of both homozygotes and heterozygotes and make it possible to determine the relationships between iron intakes, iron status and the risk of iron overload in those carrying the gene.

- Iron status of people of all ages in the UK should continue to be monitored by the Department of Health on a regular basis. In this way trends within 'at risk' groups can be followed and public health strategies to ensure adequate iron intake in 'at risk' groups developed.

- The practice of fortifying infant formulas, weaning foods and follow-on milks makes a major contribution to iron intakes in young children. There is evidence that the iron in infant formulas and follow-on milks is absorbed and that it can contribute to the maintenance of iron status.

- Current fortification of staple foodstuffs such as bread and breakfast cereals makes some contribution to the total iron intakes of adults and a greater contribution to the intakes of children. However, there is limited evidence about the bioavailability of the iron in fortified foods and therefore their contribution to iron status. There is also little evidence concerning the effects of fortified foods on the risks of iron overload in susceptible people. There are insufficient data, therefore, to recommend either the removal or the extension of iron fortification of foods in the UK.

- In general, eating meat (which is the best source of haem iron) is an effective way to ensure an adequate intake of absorbable iron. Fish and poultry provide little haem iron but they do enhance absorption of iron from other foods in the diet.

- Vegetarians need to take more care with their diet to obtain an adequate intake of iron from sources such as bread, cereals and vegetables. They need to ensure that they consume plenty of foods rich in vitamin C (such as fruits and vegetables) at the same time, in order to enhance the absorption of non-haem iron, and also minimize intakes of iron inhibitors with meals.

# GLOSSARY

**Abluminal**   Surface distal to vascular lumen.

**Achlorhydria**   Absence or near absence of hydrochloric acid in the stomach.

**Aconitase**   An iron–sulphur protein in the mitochondria that converts the aconitate intermediate in the tricarboxylic acid cycle into isocitrate.

**Aerobic**   Requiring the presence of oxygen.

**ALA-synthase**   5-aminolevulinic acid synthase — the first and rate-limiting enzyme of the haem biosynthetic pathway.

**Amenorrhoea**   The absence of menstrual periods.

**Anabolic state**   A state of metabolism during which tissue synthesis exceeds tissue breakdown and there is a net gain of body tissues over time.

**Anaemia**   A low level of haemoglobin in the blood (below that which is normal for an individual)—conventional threshold values: < 120 g/l in women and < 130 g/l in men.

**Angiodysplasia**   An abnormality of the structure of small blood vessels, sometimes leading to haemorrhage.

**Antigens**   Substances capable of inducing an immune response, i.e. specific antibody formation and specific T lymphocyte activation.

**Apoferritin**   The ferritin protein molecule not containing any iron.

**Apotransferrin**   The glycoprotein component of the transferrin molecule whose two iron-binding sites are not occupied.

**Bioavailability**   The availability of a substance from the diet for utilization in normal metabolic processes/functions.

**'Bronze diabetes'**   Hereditary haemochromatosis affecting pancreatic β-cell function (an older term now rarely used).

**Caeruloplasmin**   Copper-containing protein with a variety of oxidase activities; carrier protein of copper in the plasma.

**Catabolic state**   A state of metabolism during which body tissues are being broken down.

**Catecholamines**   A group of substances (hormones) in the body acting as neurotransmitters in the functioning of the sympathetic and central nervous systems, e.g. adrenaline.

**Chelator**   A compound that can bind with metal ions.

**Chromogen**   Substance that is converted into a pigment.

**Cristae**   The folds of the inner mitochondrial membrane.

**Crypts**   A simple glandular tube, follicle or cavity, e.g. the narrow epithelium-lined cavities in the intestinal mucosa.

**Cytochromes**   Groups of electron-transporting proteins containing a haem prosthetic group. Cytochromes are part of the respiratory electron-transport chains of mitochondria.

**Cytokines**   Substances such as growth factors that affect the proliferation and maturation of cells.

**Cytosol**   The cytoplasm within the cell other than the various membrane-bound organelles.

**Cytotoxic**   Toxic to cells.

**Diuresis**   Increased production of urine.

**Dopamine**   Tyrosine-derived neurotransmitter.

**Endocytosis**   Process by which a cell takes up material by invagination of the plasma membrane to form vesicles which enclose the material.

**Endosome**   Vesicles formed by fusion of endocytotic vesicle walls.

**Enhancers**   Substances that increase the rate or degree of an occurrence.

**Erythema**   An abnormal flushing of the skin caused by dilation of the blood capillaries.

**Erythroblasts**   Immature red blood cells (found in the bone marrow).

**Erythrocytes**   Circulating red blood cells.

**Erythrocyte protoporphyrin**   The 'free' protoporphyrin remaining in the red blood cell after the conclusion of haem synthesis. Mostly present as zinc protoporphyrin (ZPP).

**Erythroid haem**   Haem within the erythron.

**Erythron**   The whole system of red blood cells and their precursors in the body (in blood and bone marrow).

**Erythropoiesis**   Red blood cell production; this occurs in the bone marrow.

**Erythropoietin**   A glycoprotein hormone produced chiefly by the kidney and which stimulates the production of red blood cells from precursor cells.

**Ferric iron**   Iron in the trivalent (oxidized) form ($Fe^{+++}$/FeIII).

**Ferritin**   A protein shell containing up to 4000 atoms of ferric iron, found in all tissues of the body, with highest amounts in the liver and spleen.

**Ferrochelatase**   Final enzyme in haem synthesis pathway.

**Ferrous iron**   Iron in the divalent (reduced) form (Fe++/FeII).

**Ferrous sulphate (FeSO₄)**   The commonest form of iron used as the reference for the description of the relative absorption value of iron salts. Ferrous sulphate is deemed to have a value of 100 and is one of the forms used for the treatment of iron deficiency.

**Fibroblasts**   Flattened, irregularly shaped connective tissue cells, ubiquitous in fibrous connective tissue and which secrete components of the extracellular matrix, including type I collagen and hyaluronic acid.

**Fibrosis**   Thickening and scarring of tissue caused by increased synthesis of collagen.

**Fortification**   The addition of one or more nutrients to a food to increase the nutrient content of the diet.

**Free radical**   A chemically reactive molecule with an unpaired electron.

**Gastrectomy**   The surgical removal of the whole or a part of the stomach.

**Gastritis**   Inflammation of the stomach lining.

**Gene**   A stretch of DNA (or RNA for some viruses) which encodes and directs the synthesis of a protein or, in some cases, a specific RNA, e.g. t-RNA.

**Glia**   The non-neuronal cells of the brain.

**Glycoprotein**   A molecule consisting of protein linked to carbohydrate by covalent bonds.

**Haem**   An iron-containing porphyrin – this is combined with the protein globin to form haemoglobin.

**Haem iron**   Iron contained within the porphyrin ring of haem.

**Haem oxygenase**   The enzyme which breaks down haem and releases the iron.

**Haematocrit tube**   A capillary tube for estimating the packed cell volume (%) of red blood cells in blood (termed the haematocrit or PCV).

**Haematofluorometer**   A device for measuring ZPP concentrations in erythrocytes using fluorescence.

**Haemochromatosis**   A hereditary disorder in which there is excessive absorption and storage of iron: if untreated this eventually leads to tissue damage.

**Haemoglobin**   The pigment haem linked to the protein globin contained within the red blood cells. Haemoglobin can combine reversibly with oxygen and is the medium by which oxygen is transported around the body.

**Haemoglobinopathy**   A genetic defect which affects the structure, function or production of globin.

**Haemoglobinuria**   The presence of haemoglobin in the urine.

**Haemolysis**   The premature breakdown of red blood cells.

**Haemopexin**   A plasma protein; binds and transports haem to liver cells.

**Haemopoiesis**   A general term for the process of production of blood and blood cells from stem cells.

**Haemosiderin**   A degraded form of ferritin found within lysosomes in which part of the protein has been lost and the iron cores of ferritin molecules have aggregated (insoluble storage iron).

**Haptoglobin**   A plasma protein; actively binds haemoglobin released into the plasma by haemolysis.

**Hepatocyte**   Liver cell.

**Hepatoma**   A primary tumour of the liver.

**Homeostasis**   Maintenance of the constancy or balance of an environment, e.g. maintenance of equilibrium between organisms and environment.

**Hypertransfusion**   The maintenance of blood haemoglobin levels above 110 g/l by regular blood transfusion in subjects with refractory anaemias, e.g. thalassaemia.

**Hypochromic**   A pale appearance of the red blood cells on microscopy of a stained blood film due to low mean cell haemoglobin concentration. Iron deficiency is the most common cause but it also occurs in thalassaemia, a disease where iron overload may be present.

**Hypoferritinaemia**   Low serum ferritin levels.

**Hypoxia**   Low oxygen supply to blood, tissues or organs.

**Immunosuppression**   The suppression of the ability to mount an immune response; for instance, by radiation or chemotherapy which preferentially kill lymphocytes.

**Infant**   A baby less than 12 months old.

*In vitro*   Experimental observation isolated from the whole body, e.g. in isolated cell or tissue taken from the body (literally 'in glass').

*In vivo*   Experimental observation in a living body.

**Ionic iron**   An atom of iron with a charge.

**Iron carrier protein**   A protein that transports iron within the body, e.g. transferrin.

**Iron deficiency**   A general term describing an absence of storage iron (ferritin, haemosiderin) in the tissues. Anaemia is not always present.

**Iron deficiency anaemia**   Anaemia due to an inadequate supply of iron for haemopoiesis. In clinical terms, iron deficiency anaemia is one which responds to iron therapy.

**Iron deficient erythropoiesis**   Occurs when the supply of iron to the erythron is compromised, and may be present before the haemoglobin has fallen to levels defined as anaemic. This usually implies progressive negative iron balance or iron absorption that is inadequate to meet the needs of the tissues.

**Iron overload**   Excess amounts of iron in the body that under certain conditions may have harmful effects.

**Iron responsive elements** (IRE) A specific nucleotide sequence found in the mRNA coding for a number of proteins including ferritin and the transferrin receptor, involved in the regulation of translation of the mRNA sequence.

**Iron status**   A general term describing the functions of the body in relation to their dependence on iron. It is a continuous variable from severe iron deficiency anaemia through normal iron status with varying amounts of stored iron, to iron overload. Normal iron status implies erythropoiesis which is not limited by iron, and a reserve of storage iron to cope with normal physiological functions and interruptions of dietary iron supply.

**Iron–sulphur proteins**   A group of proteins containing iron complexed with either sulphur atoms derived from cysteine or a mixture of inorganic and cysteine-derived sulphur atoms.

**Iron therapy**   The giving of an oral iron compound (e.g. ferrous sulphate, ferrous fumarate) to correct iron deficiency and/or to replenish iron stores. A common dose is 100 mg elemental iron per day in an adult and 3 mg/kg body weight daily in a young child.

**Isoenzymes**   Variants of an enzyme with similar substrate-specific activity.

**Isoferritin**   One of a number of variants of the ferritin molecule.

**Isotopes**   Variants of an element having the same atomic number and identical chemical properties but differing in atomic mass.

**Lactoferrin**   A protein found in the whey fraction of milk which binds some of the iron present in milk.

**Lazeroids**   Trivial and hyperbolic term for class of iron-chelating antioxidant aminosteroid drugs.

**Leucocytes**   A group of unpigmented cells of blood, commonly known as white blood cells, which includes monocytes, granulocytes and lymphocytes. They are all derived from a common progenitor in bone marrow and are involved in the immunological protection/defence of the body.

**Ligands**   Molecules that bind to another, usually used for signalling molecules binding to receptor proteins or regulatory molecules binding to enzymes, etc.

**Lipocytes**   Cells specialized for lipid production and storage (fat cells).

**Lipofuscin**   Aged deposit of oxidized lipoproteins.

**Lochia**   Discharge from the womb after childbirth.

**Lumen**   The internal space of any hollow structure, such as the intestine.

**Luminal factors**   Factors operating within a lumen, e.g. compounds present in the gut which influence the availability of iron.

**Lysosome**   An organelle within the cell containing hydrolytic enzymes able to break down other cellular substances including potential toxins.

**Macrocytic anaemia**   – see Megaloblastic anaemia.

**Macrophages**   Types of large phagocytic mononuclear white blood cells widely distributed in tissue, which ingest invading microorganisms and also scavenge damaged cells and cellular debris.

**Megaloblast**   An abnormal form of any of the cells that are precursors of red blood cells. Megaloblasts are unusually large and have nuclei that fail to develop normally.

**Megaloblastic anaemia**   Anaemia with abnormally large erythroblasts and red blood cells (macrocytes) – usually due to a deficiency of vitamin $B_{12}$ or folate.

**Menorrhagia**   Excessive menstrual bleeding.

**Microcyte**   A red blood cell smaller than normal.

**Microcytic anaemia**   Anaemia with small red blood cells (often due to iron deficiency).

**Microglia**   Macrophage-type phagocytic glial cells of the brain.

**Microprobe technique**   Sensitive analytical method to determine elemental composition at cellular and subcellular level.

**Mitochondria**   Organelles containing ATP and the enzymes involved in several cellular metabolic activities, including haem synthesis.

**Mitogens**   Substances that cause mitosis and cell division.

**Mobilferrin**   Postulated protein carrier of iron across the gut wall.

**Monomer**   Molecule consisting of a single unit.

**Mononuclear**   Having a single nucleus.

**Mucoprotein**   Proteins found in mucus.

**Mucosal cells**   Cells forming the mucosa, e.g. the lining of the gut.

**Multipara**   Women who have had several children.

**Myelination**   The deposition of a myelin sheath around the axons (long tendrils) of nerve cells. Myelinated nervous tissues appear white rather than grey.

**Myelodysplastic syndromes**   Disorders of haemopoiesis from a genetically altered stem cell. There are characteristic morphological changes which may progress to leukaemia.

**Myoglobin**   An oxygen binding protein containing haem iron found in muscle cells.

**Neurofibrillary tangles**   Intra-neuronal deposits of aggregated neurofilament proteins present in Alzheimer brains.

**Neuromelanin**   Oxidized polymerized product of the neurotransmitter dopamine.

**Neutrophils**   White blood cells capable of ingesting bacteria, providing an important defence against infection.

**Non-haem iron**   Iron not bound to haem.

**Normochromic**   Normal microscopic appearance

of red blood cells with a normal amount of haemoglobin.

**Normocyte**   A red blood cell of normal size.

**Oligodendrocyte**   Myelin-forming cell of the brain.

**Pancreatin**   A preparation containing enzymes extracted from the pancreas.

**Parasitaemia**   Parasites in the blood stream.

**Parenchymal cells**   Cells forming the ground tissue of an organ.

**Parenteral**   Administered intramuscularly or intravenously.

**Peptides**   Chains of three or more amino acids. They can be synthesized on the *m*RNA/*t*RNA/ribosome system or can be the product of partial hydrolysis of a protein or polypeptide.

**Pernicious anaemia**   Anaemia due to a deficiency of vitamin $B_{12}$ caused by a failure of gastric mucosa to produce intrinsic factor needed to absorb the vitamin.

**Phlebotomy**   The puncture of a vein to allow controlled blood removal from the body.

**Plasma**   The non-cellular component of the blood.

**Polycythaemia**   An increased number of erythrocytes in the blood (induced at high altitude and in certain diseases).

**Polymers**   Large molecules made up of many identical subunits.

**Polymorphonuclear leucocytes**   Type of phagocytic white blood cell characterized by multipartite nucleus, which ingests invading organisms. A granulocyte.

**Polypeptides**   More than three amino acids linked by peptide bonds.

**Porphyrin**   One of a number of pigments derived from porphin.

**Primary prevention**   Measures to prevent the occurrence of disease or a disorder in a population.

**Primipara**   Women who have had one child.

**Prokaryotes**   Unicellular organisms whose small, simple cells lack a membrane-bound nucleus, mitochondria, chloroplasts and other membrane-bound organelles.

**Protoporphyrin**   The immediate precursor of haem. It is therefore a constituent of haemoglobin, myoglobin and most of the cytochromes.

**Pyelonephritis**   Bacterial infection of the kidney.

**Pyuria**   The presence of pus in the urine (due to infection).

**Pyruvate kinase deficiency**   Deficiency of the enzyme pyruvate kinase which catalyses the transfer of phosphate from phosphoenolpyruvate to ADP with formation of pyruvate and ATP (or vice versa) in glycolysis.

**Radio-isotopes**   Isotopes of elements that are radioactive (i.e. decay); some may be used as tracers in biological systems.

**Receptor**   Specific protein, glycoprotein or polysaccharide in a cell membrane which binds to a specific molecule, thus enabling the transfer of compounds or signals into the cell.

**Reducing agent**   Compound donating electron or hydrogen in metabolic reduction reactions.

**Reference Nutrient Intake (RNI)**   An intake of a nutrient sufficient to meet the requirements of almost all individuals of a healthy population.

**Refractory anaemia**   Anaemia which does not respond to usual treatments.

**Sarcoma**   Cancer of connective tissue.

**Saturation of transferrin**   The total serum iron concentration divided by the total iron binding capacity (expressed as a percentage figure).

**Secondary prevention**   Measures to detect (and then treat) early disease or disorders within a population. For example, since anaemia is so common in toddlers, some doctors arrange a haemoglobin estimation in all children they see at this age. Any anaemic children are then promptly treated. It has the advantage over primary prevention that only those requiring treatment (e.g. extra iron) are offered the intervention.

**Senescence**   Aged and dysfunctional stages of life cycle.

**Senile plaque**   Characteristic extracellular deposits of mainly aggregated β-amyloid found in high numbers in Alzheimer brains.

**Schistosomes**   Parasitic flukes of the genus *Schistosoma*, causing the human disease schistosomiasis.

**Serotonin**   Tryptophan-derived neurotransmitter.

**Serum**   Fluid component of blood, i.e. after removal of cells and clotting factors.

**Sideroblast**   Erythroblasts (precursors of red blood cells).

**Sideroblastic anaemia**   Anaemia characterized by erythropoiesis with erythroblasts containing iron-rich mitochondria (sideroblasts).

**Siderophillin**   A histochemical stain which tends to absorb iron.

**Siderophores**   Iron chelators produced by bacteria.

**Speciation**   Molecular form of element.

**Splenectomy**   The surgical removal of the spleen.

**Splenomegaly**   Enlargement of the spleen.

**Stable isotopes**   Naturally occurring isotopes of an element which are not radioactive.

**Storage iron**   Iron present in the tissues (mainly liver) as ferritin and haemosiderin which can be used for synthesis of haem and iron containing proteins when required.

**Supplements**   Concentrates of nutrients, usually in tablet or liquid form. These may be used for the prevention of nutrient deficiency diseases or symptoms. *Iron supplement* : an iron salt given to prevent or treat iron deficiency. Many people choose to take supplements irrespective of their risk of deficiency.

The dose is variable but commonly at the RNI for iron in the age group; e.g. many adult preparations provide about 14 mg daily, a preparation for children might provide 5 mg daily.

**Tachycardia** An increase in the heart rate above normal.

**Thalassaemia** An inherited abnormality of the α and β globin part of the haemoglobin molecule leading to impaired red cell haemoglobinization and anaemias. Mild 'carrier' states (heterozygotes) are very common in some parts of the world, and are usually asymptomatic. Several forms cause life-threatening anaemia and may be associated with iron overload.

**Thermogenesis** Heat generation.

**Total iron binding capacity** (TIBC) A measure of the binding capacity of transferrin for iron.

**Transcription** The copying of any DNA strand, nucleotide by nucleotide, following the base-pair rule, by an RNA polymerase to produce a complementary RNA copy.

**Transferrin** The iron binding protein of the plasma — binds two atoms of Fe per molecule and delivers iron to cells via specific receptors.

**Transit time** Time for orally ingested substances to pass through the gastrointestinal tract (from ingestion to evacuation).

**Translation** Process by which the genetic information encoded in *m*RNA directs the synthesis of specific proteins, being orderly synthesis of polypeptide chains at the ribosomes using *m*RNA as a template, by matching codons in *m*RNA with complementary codons on *t*RNAs carrying individual amino acids.

**Transport iron** Iron available for movement around the cell and systemic circulation.

**Valency** The oxidation state of an element; it gives an indication of the reactivity of an atom or a molecule.

**Varices** Dilated veins. Unless otherwise specified, these refer to the abnormal enlargement of veins in the oesophagus as a result of liver disease.

**Vegan** Someone who consumes no foods of animal origin.

**Vegetarian** Someone who consumes no fish or meat products. Milk and eggs are usually accepted.

**Xenobiotic** A non-biological chemical.

**Zinc protoporphyrin** The zinc chelate of erythrocyte protoporphyrin.

# REFERENCES

AAP (American Academy of Pediatrics Committee on Nutrition) (1985) Nutritional needs of low birth weight infants. *Pediatrics* **75**: 976–988.

AAP (American Academy of Pediatrics Committee on Nutrition) (1992) The use of whole cows milk in infancy. *Pediatrics* **89**: 1105–1109.

Abraham, R., Campbell-Brown, M., North, W.R., McFadyen, I.R. (1987) Diets of Asian pregnant women in Harrow: iron and vitamins. *Hum Nutr Appl Nutr* **41**: 164–173.

Acosta, A., Amar, M., Cornbluth-Szarfarc, S.C. *et al.* (1984) Iron absorption from typical Latin American diets. *Am J Clin Nutr* **39**: 953–962.

Adams, P.C., Zhong, R., Haist, J. *et al.* (1991) Mucosal iron in the control of iron absorption in a rat intestinal transplant model. *Gastroent* **100**: 370–374.

Addison, G.M., Beamish, M.R., Hales, C.N. *et al.* (1972) An immunoradiometric assay for ferritin in the serum of normal subjects and patients with iron deficiency and iron overload. *J Clin Pathol* **25**: 326–329.

Addy, D.P. (1986) Happiness is iron. *Brit Med J* **292**: 969–970.

Adelekan, D.A. and Thurnham, D.I. (1990) Plasma ferritin concentrations in anemic children: relative importance of malaria, riboflavin deficiency, and other infections. *Am J Clin Nutr* **51**: 453–456.

Adrian, G.S., Korinek, B.M., Yang, F. (1986) The human transferrin gene 5' region contains conserved sequences which match the control elements regulated by heavy metals glucocorticoids and acute phase reaction. *Gene* **49**: 167–175.

Aggett, P.J., Barclay, S., Whiteley, J.E. (1989) Iron for the suckling. *Acta Paed Scand* **361 (Suppl)**: 96–102.

Ahmed, F.E. (1992) Effect of nutrition on the health of the elderly. *J Am Diet Assoc* **92**: 1102–1108.

Akbar, A.N., Fitzgerald-Bocarsly, P.A., de Sousa, M. *et al.* (1986) Decreased natural killer activity in thalassemia major: a possible consequence of iron overload. *J Immunol* **136**: 1635–1640.

Alexander, D., Ball, M.J., Mann, J. (1994) Nutrient intake and haematological status of vegetarians and age–sex matched omnivores. *Eur J Clin Nutr* **48**: 538–546.

Alford, C.E., King, T.E., Campbell, P.A. (1991) Role of transferrin, transferrin receptors, and iron in macrophage listericidal activity. *J Exp Med* **174**: 459–466.

Allen, L.H. (1993) Iron-deficiency anaemia increases risk of preterm delivery. *Nutr Rev* **51**: 49–52.

Alvarez-Hernández, X., Licéaga, J., Mackay, I.C., Brock, J.H. (1989) Induction of hypoferremia and modulation of macrophage iron metabolism by tumor necrosis factor. *Lab Invest* **61**: 319–322.

Ames, B.N., Saul, R.L., Schwiers, E. *et al.* (1985) Oxidative DNA damage as related to cancer and aging: Assay of thymine glycol, thymidine glycol, and hydroxymethyluracil in human and rat urine. In *Molecular biology of aging: Gene stability and gene expression* (eds R.S. Sohal *et al.*), Raven Press, New York: 137–144.

Andelman, M.B. and Sered, B.R. (1966) Utilization of dietary iron by term infants: a study of 1048 infants from a low socioeconomic population. *Amer J Dis Child* **111**: 45–55.

Anderson, B.M., Gibson, R.S., Sabry, J.H. (1981) The iron and zinc status of long-term vegetarian women. *Am J Clin Nutr* **34**: 1042–1047.

Anonymous (1985) 'Anaemia' in athletes. *Lancet* **1**: 1490–1491.

Anonymous (1987) Iron deficiency – time for a community campaign. *Lancet* **i**: 141–142.

Anonymous (1989) Vitamin A and iron deficiency. *Nutr Rev* **47**: 119–121.

Apte, S.V. and Iyengar, L. (1970) Absorption of dietary iron in pregnancy. *Am J Clin Nutr* **23**: 73–77.

Arena, J.F., Schwartz, C., Stevenson R *et al.* (1992) Spastic paraplegia with iron deposits in the basal ganglia: a new X-linked mental retardation syndrome. *Amer J Med Genet* **43**: 479–490.

Ascherio, A., Willett, W.C., Rimm, E.B. *et al.* (1994) Dietary iron intake and risk of coronary disease among men. *Circulation* **89**: 969–974.

Aschner, M. and Aschner, J.L. (1990) Manganese transport across the blood–brain barrier: relationship to iron homeostasis. *Brain Res Bull* **24**: 857–860.

ATBC (The Alpha-tocopherol Beta-carotene Cancer Prevention Study Group) (1994) The effect of vitamin E and beta carotene on the incidence of lung cancer and other cancers in male smokers. *New Engl J Med* **330**: 1029–1035.

Aukett, M.A., Parks, Y.A., Scott, P.H., Wharton, B.A. (1986) Treatment with iron increases weight gain and psychomotor development. *Arch Dis Childh* **61**: 849–857.

Austria, J.R., Brodsky, N.L., Hurt, H. (1987) Multiply transfused preterm infants may not require iron-supplementation at recommended times. *Pediatr Res* **21**: 391A.

Avogaro, P., Bittolo Bon, G., Cazzolato, G. (1988)

Presence of a modified low density lipoprotein in humans. *Arteriosclerosis* **8**: 79–87.

Bacon, B.R., Tavill, A.S., Brittenham, G.M. *et al.* (1983) Hepatic lipid peroxidation in vivo in rats with chronic iron overload. *J Clin Invest* **71**: 429–439.

Bacon, B.R., Park, C.H., Brittenham, G.M. *et al.* (1985) Hepatic mitochondrial oxidative metabolism in rats with chronic dietary iron overload. *Hepatology* **5**: 789–797.

Baer, A.N., Dessypris, E.N., Goldwasser, E., Krantz, S.B. (1987) Blunted erythropoietin response to anemia in rheumatoid arthritis. *Brit J Haematol* **66**: 559–564.

Bagchi, K., Mohanram, M., Reddy, V. (1980) Humoral immune response in children with iron deficiency. *Br Med J* **280**: 1249–1251.

Baig, B.H., Wachsmuth, I.K., Morris, G.K. (1986) Utilization of exogenous siderophores by *Campylobacter* species. *J Clin Microbiol* **23**: 431–433.

Bailey, S., Evans, R.W., Garratt, R.C. *et al* (1988) Molecular structure of serum transferrin at 3.3-A resolution. *Biochem* **27**: 5804–5812.

Bainton, D.F. and Finch, C.A. (1964) The diagnosis of iron deficiency anemia. *Am J Med* **37**: 62–70.

Balaban, E.P., Cox, J.V., Snell, P. *et al.* (1989) The frequency of anemia and iron deficiency in the runner. *Med Sci Sports Exerc* **21**: 643–648.

Bali, P.K. and Aisen, P. (1991) Receptor-modulated iron release from transferrin: differential effects on N- and C-terminal sites. *Biochem* **30**: 9947–9952.

Ballart, I.J., Estevez, M.E., Sen, L. *et al* (1986) Progressive dysfunction of monocytes associated with iron overload and age in patients with thalassemia major. *Blood* **67**: 105–109.

Ballot, D., Baynes, R.D., Bothwell, T.H. *et al.* (1987) The effects of fruit juices and fruits on the absorption of iron from a rice meal. *Br J Nutr* **57**: 331–343.

Balmer, S.E., Scott, P.H., Wharton, B.A. (1989) Diet and faecal flora in the newborn: lactoferrin. *Arch Dis Childh* **64**: 1685–1690.

Bandi, Z.L., Schoen, I., Bee, D.E. (1985) Immunochemical methods for measurement of transferrin in serum: Effects of analytical errors and inappropriate reference intervals on diagnostic utility. *Clin Chem* **31**: 1601–1605.

Banerjee, D., Flanagan, P.R., Cluett, J., Valberg, L.S. (1986) Transferrin receptors in the human gastrointestinal tract. Relationship to body iron stores. *Gastroent* **91**: 861–869.

Barber, S.A., Bull, N.L., Buss, D.H. (1985) Low iron intakes among young women in Britain. *Br Med J* **290**: 743–744.

Barclay, R. (1985) The role of iron in infection. *Med Lab Sci* **42**: 166–177.

Barclay, S.M., Lloyd, D.J., Duffty, P., Aggett, P.J. (1989) Iron supplements for preterm or low birthweight infants. *Arch Dis Childh* **64**: 1621–1628.

Barclay, S.M., Aggett, P.J., Lloyd, D.J., Duffty, P. (1991) Reduced erythrocyte superoxide dismutase activity in low birth weight infants given iron supplements. *Pediatr Res* **29**: 297–301.

Barkai, A.I., Durkin, M., Dwork, A.J., Nelson, H.D. (1991) Autoradiographic study of iron-binding sites in the rat brain: distribution and relationship to aging. *J Neurosci Res* **29**: 390–395.

Barker, D.J., Bull, A.R., Osmond, C., Simmons, S.J. (1990) Fetal and placental size and risk of hypertension in adult life. *Br Med J* **301**: 886–891.

Barker, D.J.P., Gluckman, P.D., Godfrey, K.M. *et al.* (1993) Fetal nutrition and cardiovascular disease in adult life. *Lancet* **341**: 938–941.

Barrett, J.F.R., Whittaker, P.G., Williams, J.G., Lind, T. (1992) Absorption of non-haem iron in normal women measured by the incorporation of two stable isotopes into erythrocytes. *Clin Sci* **83**: 213–219.

Barrett, J.F.R., Whittaker, P.G., Williams, J.G., Lind, T. (1994) Absorption of non-haem iron from food during normal pregnancy. *Br Med J* **309**: 79–82.

Barry, D.M.J. and Reeve, A.W. (1976) Iron and neonatal infection – the Hawkes Bay experience. *NZ Med J* **84**: 287.

Barry, D.M.J. and Reeve, A.W. (1988) Iron and infection. *Br Med J* **296**: 1736.

Bassett, M.L., Halliday, J.W., Powell, L.W. (1986) Value of hepatic iron measurements in early haemochromatosis and determination of critical iron concentration associated with fibrosis. *Hepatology* **6**: 24–29.

Basta, S.S., Soekirman, M.S., Karyadi, D., Scrimshaw, N.S. (1979) Iron deficiency anemia and the productivity of adult males in Indonesia. *Am J Clin Nutr* **32**: 916–925.

Batey, R.G., Hussein, S., Sherlock, S., Hoffbrand, A.V. (1978) The role of serum ferritin in the management of idiopathic haemochromatosis. *Scand J Gastroenterol* **13**: 953–957.

Baynes, R.D. and Bothwell, T.H. (1990) Iron deficiency. *Annu Rev Nutr* **10**: 133–148.

Baynes, R.D., Bothwell, T.H., Bezwoda, W.R. *et al.* (1987) Relationship between absorption of inorganic and food iron in field studies. *Ann Nutr Metab* **31**: 109–116.

Baynes, R.D., Macfarlane, B.J., Bothwell, T.H. *et al.* (1990) The promotive effect of soy sauce on iron absorption in human subjects. *Eur J Clin Nutr* **44**: 419–424.

Beard, J.L. (1994) Iron deficiency: assessment during pregnancy and its importance in pregnant adolescents. *Am J Clin Nutr* **59 (Suppl)**: 502S–510S.

Beard, J.L., Tobin, B.W., Smith, S.M. (1988a) Norepinephrine turnover in iron deficiency at three environmental temperatures. *Am J Physiol* **255**: R90–R96.

Beard, J.L., Weaver, C.M., Lynch, S.R. *et al.* (1988b) The effect of soybean phosphate and phytate content on iron bioavailability. *Nutr Res* **8**: 345–352.

Beard, J., Tobin, B., Green, W. (1989) Evidence for thyroid hormone deficiency in iron deficient anemic rats. *J Nutr* **119**: 772–778.

Beard, J.L., Borel, M.J., Derr, J. (1990a) Impaired thermoregulation and thyroid function in iron-deficiency anemia. *Am J Clin Nutr* **52**: 813–819.

Beard, J.L., Tobin, B.W., Smith, S.M. (1990b) Effects of iron repletion and correction of anemia on norepinephrine turnover and thyroid metabolism in iron deficiency. *Proc Soc Exp Biol Med* **193**: 306–312.

Beard, J.L., Connor, J.R., Jones, B.C. (1993) Iron in the brain. *Nutr Rev* **51**: 157–170.

Beasley, R.P., Lin, C.C., Hwang, L.Y. *et al* (1981)

Hepatocellular carcinoma and hepatitis B virus. *Lancet* **2**: 1129–1132.

Beaton, G.H., Corey, P.N., Steele, C. (1989) Conceptual and methodological issues regarding the epidemiology of iron deficiency and their implications for studies of the functional consequences of iron deficiency. *Am J Clin Nutr* **50**: 575–88.

Beckman, J.S., Beckman, T.W., Chen, J. *et al.* (1990) Apparent hydroxyl radical production by peroxynitrite: implication for the endothelial injury from nitric oxide and superoxide. *Proc Natl Acad Sci USA* **87**: 1620–1624.

Beckman, L. and Beckman, G. (1986) Decrease of transferrin C2 frequency with age. *Hum Hered* **36**: 254–255.

Becroft, D.M.O., Dix, M.R., Farmer, K. (1977) Intramuscular iron dextran and susceptibility of neonates to bacterial infections. *Arch Dis Childh* **52**: 778–781.

Beguin, Y., Lipscei, G., Thoumsin, H., Fillet, G. (1991) Blunted erythropoietin production and decreased erythropoiesis in early pregnancy. *Blood* **78**: 89–93.

Ben-Shachar, D. and Youdim, M.B.H. (1993) Iron, melanin and dopamine interaction: relevance to Parkinson's disease. *Prog Neuro Psychopharmacol & Biol Psychiat* **17**: 139–150.

Bender, A.E. (1993) Breakfast – Role in the diet. In *Encyclopaedia of Food Science, Food Technology and Nutrition* (eds R. Macrae, R.K. Robinson, M.J. Sadler), Academic Press, London: 488–490.

Benito, E., Stiggelbout, A., Bosch, F.X. *et al.* (1991) Nutritional factors in colorectal cancer risk: a case-control study in Majorca. *Int J Cancer* **49**: 161–167.

Benkovic, S.A. and Connor, J.R. (1993) Ferritin, transferrin, and iron in selected regions of the adult and aged rat brain. *J Compar Neurol* **338**: 97–113.

Berger, J., Schneider, D., Dyck, J.L. *et al.* (1992) Iron deficiency, cell-mediated immunity and infection among 6–36 month old children living in rural Togo. *Nutr Res* **12**: 39–49.

Bergeron, R.J.and Ingeno, M.J. (1987) Microbial iron chelator-induced cell cycle synchronization in L1210 cells: potential in combination therapy. *Cancer Res* **47**: 6010–6016.

Bergstrom, E., Hernell, O., Lonnerdal, B., Persson, L.A. (1995) Sex differences in iron stores of adolescents – what is normal? *J Paediatr Gastroenterol Nutr* **20**: 215–224.

Bessman, J.D., Gilmer, P.R., Gardner, F.H. (1983) Improved classification of anemias by MCV and RDW. *Am J Clin Pathol* **80**: 322–236.

Beutler, E., Larsh, S.E., Gurney, C.W. (1960) Iron therapy in chronically fatigued, nonanemic women: a double-blind study. *Ann Intern Med* **52**: 378–394.

Beveridge, B.R., Bannerman, R.M., Evanson, J.M., Witts, L.J. (1965) Hypochromic anaemia. A retrospective study and follow-up of 378 in-patients. *Q J Med* **34**: 145–161.

Bezwoda, W.R., Disler, P.B., Lynch, S.R. *et al.* (1976) Patterns of food iron absorption in iron-deficient white and Indian subjects and in venesected haemochromatotic patients. *Br J Haem* **33**: 265–276.

Bezwoda, W., Charlton, R., Bothwell, T. *et al.* (1978) Gastric hydrochloric acid and iron absorption. *J Lab Clin Med* **92**: 108–116.

Bhaskaram, P., Sharada, K., Sivakumar, B. *et al.* (1989) Effect of iron and vitamin A deficiencies on macrophage function in children. *Nutr Res* **9**: 35–45.

Bidoli, E., Franceschi, S., Talamini, R. *et al.* (1992) Food consumption and cancer of the colon and rectum in north-eastern Italy. *Int J Cancer* 1992; **50**: 223–229.

Bindra, G.S., Gibson, R.S. (1986) Fe status of predominantly lacto-ovo-vegetarian East Indian immigrants to Canada: a model approach. *Am J Clin Nutr* **44**: 643–652.

Bingol, A.C., Altay, C., Say, B., Donmez, S. (1975) Plasma, erythrocyte and leucocyte ascorbic acid concentrations in children with iron deficiency anaemia. *J Pediatr* **86**: 902–904.

Bishop, D.F. (1990) Two different genes encode delta-aminolevulinate synthase in humans: Nucleotide sequence of cDNA for the housekeeping and erythroid genes. *Nucleic Acid Res* **18**: 7187.

Bissenden, J.G., Scott, P.H., Hallum, J. *et al.* (1981) Racial variations in tests of placental function. *Br J Obstet Gynaecol* **88**: 109–114.

Bjorn-Rasmussen, E. (1974) Iron absorption from wheat bread. *Nutr Metab* **16**: 101–110.

Blake, D.R., Lunec, J., Ahern, M. *et al.* (1985) Effect of intravenous iron dextran on rheumatoid synovitis. *Ann Rheum Dis* **44**: 183–188.

Bloxam, D.L., Williams, N.R., Waskett, R.J. *et al.* (1989) Maternal zinc during oral iron supplementation in pregnancy: a preliminary study. *Clin Sci* **76**: 59–65.

Blum, S.M., Sherman, A.R., Boileau, R.A. (1986) The effects of fitness-type exercise on iron status in adult women. *Am J Clin Nutr* **43**: 456–463.

Bodley, J.L., Hanley, W.B., Clarke, J.T., Zlotkin, S. (1993) Low iron stores in infants and children with treated phenylketonuria: a population at risk for iron deficiency and associated cognitive deficits. *Eur J Pediatr* **15**: 140–143.

Bomford, A. and Williams, R. (1976) Long term results of venesection therapy in idiopathic haemochromatosis. *Q J Med* **45**: 611–623.

Bondestam, M., Foucard, T., Gebre-Medhin, M. (1985) Subclinical trace element deficiency in children with undue susceptibility to infection. *Acta Paediatr Scand* **74**: 515–520.

Borel, M.J., Smith, S.M., Derr, J., Beard, J.L. (1991) Day-to-day variation in iron-status indices in healthy men and women. *Am J Clin Nutr* **54**: 729–735.

Bornstein, M.H. and Sigman, M.D. (1986) Continuity in mental development from infancy. *Child Development* **57**: 251–274.

Bothwell, T.H., Charlton, R.W., Cook, J.D., Finch, C.A. (1979) *Iron metabolism in man*, Blackwell Scientific Publications, Oxford.

Bottiger, L.-E. and Carlson, L.A. (1980) Risk factors for ischaemic vascular death for men in the Stockholm prospective study. *Atherosclerosis* **36**: 389–408.

Bottomley, S.S. and Muller-Eberhard, U. (1988) Pathophysiology of heme synthesis. *Seminars in Hematology* **25**: 282–302.

Bowman, B.H., Yang, F., Adrian, G.S. (1988) *Adv Genet* **25**: 1–8.

Boxer, L.A., Coates, T.D., Haak RA *et al.* (1982) Lactoferrin deficiency associated with altered granulocyte function. *New Engl J Med* **303**: 404–410.

Brabin, L. and Brabin, B.J. (1992) The cost of successful adolescent growth and development in girls in relation to iron and vitamin A status. *Am J Clin Nutr* **55**: 955–958.

Bradley, C.K., Hillman, L., Sherman, A.R. (1993) Evaluation of two iron fortified, milk-based formulas during infancy. *Pediatrics* **91**: 908–914.

Brieland, J.K., Clarke, S.J., Karmiol, S. *et al.* (1992) Transferrin: a potential source of iron for oxygen free radical-mediated endothelial cell injury. *Arch Biochem Biophys* **294**: 265–270.

Brigden, M.L. (1993) Iron deficiency anaemia. Every case is instructive. *Postgrad Med* **93**: 181–182, 185–192.

Brink, B., Disler, P., Lynch, S. *et al.* (1976) Patterns of iron storage in dietary iron overload and idiopathic hemochromatosis. *J Lab Clin Med* **88**: 725–731.

Brise, H. and Hallberg, L. (1962) A method for comparative studies on iron absorption in man using two radio-iron isotopes. *Acta Med Scand* **376 (Suppl 59)**: 7–21.

Brittenham, G.M. (1992) Development of iron-chelating agents for clinical use. *Blood* **80**: 569–574.

Brittenham, G.M., Griffith, P.M., Nienhuis, A.W. *et al* (1994) Efficacy of deferoxamine in preventing complications of iron overload in patients with thalassemia major. *N Engl J Med* **331**: 567–573.

Britton, R.S., O'Neill, R., Bacon, B.R. (1990) Hepatic mitochondrial malondialdehyde metabolism in rats with chronic iron overload. *Hepatology* **11**: 93–97.

Britton, R.S., Li, S.C.Y., O'Neill, R. *et al.* (1993) Chronic iron overload in rats results in activation of hepatic lipocytes [Abstract]. In *Proceedings of the 4th International Conference on Haemochromatosis and Clinical Problems in Iron Metabolism.* Jerusalem: 32.

Brock, J.H. (1981) The effect of iron and transferrin on the response of serum-free cultures of mouse lymphocytes to concanavalin A and lipopolysaccharide. *Immunology* **43**: 387–392.

Brock, J.H. (1993) Iron and immunity. *J Nutr Immunol* **2**: 47–106.

Brock, J.H. and Ng, J. (1983) The effect of desferrioxamine on the growth of *Staphylococcus aureus, Yersinia enterocolitica* and *Streptococcus faecalis* in human serum: uptake of desferrioxamine-bound iron. *FEMS Microbiol Lett* **20**: 439–442.

Brock, J.H., Licéaga, J., Kontoghiorghes, G.J. (1988) The effect of synthetic iron chelators on bacterial growth in human serum. *FEMS Microbiol Immunol* **47**: 55–60.

Brock, J.H., Djeha, A., Ismail, M. *et al.* (1993) Cellular responses to iron and iron compounds. In *Progress in iron research* (ed. C. Hershko), Plenum Press, New York.

Broxmeyer, H.E. (1992) H-ferritin: a regulatory cytokine that down-modulates cell proliferation. *J Lab Clin Med* **120**: 367–370.

Brune, M., Rossander, L., Hallberg, L. (1989) Iron absorption and phenolic compounds: importance of different phenolic structures. *Eur J Clin Nutr* **43**: 547–558.

Bryan, C.F., Leech, S.H., Ducos, R. *et al.* (1984) Thermostable erythrocyte rosette-forming lymphocytes in hereditary hemochromatosis. I. Identification in peripheral blood. *J Clin Immunol* **4**: 134–142.

Bryan, C.F., Leech, S.H., Kumar, P. *et al.* (1991) The immune system in hereditary hemochromatosis: a quantitative and functional assessment of the cellular arm. *Amer J Med Sci* **301**: 55–61.

Buick, F.J., Gledhill, N., Froese, A.B. *et al.* (1980) Effect of induced erythrocythemia on aerobic work capacity. *J Appl Physiol* **48**: 636–642.

Bull, N.L. (1985) Dietary habits of 15 to 25-year-olds. *Hum Nutr: Appl Nutr* **39A (Suppl 1)**: 1–68.

Bull, N.L. and Buss, D.H. (1980) Haem and non-haem iron in British household diets. *J Hum Nutr* **34**: 141–145.

Bullen, J.J. and Joyce, P.R. (1982) Abolition of the bacterial function of polymorphs by ferritin–antiferritin complexes. *Immuonology* **46**: 497–505.

Bunin, G.R., Kuijten, R.R., Buckley, J.D. *et al.* (1993) Relation between maternal diet and subsequent primitive neuroectodermal brain tumors in young children. *N Engl J Med* **329**: 536–541.

Burman, D. (1972) Haemoglobin levels in normal infants aged 3–24 months and the effect of iron. *Arch Dis Childh* **47**: 261–271.

Burr, M.L., Sweetham, P.M., Hurley, R.J., Powell, G.H. (1974) Effects of age and intake on plasma ascorbic acid levels. *Lancet* **i**: 163–164.

Burt, M.T., Halliday, J.W., Powell, L.W. (1993) Iron and coronary heart disease. *Br Med J* **307**: 575–576.

Calkins, B.M., Whittaker, D.J., Nair, P.P. *et al.* (1984) Diet, nutrition intake and metabolism in populations at high and low risk for colon cancer. *Am J Clin Nutr* **40**: 896–905.

Cambell, G.D., Steinberg, M.H., Bower, J.D. (1975) Ascorbic acid induced haemolysis in G-6-PD deficiency. *Ann Intern Med* **82**: 810.

Cannata, J.B., Fernandez-Soto, I., Fernandez-Menendez, M.J. *et al.* (1991) Role of iron metabolism in absorption and cellular uptake of aluminum. *Kidney Int* **39**: 799–803.

Cantinieaux, B., Hariga, C., Ferster, A. *et al.* (1987) Neutrophil dysfunctions in thalassaemia major: the role of cell iron overload. *Eur J Haematol* **39**: 28–34.

Cantinieaux, B., Hariga, C., Ferster, A. *et al.* (1990) Desferrioxamine improves neutrophil phagocytosis in thalassemia major. *Am J Hematol* **35**: 13–17.

Cantwell, R.J. (1974) The long term neurological sequelae of anemia in infancy. *Pediatr Res* **8**: 342.

Carriaga, M.T., Skikne, B.S., Finley, B. *et al.* (1991) Serum transferrin receptor for the detection of iron deficiency in pregnancy. *Am J Clin Nutr* **54**: 1077–1081.

Cartier, L.-J., Ohira, Y., Chen, M. *et al.* (1986) Perturbation of mitochondrial composition in muscle by iron deficiency. *J Biol Chem* **261**: 13827–13832.

Cartwright, G.E. (1966) The anaemia of chronic disorders. *Semin Haem* **3**: 351–375.

Cavanaugh, P.G. and Nicolson, G.L. (1991) Lung-derived growth factor that stimulates the growth of lung-metastasizing tumor cells: identification as transferrin. *J Cell Biochem* **47**: 261–271.

Cavell, P.A. and Widdowson, E.M. (1964) Intakes and excretion of iron, copper and zinc in the neonatal period. *Arch Dis Childh* **39**: 496–501.

Cavill, I., Worwood, M., Jacobs, A. (1975) Internal regulation of iron absorption. *Nature* **256**: 328–329.

Cavill, I., Jacobs, A., Worwood, M. (1986) Diagnostic methods for iron status. *Ann Clin Biochem* **23**: 168–171.

Cazzola, M. and Ascari, E. (1986) Red cell ferritin as a diagnostic tool [Annotation]. *Brit J Haematol* **62**: 209–213.

Cazzola, M. and Beguin, Y. (1992) New tools for clinical evaluation of erythron function in man [Annotation]. *Brit J Haematol* **80**: 278–284.

Cazzola, M., Dezza, L., Bergamaschi, G. *et al.* (1983) Biologic and clinical significance of red cell ferritin. *Blood* **62**: 1078–1087.

Cazzola, M., Bergamaschi, G., Dezza, L., Arosio, P. (1990) Manipulations of cellular iron metabolism for modulating normal and malignant cell proliferation: achievements and prospects. *Blood* **75**: 1903–1919.

Celsing, F., Blomstrand, E., Werner, B. *et al* (1986) Effects of iron deficiency on endurance and muscle enzyme activity in man. *Med Sci Sports Exerc* **18**: 156–161.

Chanarin, I., McFadyen, I.R., Kyle, R. (1977) The physiological macrocytosis of pregnancy. *Br J Obstet Gynecol* **84**: 504–508.

Chandra, R.K. (1975) Impaired immunocompetence associated with iron deficiency. *J Pediatr* **86**: 899–902.

Chang, L.L. (1973) Storage iron in foetal livers. *Acta Paediat Scand* **62**: 173–175.

Chapman, D.E., Good, M.F., Powell, L.W., Halliday, J.W. (1988) The effect of iron, iron-binding proteins and iron-overload on human natural killer cell activity. *J Gastroenterol Hepatol* **3**: 9–17.

Chapman, R.W.G., Gorman, A., Laulight, M. *et al.* (1982a) Binding of serum ferritin to concanavalin A in patients with iron overload and with chronic liver disease. *J Clin Pathol* **35**: 481–486.

Chapman, R.W.G., Hussain, M.A.M., Gorman, A. *et al.* (1982b) Effect of ascorbic acid deficiency on serum ferritin concentration in patients with β-thalassaemia major and iron overload. *J Clin Pathol* **35**: 487–491.

Charlton, R.W., Jacobs, P., Torrance, J.D., Bothwell, T.H. (1965) The role of the intestinal mucosa in iron absorption. *J Clin Invest* **44**: 543–554.

Charlton, R.W., Hawkins, D.M., Mavor, W.O., Bothwell, T.H. (1970) Hepatic iron storage concentrations in different population groups. *Am J Clin Nutr* **23**: 358–371.

Charlton, R.W., Derman, D., Skikne, B. *et al.* (1977) Anaemia, iron deficiency and exercise: extended studies in human subjects. *Clin Sci Mol Med* **53**: 537–541.

Charoenlarp, P., Dhanamitta, S., Kaewvichit, R. *et al.* (1988) A WHO collaborative study on iron supplementation in Burma and in Thailand. *Am J Clin Nutr* **47**: 280–297.

Cherici, R., Sawatzki, G., Tamisari, L. *et al.* (1992) Supplementation of an adapted formula with bovine lactoferrin 2. Effects on serum iron, ferritin and zinc levels. *Acta Paediatrica* **81**: 475–479.

Chitambar, C. and Seligman, P. (1986) Effect of different transferrin forms on transferrin receptor expression, iron uptake, and cellular proliferation of human leukemic HL60 cells: mechanisms responsible for the specific cytotoxicity of transferrin-gallium. *J Clin Invest* **78**: 1538–1546.

Chow, C.K. (1976) Biochemical responses in lungs of ozone-tolerant rats. *Nature* 1976; **260**: 721–722.

Clark, I.A., Cowden, W.B., Rockett, K.A. (1993) Nitric oxide and cerebral malaria. *Lancet* **341**: 632–633.

Coenen, J.L.L.M., van Dieijen-Visser, M.P., van Pelt, J. *et al.* (1991) Measurements of serum ferritin used to predict concentrations of iron in bone marrow in anemia of chronic disease. *Clin Chem* **37**: 560–563.

Cohen, B.J.B. and Gibor, Y. (1980) Anaemia and menstrual blood loss. *Obstet Gynecol Surv* **35**: 597–618.

Cole, P. and Goldman, M.B. (1975) Occupation. In *Persons at High Risk of Cancer: An Approach to Cancer, Etiology and Control* (ed. J.F. Fraumeni Jr), Academic Press, London.

Commission of the European Communities (EC) (1993) *Nutrient and Energy intakes for the European Community*. Reports of the Scientific Committee for Food, 31st series. Office for Official Publications of the European Communities, Luxembourg.

Connor, J.R., Menzies, S.L., St Martin, S.M., Mufson, E.J. (1990) Cellular distribution of transferrin, ferritin, and iron in normal and aged human brains. *J Neurosci Res* **27**: 595–611.

Connor, J.R., Menzies, S.L., St Martin, S.M., Mufson, E.J. (1992) A histochemical study of iron, transferrin and ferritin in Alzheimer's disease brains. *J Neurosci Res* **31**: 75–83.

Connor, J.R., Boeshore, K.L., Benkovic, S.A., Menzies, S.L. (1994) Isoforms of ferritin have a specific cellular distribution in the brain. *J Neurosci Res* **37**: 461–465.

Conrad, M.E. and Barton, J.C. (1980) Anaemia and iron kinetics in anaemia. *Semin Hematol* **17**: 149–163.

Conrad, M.E. and Crosby, W.H. (1963) Intestinal mucosal mechanisms controlling iron absorption. *Blood* **22**: 406–415.

Conrad, M.E., Umbreit, J.N., Moore, E.G. *et al.* (1990) A newly identified iron binding protein in duodenal mucosa of rats. *J Biol Chem* **265**: 5273.

Conrad, M.E., Umbreit, J.N., Moore, E.G., Rodning, C.R. (1992) Newly identified iron-binding protein in human duodenal mucosa. *Blood* 1992; **79**: 244–247.

Cook, J.D. (1982) Clinical evaluation of iron deficiency. *Semin Hematol* **19**: 6–18.

Cook, J.D. (1990) Adaptation in iron metabolism. *Am J Clin Nutr* **51**: 301–308.

Cook, J.D. (1994) The effect of endurance training on iron metabolism. *Semin Hematol* **31**: 146–154.

Cook, J.D. and Monsen, E.R. (1976a) Food iron absorption in man. II. The effect of EDTA on absorption of dietary non-heme iron. *Am J Clin Nutr* **29**: 614–620.

Cook, J.D. and Monsen, E.R. (1976b) Food iron absorption in man. III. Comparison of the effect of animal proteins on non-heme iron absorption. *Am J Clin Nutr* **29**: 859–867.

Cook, J.D. and Skikne, B.S. (1989) Iron deficiency: definition and diagnosis. *J Intern Med* **226**: 349–355.

Cook, J.D., Lipschitz, D.A., Miles, L.E.M., Finch, C.A.

(1974) Serum ferritin as a measure of iron stores in normal subjects. *Am J Clin Nutr* **27**: 681–687.

Cook, J.D., Finch, C.A., Smith, N.J. (1976) Evaluation of the iron status of a population. *Blood* **48**: 449–455.

Cook, J.D., Morck, T.A., Lynch, S.R. (1981) The inhibitory effect of soy proteins on nonheme iron absorption in man. *Am J Clin Nutr* **34**: 2622–2629.

Cook, J.D., Skikne, B.S., Lynch, S.R., Reusser, M.E. (1986) Estimates of iron sufficiency in the US population. *Blood* **68**: 726–731.

Cook, J.D., Dassenko, S.A., Lynch, S.R. (1991) Assessment of the role of non heme-iron availability in iron balance. *Am J Clin Nutr* **54**: 717–722.

Cook, J.D., Skikne, B., Baynes, R.D. (1993) Serum transferrin receptors. *Ann Rev Med* **44**: 63–74.

Cooper, R.S. and Liao, Y. (1993) Iron stores and coronary heart disease: negative findings in the NHANES I epidemiologic follow-up study [Abstract]. *Circulation* **87**: 686.

Cory, J.G., Lasater, L., Sato, A. (1981) Effect of iron-chelating agents on inhibitors of ribonucleotide reductase. *Biochem Pharmacol* **30**: 979–984.

Cotter, P.D., Willard, H.F., Gorski, J.L., Bishop, D.F. (1992) Assignment of human erythroid delta-amino-levulinate synthase (ALAS2) to a distal subregion of band Xp11.21 by PCR analysis of somatic cell hybrids containing Xtosome translocations. *Genomics* **13**: 211–212.

Cox, L.A. and Adrian, G.S. (1993) Post-transcriptional regulation of chimeric human transferrin genes by iron. *Biochem* **32**: 4738–4745.

Cox, T.M., Mazurier, J., Spik, G., Montreuil, J. and Peters, T.J. (1979) Iron binding proteins and influx of iron across the duodenal brush border: evidence for specific lactotransferrin receptors in the human intestine. *Biochem Biophys Acta* **588**: 120–128.

Cox, T.M. and Peters, T.J. (1979) The kinetics of iron uptake *in vitro* by human duodenal mucosa: studies in normal subjects. *J Physiol* **289**: 469–478.

Crandon, J.H., Lennihen, R., Mikal, S., Reif, A.E. (1961) Ascorbic acid economy in surgical patients. *Ann N Y Acad Sci* **92**: 246–267.

Craven, C.M., Alexander, J., Eldridge, M. *et al.* (1987) Tissue distribution and clearance kinetics of non-transferrin-bound iron in the hypotransferrinemic mouse: a rodent model for hemochromatosis. *Proc Natl Acad Sci USA* **84**: 3457–3461.

Cravioto, J. and Delicardi, E.R. (1972) Environmental correlates of severe clinical malnutrition and language development in survivors from kwashiorkor and marasmus. In *Nutrition, the Nervous System and Behaviour*, PAHO Scientific Pubn 251, Pan American Health Organization, Washington DC: 73–94.

Crawley, H. (1988) *Food Portion Sizes*, HMSO, London.

Crichton, R.R. and Ward, R.J. (1992) Iron metabolism – perspectives in view. *Biochemistry* **31**: 11254–11264.

Cullen, K.J., Stenhouse, N.S., Wearne, K.L. (1981) Raised haemoglobin and risk of cardiovascular disease. *Lancet* **ii**: 1288–1289.

Cummins, A.G., Duncombe, V.M., Bolin, T.D. *et al.* (1978) Suppression of rejection of *Nippostrongylus brasiliensis* in iron and protein deficient rats:

effect of syngeneic lymphocyte transfer. *Gut* **19**: 823–826.

Cunningham, J.J., Ellis, S.L., McVeigh, K.L. *et al.* (1991) Reduced mononuclear leucocyte ascorbic acid content in adults with insulin-dependent diabetes mellitus consuming adequate dietary vitamin C. *Metabol* **40**: 146–149.

Dacie, Sir J.V. and Lewis, S.M. (1991) *Practical Haematology*, Churchill Livingstone, Edinburgh.

Dagnelie, P.C., van Staveren, W.A., Vergote, F.J.V.R.A. *et al.* (1989) Increased risk of vitamin $B_{12}$ and iron deficiency in infants on macrobiotic diets. *Am J Clin Nutr* **50**: 818–824.

Dagnelie, P.C., van Staveren, W.A., Hautvast, J.G. (1991) Stunting and nutrient deficiencies in children on alternative diets. *Acta Paed Scand* **374 (Suppl)**: 111–118.

Dalhøj, J., Kiaer, H., Wiggers, P. *et al.* (1990) Iron storage disease in parents and sibs of infants with neonatal hemochromatosis: 30-year follow-up [Review]. *Am J Med Genet* **37 (3)**: 342–345.

Dallman, P.R. (1984) Diagnosis of anemia and iron deficiency: analytic and biological variations of laboratory tests. *Am J Clin Nutr* **39**: 937–941.

Dallman, P.R. (1986) Iron deficiency in the weanling: a nutritional problem on the way to resolution. *Acta Paed Scand* **323 (Suppl)**: 59–67.

Dallman, P.R. (1987) Iron deficiency and the immune response. *Am J Clin Nutr* **46**: 329–334.

Dallman, P.R. (1989a) Iron deficiency: does it matter? *J Intern Med* **226**: 367–372.

Dallman, P.R. (1989b) Review of iron metabolism In *Dietary iron: birth to two years* (ed. L.J. Filer), Raven Press, New York: 1–18.

Dallman, P.R. (1990) Progress in the prevention of iron deficiency in infants. *Acta Paed Scand* **365 (Suppl)**: 28–37.

Dallman, P.R. and Spirito, R.A. (1977) Brain iron in the rat: extremely slow turnover in normal rats may explain long-lasting effects of early iron deficiency. *J Nutr* **107**: 1075–1081.

Dallman, P.R., Refino, C.A., Dallman, M.F. (1984a) The pituitary–adrenal response to stress in the iron-deficient rat. *J Nutr* **114**: 1747–1753.

Dallman, P.R., Yip, R., Johnson, C. (1984b) Prevalence and causes of anemia in the United States, 1976 to 1980. *Am J Clin Nutr* **39**: 437–445.

Dauncy, M.J., Davies C.G., Shaw, J.C., Urman, J. (1978) The effect of iron supplements and blood transfusion on iron absorption by low birthweight infants fed on pasteurised human breast milk. *Pediatr Res* **12**: 899–904.

Dautry-Varsat, A., Ciechanover, A., Lodish, H.F. (1983) pH and the recycling of transferrin during receptor-mediated endocytosis. *Proc Natl Acad Sci USA* **80**: 2258–2262.

Davidson, L., Kastenmayer, P., Yuen, M. *et al.* (1994) Influence of lactoferrin on iron absorption from human milk in infants. *Pediatr Res* **35**: 117–124.

Dawkins, S.J., Cavill, I., Ricketts, C., Worwood, M. (1979) Variability of serum ferritin concentration in normal subjects. *Clin Lab Haematol* **1**: 41–46.

De Belder, A.J., Radomski, M.W., Why, H.J.F. *et al.*

(1993) Nitric oxide synthase activities in human myocardium. *Lancet* **341**: 84–85.

de Jong, G. and van Eijk, G.H. (1989) Functional properties of the carbohydrate moiety of human transferrin. *Int J Biochem* **21**: 252–263.

de Jong, G., van Dijk, J.P., van Eijk, H.G. (1990) The biology of transferrin. *Clin Chim Acta* **190**: 1–46.

de Jong, G., van Noort, W.L., Feelders, R.A. *et al.* (1992) Adaptation of transferrin protein and glycan synthesis. *Clin Chim Acta* 1992; **212**: 27–45.

De Maeyer, E. and Adiels-Tegman, M. (1985) The prevalence of anaemia in the world. *Wld Hlth Statist Quart* **38**: 302–316.

De Sousa, M. (1989) Immune cell function in iron overload. *Clin Exp Immunol* **75**: 1–6.

de Vizia, B., Poggi, V., Conenna, R. *et al.* (1992) Iron absorption and iron deficiency in infants and children with gastrointestinal disease. *J Ped Gastro Nutr* **14**: 21–6.

Decarli, A. and La Vecchia, C. (1986) Environmental factors and cancer mortality in Italy: correlation with exercise. *Oncology* **43**: 116–126.

Deehr, M.S., Dallal, G.E., Smith, K.T. *et al.* (1990) Effects of different calcium sources on iron absorption in postmenopausal women. *Am J Clin Nutr* **51**: 95–99.

Deinard, A.S., Schwartz, S., Yip, R. (1983) Developmental changes in serum ferritin and erythrocyte protoporphyrin in normal (nonanemic) children. *Am J Clin Nutr* **38**: 71–76.

Deinhard, A., Gilbert, A., Dodds, M., Egeland, B. (1981) Iron deficiency and behavioral deficits. *J Pediatr* **68**: 828–832.

Deinhard, A., List, B.A., Lindgren, B. *et al.* (1986) Cognitive deficits in iron deficient and iron deficient anaemic children. *J Pediatr* **108**: 681–689.

Denham, M.J. and Chanarin, I. (eds) (1985) *Blood disorders in the elderly*, Churchill Livingston, Edinburgh.

Derman, D., Sayers, M., Lynch, S.R. *et al.* (1977) Iron absorption from a cereal-based meal containing cane sugar fortified with ascorbic acid. *Br J Nutr* **38**: 261–269.

Derman, D.P., Bothwell, T.H., Torrance, J.D. *et al.* (1980) Iron absorption from maize (*Zea mays*) and sorghum (*Sorghum vulgare*) beer. *Br J Nutr* **43**: 271–279.

Dexter, D.T., Carayon, A., Javoy-Agid, F. *et al.* (1991) Alterations in the levels of iron, ferritin and other trace metals in Parkinson's disease and other neurodegenerative diseases affecting the basal ganglia. *Brain* **114**: 1953–1975.

DH (Department of Health) (1989) *The Diets of British Schoolchildren*. Report on Health and Social Subjects No. 36. HMSO, London.

DH (1991) *Dietary reference values for food energy and nutrients for the United Kingdom*. Report on Health and Social Subjects No. 41. HMSO, London.

DH (1994) *Weaning and the weaning diet*. Report on Health & Social Subjects No. 45. HMSO, London.

DHSS (Department of Health and Social Security) (1972) *A nutrition survey of the elderly*. Report on Health and Social Subjects No. 3. HMSO, London.

DHSS (1979) *Nutrition and health in old age*. Report on Health and Social Subjects No. 16. HMSO, London.

DHSS (1980) *Artificial feeds for the young infant*. Report on Health and Social Subjects No. 18. HMSO, London.

DHSS (1981) *Nutritional aspects of bread and flour*. Report on Health and Social Subjects No. 23. HMSO, London.

DHSS (1988) *Present day practice in infant feeding: third report*. Report on Health and Social Subjects No. 32. HMSO, London.

Dhur, A., Galán, P., Hannoun, C. *et al.* (1990) Effects of iron deficiency upon the antibody response to influenza virus in rats. *J Nutr Biochem* **1**: 629–634.

Diehl, D.M., Lohman, T.G., Smith, S.C., Kertzer, R.J. (1986) The effects of physical training on iron status of female field hockey players. *Int J Sports Med* **7**: 264–270.

Dillmann, E., Johnson, D.G., Martin, J. *et al.* (1979) Catecholamine elevation in iron deficiency. *Am J Physiol* **237**: R297–R300.

Dillmann, E., Gale, C., Green, W. *et al.* (1980) Hypothermia in iron deficiency due to altered triiodothyronine metabolism. *Am J Physiol* **239**: R377–R381.

Dinarello, C.A. (1984) Interleukin-1 and the pathogenesis of the acute phase response. *New Eng J Med* **311**: 1413–1418.

Dirren, H., Decarli, B., Lesourd, B. *et al.* (1991) Nutritional status: haematology and albumin. *Eur J Clin Nutr* **45 (Suppl 3)**: 43–52.

Disler, P.B., Lynch, S.R., Charlton, R.W. *et al.* (1975) The effect of tea on iron absorption. *Gut* **16**: 193–200.

Djeha, A. and Brock, J.H. (1992a) Uptake and intracellular handling of iron from transferrin and iron chelates by mitogen stimulated mouse lymphocytes. *Biochem Biophys Acta* **1133**: 147–152.

Djeha, A. and Brock, J.H. (1992b) Effect of transferrin, lactoferrin and chelated iron on human T-lymphocytes. *Br J Haematol* **80**: 235–241.

Dobbing, J. (1987) *Early Nutrition and Later Achievement*, Academic Press, London.

Dommergues, J.P., Arachambeaud, M.P., Ducot, B. *et al.* (1989) Carence en fer et tests de développement psychomoteur. *Arch Fr Pediatr* **46**: 487–490.

Dormandy, T.L. (1969) Biological rancidification. *Lancet* **ii**: 684–688.

Dormandy, T.L. (1983) An approach to free radicals. *Lancet* **ii**: 1010–1014.

Doyle, J.J. and Zipursky, A. (1992) Neonatal blood disorders. In *Effective care of the newborn infant* (eds J.L. Sinclair, M.B. Bracken), Oxford University Press, Oxford: 426–453.

Doyle, W., Jenkins, S., Crawford, M.A., Puvandendran, K. (1994) Nutritional status of schoolchildren in an inner city area. *Arch Dis Childh* **70**: 376–381.

Draper, A., Lewis, J., Malhotra, N., Wheeler, E. (1993) The energy and nutrient intake of different types of vegetarian: a case for supplements? *Br J Nutr* **69**: 3–19.

Dressendorfer, R.H., Wade, C.E., Amsterdam, R.A. (1981) Development of pseudoanemia in marathon runners during a 20-day road race. *JAMA* **248**: 1215–1219.

Driva, A., Kafatos, A., Salman, M. (1985) Iron deficiency

and the cognitive and psychomotor development of children: A pilot study with institutionalised children. *Early Child Devel and Care* **22**: 73–82.

Dufaux, B., Hoederath, A., Streetberger, I. *et al.* (1981) Serum ferritin, transferrin, haptoglobin, and iron in middle- and long-distance runners, elite rowers and professional racing cyclists. *Int J Sports Med* **2**: 43–46.

Duflou, H., Maenhaut, W., De Reuck, J. (1989) Regional distribution of two minor and six trace elements in normal human brain. In *Current trends in trace elements research* (eds G. Chazot, M. Abdulla, P. Arnaud), Smith-Gordon, London: 135–140.

Duggan, M.B., Steel, G., Elwys, G. *et al.* (1991) Iron status, energy intake, and nutritional status of healthy young Asian children. *Arch Dis Childh* **66**: 1386–1389.

Duggan, M.B., Harbottle, L., Noble, C. (1992) The weaning diet of healthy Asian children living in Sheffield. *J Hum Nutr Dietet* **5**: 189–200.

Dwyer, J., Wood, C., McNamara, J. *et al.* (1987) Abnormalities in the immune system of children with beta-thalassaemia. *Clin Exp Immunol* **63**: 621–629.

Dyrks, T., Dyrks, E., Hartmann, T. *et al.* (1992) Amyloidogenicity of βA4 and βA4-bearing amyloid protein precursor fragments by metal-catalysed oxidation. *J Biol Chem* **267**: 18210–18217.

Earley, A., Valman, H.B., Altman, D.G., Pippard, M.J. (1990) Microcytosis, iron deficiency, and thalassaemia in preschool children. *Arch Dis Child* **65**: 610–614.

Eaton, P.M., Wharton, P.A., Wharton, B.A. (1984) Nutrient intake of pregnant women at Sorrento Maternity Hospital, Birmingham. *Br J Nutr* **52**: 457–468.

Eaton, S.B. and Konner, M. (1985) Paleolithic nutrition: a consideration of its nature and current implications. *N Eng J Med* **312**: 283–289.

Edgerton, V.R., Bryant, S.L., Gillespie, C.A., Gardner, G.W. (1972) Iron deficiency and physical performance and activity of rats. *J Nutr* **102**: 384–400.

Edgerton, V.R., Gardner, G.W., Ohira, Y. *et al.* (1979) Iron-deficiency anaemia and its effect on worker productivity and activity patterns. *Br Med J* **2**: 1546–1549.

Edling, J.E., Britton, R.S., Grisham, M.B., Bacon, B.R. (1990) Increased unwinding of hepatic double-stranded DNA (dsDNA) in rats with chronic dietary overload. *Gastroent* **98**: A585.

EEC (European Economic Community) (1991) EEC Commission Directive on Infant Formulae and Follow-on Formulae No 91/321/EEC. *Off J Eur Commun* **L175**: 35–40.

Ehmann, W.D., Markesbery, W.R., Alauddin, M. *et al.* (1986) Brain trace elements in Alzheimer's disease. *Neurotoxicol* **7**: 197–206.

Ehn, L., Carlmark, B., Hoglund, V. (1980) Iron status in athletes involved in intense physical activity. *Med Sci Sports Exerc* **12**: 61–64.

Ehrenkranz, R.A. (1992) Iron, folic acid and vitamin B12. In *Nutritional Needs of the Preterm Infant* (eds R.A. Tsang, A. Lucas, R. Uauy, S. Zlotkin), Williams & Wilkins, New York: 177–194.

Ehrenkranz, R.A., Gettner, P.A., Nelli, C. (1992) Iron absorption and incorporation into red blood cells by very low birthweight infants: studies with the stable isotope $^{58}$Fe. *J Pediatr Gastro Nutr* **15**: 270–278.

Ehrhardt, P. (1986) Iron deficiency in young Bradford children from different ethnic groups. *Br Med J* **292**: 90–93.

Elsborg, L., Lund, V., Bastrup-Madsen, P. (1976) Serum vitamin B$_{12}$ levels in the aged. *Acta Med Scand* **200**: 309–314.

Elwood, P.C., Shinton, N.K., Wilson, C.I.D. *et al.* (1971) Haemoglobin, vitamin B$_{12}$, and folate levels in the elderly. *Brit J Haematol* **21**: 557.

Elwood, P.C., Burr, M.L., Hole, D. *et al.* (1972) Nutritional state of elderly Asian and English subjects in Coventry. *Lancet* **i**: 1224–1227.

Elwood, P.C., Walters, W.E., Benjamin, I.T., Sweetnam, P.M. (1974) Mortality and anaemia in women. *Lancet* **i**: 891–894.

Emond, A.M., Hawkins, N., Pennock, C., Golding, J. (1995) [Personal communication].

English, R.M. and Bennett, S.A. (1991) Iron status of Australian children. *Med J of Australia* **152**: 582–586.

Enstrom, J.E., Kanim, L.E., Klein, M.A. (1992) Vitamin C intake and mortality among a sample of the United States population. *Epidemiol* **3**: 194–202.

Eraklis, A.J. and Filler, R.M. (1972) Splenectomy in childhood: a review of 1413 cases. *J Pediat Surg* **7**: 382–388.

Eschbach, J.W., Egrie, J.C., Downing, M.R. *et al.* (1987) Correction of the anemia of end-stage renal disease with recombinant human erythropoietin. Results of a combined phase I and II clinical trial. *N Eng J Med* **316**: 73–78.

ESPGAN (European Society for Paediatric Gastroenterology and Nutrition) (1977) Guidelines on infant nutrition. I. Recommendations for the composition of an adapted formula. *Acta Paed Scand* **262 (Suppl)**.

ESPGAN (1987) Nutrition and feeding of preterm infants. *Acta Paed Scand* **336 (Suppl)**: 6–7.

Evans, P.H. (1993) Free radicals in brain metabolism and pathology. *Br Med Bull* **49**: 577–587.

Evans, P.H. (1994) Nutrient and toxin interactions in neurodegenerative disease. *Proc Nutr Soc* **53**: 431–442.

Evans, P.H., Peterhans, E., Bürge, T., Klinowski, J. (1992) Aluminosilicate-induced free radical generation by murine brain glial cells in vitro: potential significance in the aetiopathogenesis of Alzheimer's dementia. *Dementia* **3**: 1–6.

Evans, R.M., Currie, L., Campbell, A. (1982) The distribution of ascorbic acid between various cellular components of the blood in normal individuals, and its relation to the plasma concentration. *Br J Nutr* **47**: 473–482.

Ewertz, M. and Gill, C. (1990) Dietary factors and breast cancer risk in Denmark. *Int J Cancer* **46**: 779–784.

Fairley, P.C. and Foland, J. (1990) Iron deficiency anemia: how to diagnose and correct. *Postgrad Med* **87**: 89–101.

Fairweather-Tait, S.J. (1986) Iron availability – the implications of short-term regulation. *BNF Nutr Bull* **11**: 174–180.

Fairweather-Tait, S.J. (1989) Iron in food and its bioavailability. *Acta Paed Scand* **361 (Suppl)**: 12–20.

Fairweather-Tait, S.J. (1992a) Bioavailability of trace elements. *Food Chem* **43**: 213–217.

Fairweather-Tait, S.J. (1992b) The metabolism of iron

and its bioavailability in foods. In *The contribution of nutrition to human and animal health* (eds E.M. Widdowson, E.C. Mathers), Cambridge University Press, Cambridge: 151–161.

Fairweather-Tait, S.J. (1993) Iron [Flair Concerted Action No 10 Status Papers]. *Int J Vit Nutr Res* **63**: 296–301.

Fairweather-Tait, S.J. and Minski, M.J. (1986) Studies on iron availability in man, using stable isotope techniques. *Br J Nutr* **55**: 279–285.

Fairweather-Tait, S.J. and Wright, A.J.A. (1991) Small intestinal transit time and iron absorption. *Nutr Res* **11**: 1465–1468.

Fairweather-Tait, S.J., Swindell, T.E., Wright, A.J.A. (1985) Further studies in rats on the influence of previous iron intake on the estimation of bioavailability of iron. *Br J Nutr* **54**: 79–86.

Fairweather-Tait, S.J., Balmer, S.E., Scott, P.H., Minski, M.J. (1987) Lactoferrin and iron absorption in newborn infants. *Pediatr Res* **22**: 651–654.

Fairweather-Tait, S.J., Powers, H.J., Minski, M.J. *et al.* (1992) Riboflavin deficiency and iron absorption in adult Gambian men. *Ann Nutr Metab* **36**: 34–40.

Fairweather-Tait, S.J., Fox, T.E., Wharf, S.G., Eagles, J. (1995) The bioavailability of iron in different weaning foods and the enhancing effects of a fruit drink containing ascorbic acid. *Ped Res* **37**: 389–394.

FAO (Food and Agriculture Organisation) (1990) *FAO Production Yearbook*, vol. 44, Food and Agriculture Organisation, Rome.

FAO/WHO (Food and Agriculture Organisation/ World Health Organisation) (1988) *Requirements of vitamin A, iron, folate and vitamin B$_{12}$*. Joint Expert Consultation Report. FAO Food and Nutrition Series 23, FAO, Rome.

Fargion, S., Arosio, P., Fracanzani, A.L. *et al.* (1988) Characteristics and expression of binding sites specific for ferritin H-chain on human cell lines. *Blood* **71**: 753–757.

Fargion, S., Mandelli, C., Piperno, A. *et al.* (1992) Survival and prognostic factors in 212 Italian patients with genetic hemochromatosis. *Hepatology* **15**: 655–659.

Ferguson, B.J., Skikne, B.S., Simpson, K.M. *et al.* (1992) Serum transferrin receptor distinguishes the anemia of chronic disease from iron deficiency anemia. *J Lab Clin Med* **19**: 385–390.

Finch, C.A., Miller, L.R., Inandar, A.R. *et al.* (1976) Iron deficiency in the rat. Physiological and biochemical studies of muscle dysfunction. *J Clin Invest* **58**: 447–453.

Finch, C.A., Cook, J.D., Labbe, R.F., Culala, M. (1977) Effect of blood donation on iron stores as evaluated by serum ferritin. *Blood* **50**: 441–447.

Finch, C.A., Gollnick, P.D., Hlastala, M.P. *et al.* (1979) Lactic acidosis as a result of iron deficiency. *J Clin Invest* **61**: 129–137.

Flament, J., Goldman, M., Waterlot, Y. *et al.* (1986) Impairment of phagocytic oxidative metabolism in hemodialyzed patients with iron overload. *Clin Nephrol* 1986; **25**: 227–230.

Florence, T.M. and Stauber, J.L. (1989) Manganese catalysis of dopamine oxidation. *Sci Total Environ* **78**: 233–240.

Flowers, C.H., Skikne, B.S., Covell, A.M., Cook, J.D. (1989) The clinical measurement of serum transferrin receptor. *J Lab Clin Med* **114**: 368.

Fomon, S.J. (1987) Bioavailability of supplemental iron in commercially prepared dry infant cereals. *J Pediatr* **110**: 660–662.

Fomon, S.J., Zeigler, E.E., Nelson, S.E., Edwards, B.B. (1981) Cow milk feeding in infancy: gastrointestinal blood loss and iron nutritional status. *J Pediatr* **98**: 540–545.

Fomon, S.J., Eckhard, E., Ziegler, E.E. *et al.* (1989) Iron absorption from infant foods. *Ped Res* **26**: 250–254.

Fomon, S.J., Zeigler, E.E., Nelson, S.E. (1993) Erythrocyte incorporation of ingested 58Fe by 56 day old breast-fed and formula-fed infants. *Pediatr Res* **33**: 573–576.

Fontecave, M. and Pierre, J.L. (1991) Iron metabolism: the low-molecular-mass iron pool. *Biol Metals* **4**: 133–135.

Fordy, J. and Benton, D. (1994) Does low iron status influence psychological functioning? *J Hum Nutr Diet* **7**: 127–133.

Frei, B. (1991) Ascorbic acid protects lipids in human plasma and low density lipoprotein against oxidative damage. *Am J Clin Nutr* **54**: 1113S–1118S.

Fridovich, I. (1983) Superoxide radical: an endogenous toxicant. *Ann Rev Pharmacol Toxicol* 1983; **23**: 239–257.

Fuchs, G., DeWeir, M., Hutchinson, S. *et al.* (1993a) Gastrointestinal blood loss in older infants: impact of cows milk versus formula. *J Pediatr Gastro Nutr* **16**: 4–9.

Fuchs, G.J., Farris, R., DeWeir, M. *et al.* (1993b) Iron status and intake of older infants fed formula vs cow milk with cereal. *Am J Clin Nutr* **58**: 343–348.

Galanello, R., Turco, M.P., Barellas, S. *et al.* (1990) Iron stores and iron deficiency anaemia in children heterozygous for beta-thalassaemia. *Haematologica* **75**: 319–22.

Gale, E., Torrance, J., Bothwell, T. (1963) The quantitative estimation of total iron stores in human bone marrow. *J Clin Invest* **42**: 1076–1082.

Gallagher, S.A., Johnson, L.K., Milne, D.H. (1989) Short-term and long-term variability of indices related to nutritional status I: Ca, Cu, Fe, Mg, and Zn. *Clin Chem* **35**: 369–373.

Garby, L., Sjolin, S., Vuille, J.-C. (1963) Studies on erythrokinetics in infancy III: disappearance from plasma and red cell uptake of radio-active iron injected intravenously. *Acta Paed Scand* **52**: 537–553.

Garby, L., Irnell, L., Werner, I. (1969) Iron deficiency in women of fertile age in a Swedish community. III. Estimation of prevalence based on response to iron supplementation. *Acta Med Scand* **185**: 113–117.

Gardner, G.W., Edgerton, V.R., Senewiratne, B. *et al.* (1977) Physical work capacity and metabolic stress in subjects with iron deficiency anemia. *Am J Clin Nutr* **30**: 910–917.

Gardner Merchant Educational Services (1994) *The Gardner Merchant School Meals Survey: 'What are our children eating?'*, Gardner Merchant Ltd, Kenley.

Garrett, S. and Worwood, M. (1994) Zinc protoporphyrin and iron deficient erythropoiesis. *Acta Haematol* **91**: 21–25.

Garry, P.J., Owen, G.M., Hopper, E.M., Gilbert, B.A. (1981) Iron absorption from human milk and formula with and without iron supplementation. *Pediatr Res* **15**: 822–828.

Gascón, P., Zoumbos, N.C., Young, N.S. (1984) Immunological abnormalities in patients receiving multiple blood transfusions. *Ann Intern Med* **100**: 173–177.

Gear, J.S., Mann, J.I., Thorogood, M. *et al.* (1980) Biochemical and haematological variables in vegetarians (Letter). *Br Med J* **280**: 1415.

Gelman, B.B., Rodriguez-Wolf, M.G., Wen, J. *et al.* (1992) Siderotic cerebral macrophages in the acquired immunodeficiency syndrome. *Arch Pathol Lab Med* **116**: 509–516.

Gey, K.F. (1993) Prospects for the prevention of free radical disease, regarding cancer and cardiovascular disease. *Brit Med Bull* **49**: 679–699.

Ghersi-Egea, J.F. and Livertoux, M.-H. (1992) Evidence for drug metabolism as a source of reactive species in the brain. In *Free radicals and aging* (eds I. Emerit, B. Chance), Birkhäuser Verlag, Basel: 219–226.

Gillooly, M., Bothwell, T.H., Torrance, J.D. *et al.* (1983) The effects of organic acids, phytates and polyphenols on the absorption of iron from vegetables. *Br J Nutr* **49**: 331–342.

Gillooly, J.D., Torrance, J.D., Bothwell, T.H. *et al.* (1984) The relative effect of ascorbic acid on iron absorption from soy based and milk based infant formulas. *Am J Clin Nutr* **40**: 522–527.

Gillum, R.F. and Makuc, D.M. (1992) Serum albumin, coronary heart disease, and death. *Am Heart J* **123**: 507–513.

Gleerup, A., Rossander-Hulthen, L., Gramatkovski, E., Hallberg, L. (1995) Iron absorption from the whole diet: comparison of the effect of two different distributions of daily calcium intake. *Am J Clin Nutr* **61 (1)**: 97–104.

Godfrey, K.M., Redman, C.W.G., Barker, D.J.P., Osmond, C. (1991) The effect of maternal anaemia and iron deficiency in the ratio of fetal weight to placental weight. *Br J Obstet Gynaecol* **98**: 886–891.

Goldbloom, D.S. and Garfinkel, P.E. (1993) Anorexia Nervosa; Bulimia Nervosa. In *Encyclopaedia of Food Science, Food Technology and Nutrition* (eds R. Macrae, R.K. Robinson, M.J. Sadler), Academic Press, London: 200–203, 533–536.

Golden, M.H.N. and Ramdath, D.D. (1987) Free radicals in the pathogenesis of kwashiorkor. *Proc Nutr Soc* **46**: 53–68.

Golden, M.H.N., Golden, B.E., Bennett, F.I. (1985) Relationship of trace element deficiencies to malnutrition. In *Trace elements in nutrition of children* (ed. R.K. Chandra), Nestle Nutrition, Vevey/Raven Press, New York: 185–207.

Gonzalez de Aledo Linos, A., Rollan Rollan, A., Bonilla Miera, C. (1990) Prospective study of the prevalence of iron deficiency in breast fed infants in Cantabria, its relation to the introduction of cow's milk and psychomotor development. *An Esp Pediatr* **32**: 24–27.

Good, P.F., Olanow, C.W., Perl, D.P. (1992) Neuromelanin-containing neurons of the substantia nigra accumulate iron and aluminum in Parkinson's disease: a LAMMA study. *Brain Res* **593**: 343–346.

Goodman, M.T., Nomura, A.M.Y., Wilkens, L.R., Hankin, J. (1992) The association of diet, obesity, and breast cancer in Hawaii. *Cancer Epid Biomark Prevent* **1**: 269–275.

Gordeuk, V.R., Brittenham, G.M., McLaren, G.D., Spagnuolo, P.J. (1986) Hyperferremia in immunosuppressed patients with acute nonlymphocytic leukemia and the risk of infection. *J Lab Clin Med* **108**: 466–472.

Gordeuk, V.R., Brittenham, G.M., Hughes, M. *et al.* (1987) High-dose carbonyl iron for iron deficiency anemia: a randomized double-blind trial. *Am J Clin Nutr* **46**: 1029–1034.

Gordeuk, V.R., Prithviraj, J., Dolinar, T., Brittenham, G.M. (1988) Interleukin 1 administration in mice produces hypoferremia despite neutropenia. *J Clin Invest* **82**: 1934–1938.

Gordeuk, V., Mukiibi, J., Hasstedt, S.J. *et al* (1992) Iron overload in Africa. Interaction between a gene and dietary iron content. *N Engl J Med* **326**: 95–100.

Gorton, M.K., Hepner, R., Workman, J. (1963) Iron metabolism in premature infants. I: absorption and utilisation of iron as measured by isotope studies. *J Pediatr* 1963; **63**: 1063–1071.

Gould, S. (1981) *The Mismeasure of Man*, W.W. Norton, New York.

Graf, E., Mahoney, J.R., Bryant, R.G., Eaton, J.W. (1984) Iron catalyzed free radical formation. Stringent requirement for free iron coordination site. *J Biol Chem* **259**: 3620–3624.

Granick, S. (1949) Iron metabolism and hemochromatosis. *Bull NY Acad Med* **25**: 403–428.

Grasbeck, R., Kuovonen, I., Lundberg, M., Tenhunen, R. (1979) An intestinal receptor for heme. *Scand J Haematol* **23**: 5–9.

Green, L., Baker, B.A., Lucas, A. *et al.* (1991/2) [Unpublished data].

Green, R., Esparza, I., Schreiber, R. (1988) Iron inhibits the non-specific tumoricidal activity of macrophages: a possible contributory mechanism for neoplasia in hemochromatosis. *Ann NY Acad Sci* **526**: 301–309.

Greenwood, C.T. and Richardson, D.P. (1979) Nutrition during adolescence. *World Rev Nutr & Diet* **33**: 1–41.

Gregory, J., Foster, K., Tyler, H., Wiseman, M. (1990) *The Dietary and Nutritional Survey of British Adults*, HMSO, London.

Gregory, J.R., Collins, D.L., Davies, P.S.W. *et al* (1995) *National diet and nutrition survey: children aged $1\frac{1}{2}$ to $4\frac{1}{2}$ years. Volume 1: Report of the diet and nutrition survey*, HMSO, London.

Griffiths, E. (1987) The iron uptake systems of pathogenic bacteria. In *Iron and Infection* (eds J.J. Bullen, E. Griffiths), John Wiley, Chichester: 69–137.

Griffiths, E. and Humphreys, J. (1977) Bacteriostatic effect of human milk and bovine colostrum on *Escherichia coli*: importance of bicarbonate. *Infect immunity* **15**: 396–401.

Grindulis, H., Scott, P.H., Belton, N.R., Wharton, B.A. (1986) Combined deficiency of iron and vitamin D in Asian toddlers. *Arch Dis Childh* **61**: 843–848.

Guglielmo, P., Cunsolo, F., Lombardo, T. *et al.* (1984) T-subset abnormalities in thalassaemia intermedia: possible evidence for a thymus functional deficiency. *Acta Haematol* **72**: 361–367.

Guindi, M.E., Lynch, S.R., Cook, J.D. (1988) Iron absorption from fortified flat breads. *Br J Nutr* **59**: 205–213.

Gutteridge, J.M.C. (1986) Antioxidant properties of the proteins ceruloplasmin, albumin, and transferrin. A study of their activity in serum and synovial fluid from patients with rheumatoid arthritis. *Biochim Biophys Acta* **869**: 119–127.

Gutteridge, J.M.C. (1987) The antioxidant activity of haptoglobin towards haemoglobin-stimulated lipid peroxidation. *Biochem Biophys Acta* **917**: 219–223.

Gutteridge, J.M.C., Rowley, D.A., Halliwell, B., Westermarck, T. (1982) Increased non-protein bound iron and decreased protection against superoxide-radical damage in cerebrospinal fluid from patients with neuronal ceroid lipofuscinoses. *Lancet* **ii**: 459–461.

Gutteridge, J.M.C., Halliwell, B., Treffry, A. *et al.* (1983) Effect of ferritin-containing fractions with different iron loading on lipid peroxidation. *Biochem J* **209**: 557–560.

Gutteridge, J.M.C., Quinlan, G.J., Clark, I., Halliwell, B. (1985) Aluminium salts accelerate peroxidation of membrane lipids stimulated by iron salts. *Biochim Biophys Acta* **835**: 441–447.

Gutteridge, J.M.C., Cao, W., Chevion, M. (1991) Bleomycin-detectable iron in brain tissue. *Free Rad Res Comms* **11**: 317–320.

Hall, D.M.P. (1989) *Health for all children: a programme for child health surveillance*, Oxford Medical Publishers, Oxford: 34–36.

Hall, E.D. and McCall, J.M. (1993) Lazaroids: potent inhibitors of iron-dependent lipid peroxidation for neurodegenerative disorders. In *Iron in central nervous system disorders* (eds P. Riederer, M.B.H. Youdim), Springer-Verlag, Vienna: 173–188.

Hall, R.T., Wheeler, R.E., Benson, J. *et al.* (1993) Feeding iron-fortified formula during initial hospitalisation to infants less than 1800g weight. *Pediatrics* **92**: 409–414.

Hallberg, L. (1981) Bioavailability of dietary iron in man. *Ann Rev Nutr* **1**: 123–147.

Hallberg, L. and Magnusson, B. (1984) The etiology of 'Sports Anaemia'. *Acta Med Scand* **216**: 145–148.

Hallberg, L. and Rossander, L. (1982) Effect of different drinks on the absorption of non-heme iron from composite meals. *Hum Nutr: Appl Nutr* **36A**: 116–123.

Hallberg, L. and Rossander-Hulten, L. (1991) Iron requirements in menstruating women. *Am J Clin Nutr* **54**: 1047–1058.

Hallberg, L., Högdahl, A.-M., Nilsson, L., Rybo, G. (1966) Menstrual blood loss – a population study; variation at different ages and attempts to define normality. *Acta Obstet Gynaecol Scand* **45**: 320–351.

Hallberg, L., Brune, M., Rossander, L. (1986) Effect of ascorbic acid on iron absorption from different types of meals. *Hum Nutr: Appl Nutr* **40A**: 97–113.

Hallberg, L., Brune, M., Erlandsson, M. *et al* (1991) Calcium: effects of different amounts of non-heme and heme-iron absorption. *Am J Clin Nutr* **53**: 112–119.

Hallberg, L., Rossander-Hulten, L., Brune, M., Gleerup, A. (1992) Calcium and iron absorption: mechanism of action and nutritional importance. *Eur J Clin Nutr* **46**: 317–327.

Hallberg, L., Bengtsson, C., Lapidus, L. *et al.* (1993a) Screening for iron deficiency: an analysis based on bone-marrow examination and serum ferritin determination of a population sample of women. *Brit J Haematol* **85**: 787–798.

Hallberg, L., Hulten, L., Lindstedt, G. *et al.* (1993b) Prevalence of iron deficiency in Swedish adolescents. *Pediatr Res* **34**: 680–687.

Hallgren, B. and Sourander, P. (1958) The effect of age on the non-haemin iron in the human brain. *J Neurochem* **3**: 41–51.

Hallgren, B. and Sourander, P. (1960) The non-haem iron in the cerebral cortex in Alzheimer's disease. *J Neurochem* **5**: 307–310.

Halliwell, B. (1989) Free radicals, reactive oxygen species and human disease: a critical review with special reference to atherosclerosis. *Brit J Exp Path* **70**: 737–757.

Halliwell, B. (1991) Drug antioxidant effects. *Drugs* **42**: 569–605.

Halliwell, B. and Aruoma, O.I. (1991) DNA damage by oxygen-derived species. *FEBS Lett* **281**: 9–19.

Halliwell, B. and Gutteridge, J.M.C. (1984) Oxygen toxicity, oxygen radicals, transition metals and disease. *Biochem J* **219**: 1–14.

Halliwell, B. and Gutteridge, J.M.C. (1985a) *Free radicals in biology and medicine*, Clarendon Press, Oxford.

Halliwell, B. and Gutteridge, J.M.C. (1985b) Oxygen radicals and the nervous system. *Trends Neurosci* **8**: 22–26.

Halliwell, B. and Gutteridge, J.M.C. (1990) Role of free radicals and catalytic metal ions in human disease: an overview. *Meth Enzymol* **186**: 1–85.

Hallquist, N.A. and Sherman, A.R. (1989) Effect of iron deficiency on the stimulation of natural killer cells by macrophage-produced interferon. *Nutr Res* **9**: 283–292.

Hallquist, N.A., McNeil, L., Lockwood, J.F., Sherman, A.R. (1992) Maternal-iron-deficiency effect on peritoneal macrophage and peritoneal natural-killer-cell cytotoxicity in rat pups. *Am J Clin Nutr* **55**: 741–746.

Hambidge, K.M., Krebs, N.F., Sibley, L., English, J. (1987) Acute effects of iron therapy on zinc status during pregnancy. *Obstet Gynecol* **70**: 593–596.

Hankes, L.V., Jansen, C.R., Schmaeler, M. (1974) Ascorbic acid catabolism in Bantu with haemo-siderosis (scurvy). *Biomed Med* **9**: 244–255.

Harada, T., Baba, M., Torii, I., Morikawa, S. (1987) Ferritin selectively suppresses delayed-type hypersensitivity responses at induction or effector phase. *Cell Immunol* **109**: 75–88.

Harju, E. (1988) Empty iron stores as a significant risk factor in abdominal surgery. *J Parenteral Enteral Nutr* **12**: 282–285.

Harris, E.D. (1992) New insights into placental iron transport. *Nutr Rev* **50**: 329–331.

Harris, R.J., Armstrong, D., Ali, R., Loynes, A. (1983) Nutritional survey of Bangladeshi children aged under 5 years in the London borough of Tower Hamlets. *Arch Dis Childh* **58**: 428–432.

Harrison, M.J.G., Pollock, S., Kendall, B.E., Marshall, J. (1981) Effect of haematocrit on carotid stenosis and cerebral infarction. *Lancet* **ii**: 114–115.

Harrison, P.M. (1986) The structure and function of ferritin. *Biochem Ed* **14**: 153–162.

Hartley, J.A., Gibson, N.W., Zwelling, L.A., Yuspa, S.H. (1985) Association of DNA strand breaks with accelerated terminal differentiation in mouse epidermal cells exposed to tumor promotors. *Cancer Res* **45**: 4864–4870.

Harvey, P.W.J., Heywood, P.F., Nesheim, M.C. *et al.* (1989) The effect of iron therapy on malaria infection in Papua New Guinea school children. *Am J Trop Med Hyg* **40**: 12–18.

Haschke, F., Vanura, H., Male, C. *et al.* (1993) Iron nutrition and growth of breast- and formula-fed infants during the first 9 months of life. *J Pediatr Gastro Nutr* **16**: 151–6.

Hastka, J., Lasserre, J.J., Schwarzbeck, A. *et al.* (1993) Zinc protoporphyrin in anemia of chronic disorders. *Blood* **81**: 1200–1204.

Haurani, F.I. (1989) The effects of free radicals on cobalamin and iron. *Free Rad Res Comms* **7**: 241–243.

Heilbrun, L.K., Nomura, A., Hankin, J.H., Stemmerman, G.N. (1989) Diet and colorectal cancer with special reference to fibre intake. *Int J Cancer* **44**: 1–6.

Helman, A.D. and Darton-Hill, I. (1987) Vitamin and iron status in new vegetarians. *Am J Clin Nutr* **45**: 785–789.

Helyar, L. and Sherman, A.R. (1987) Iron deficiency and interleukin 1 production by rat leukocytes. *Am J Clin Nutr* **46**: 346–352.

Heresi, G. (1986) Trace elements and immunity. In *Proceedings of the XIII International Congress of Nutrition* (eds T.G. Taylor, N.K. Jenkins), Pergamon Press, London: 729–733.

Hershko, C. (1987) Non-transferrin plasma iron. *Br J Haematol* **66**: 149–151.

Hershko, C. (1994) Iron chelators. In *Iron metabolism in health and disease* (eds J.H. Brock, J.W. Halliday, M.J. Pippard, L.W. Powell), W.B. Saunders, London: 391–426.

Hershko, C., Bar-Or, D., Gaziel, Y. *et al.* (1981) Diagnosis of iron deficiency anemia in a rural population of children. Relative usefulness of serum ferritin, red cell protoporphyrin, red cell indices, and transferrin saturation determinations. *Am J Clin Nutr* **34**: 1600–1610.

Hershko, C., Peto, T.E.A., Weatherall, D.J. (1988) Iron and infection. *Br Med J* **296**: 660–664.

Hertog, M.G.L., Feskens, E.J.M., Hollman, P.C.H. *et al.* (1993) Dietary antioxidant flavonoids and risk of coronary heart disease: the Zutphen Elderly Study. *Lancet* **342**: 1007–1011.

Hertog, M.G.L., Kromhout, D., Heys, A. *et al.* (1995) Flavonoid intake and risk of coronary heart disease and cancer mortality in the seven countries study. *Am J Epidemiol* (in press).

Hertrampf, E., Cayazzo, M., Pizarro, F., Stekel, A. (1986) Bioavailability of iron in soy-based formula and its effect on iron nutriture in infancy. *Pediatr* **78**: 640–645.

Heywood, A., Oppenheimer, S., Heywood, P., Jolley, D. (1989) Behavioral effects of iron supplementation in infants in Madang, Papua New Guinea. *Am J Clin Nutr* **50**: 630–637.

Hibbard, B.M. (1988) Iron and folate supplements during pregnancy: supplementation is valuable only in selected patients. *Br Med J* **297**: 1324–1326.

Hibbs, J.B., Taintor, R.R., Vavrin, Z. (1984) Iron depletion: possible cause of tumor cell cytotoxicity induced by activated macrophages. *Biochem Biophys Res Comm* **123**: 716–723.

Hider, R.C. (1984) Siderophore mediated absorption of iron. *Struct Bonding* **58**: 25–87.

Higgins, A.C., Pencharz, P.B., Strawbridge, J.E. *et al.* (1982) Maternal haemoglobin changes and their relationship to infant birth weight in mothers receiving a programme of nutritional assessment and rehabilitation. *Nutr Res* **2**: 641–9.

Higginson, J., Gerritsen, T., Walker, A.R.P. (1953) Siderosis in the Bantu of South Africa. *Am J Path* **29**: 779–815.

Higgs, J.M. and Wells, R.S. (1972) Chronic mucocutaneous candidiasis: associated abnormalities of iron metabolism. *Br J Dermatol* **86 (Suppl)**: 88–102.

Hill, J.M. and Switzer, R.C. (1984) The regional distribution and cellular localization of iron in the rat brain. *Neuroscience* **11**: 595–603.

Hirayama, T. (1990) Diet and mortality. In *Lifestyle and mortality* (ed. T. Hirayama), Karger, Basel: 73–95.

Hislop, T.G., Coldman, A.J., Elwood, J.M. *et al.* (1986) Childhood and recent eating patterns and risk of breast cancer. *Cancer Detect Prev* **9**: 47–58.

Hislop, T.G., Kan, L., Coldman, A.J. *et al* (1988) Influence of oestrogen receptor status on dietary risk factors for breast cancer. *Can Med Assoc J* **138**: 424–430.

Hodgetts, J., Peters, S.W., Hoy, T.G., Jacobs, A. (1986) The ferritin content of normoblasts and megaloblasts from human bone marrow. *Clin Sci* **70**: 47–51.

Hodkinson, H.M. (1985) Screening for anaemia and its prevention. In *Blood disorders in the elderly* (eds M.J. Denham, I. Chanarin), Churchill Livingstone, Edinburgh: 100–108.

Hoepelman, I.M., Bezmer, W.A., Van Doornmalen, E. *et al.* (1989) Lipid peroxidation of human granulocytes (PMN) and monocytes by iron complexes. *Br J Haematol* **72**; 584–588.

Hoffbrand, A.V., Ganeshaguru, K., Hooton, J.W.L., Tattersall, M.H.N. (1976) Effect of iron deficiency and desferrioxamine on DNA synthesis in human cells. *Br J Haematol* **33**: 517–526.

Holland, B., Welch, A.A., Unwin, I.D. *et al.* (1991) *The Composition of Foods* (5th edn), Royal Society of Chemistry, Cambridge.

Holland, J., Langford, P.R., Towner, K.J., Williams, P. (1992) Evidence for in vivo expression of transferrin-binding proteins in *Haemophilus influenzae* type b. *Infect Immunity* **60**: 2986–2991.

Hoogstraten, J., De Sa, D.J., Knisely, A.S. (1990) Fetal liver disease may precede extrahepatic siderosis in neonatal hemochromatosis. *Gastroent* **98 (6)**: 1699–1701.

Horn, E. (1988) Iron and folate supplements during pregnancy: supplementing everyone treats those at risk and is cost effective. *Br Med J* **297**: 1325–1327.

Horowitz, F.D. (1989) Using developmental theory to guide the search for the effects of biological risk factors on the development of children. *Am J Clin Nutr* **50**: 509–597.

Howarth, J.M., Gliner, J.A., Folinsbee, L.J. (1981)

Adaptation to ozone, duration of effect. *Am Rev Respir Dis* **123**: 496–499.

Howell, D. (1971) Significance of iron deficiencies. Consequences of mild deficiency in children. In *Extent and Meaning of Iron Deficiency in the US*. Summary proceedings of a workshop of the Food and Nutrition Board. National Academy of Sciences, Washington DC.

Hruszkewycz, A.M. (1988) Evidence for mitochondrial DNA damage by lipid peroxidation. *Biochem Biophys Res Comm* **153**: 191–197.

Hu, J., Liu, Y., Yu, T. *et al*. (1991) Diet and cancer of the colon and rectum. A case-control study in China. *Int J Epidemiol* **20**: 362–367.

Huch, R. and Huch, A. (1993) Maternal and fetal erythropoietin: physiological aspects and clinical significance. *Ann Med* **25**: 289–293.

Huebers, H.A. (1986) Iron absorption: molecular aspects and its regulation. *Acta Haematologica Japonica* **49**: 1528–1535.

Huebers, H.A. and Finch, C.A. (1987) The physiology of transferrin and transferrin receptors. *Physiol Res* **67**: 520–582.

Huebers, H.A., Josephson, B., Huebers, E. *et al*. (1984) Occupancy of the iron binding sites of human transferrin. *Proc Natl Acad Sci USA* **81**: 4326–4330.

Huebers, H., Csiba, E., Huebers, E., Finch, C.A. (1985) Molecular advantage of diferric transferrin in delivering iron to reticulocytes: a comparative study. *Proc Soc Exp Med* **179**: 222–226.

Huebers, H.A., Eng, M.J., Josephson, B.M. *et al*. (1987) Plasma iron and transferrin iron-binding capacity evaluated by colorimetric and immunoprecipitation methods. *Clin Chem* **33**: 273–277.

Hume, R. and Weyers, E. (1973) Changes in leucocyte ascorbic acid during the common cold. *Scot Med J* **18**: 3–7.

Hunding, A., Jordal, R., Paulev, P.-E. (1981) Runner's anaemia and iron deficiency. *Acta Med Scand* **209**: 315–318.

Hurrell, R.F., Lynch, S.R., Trinidad, T.P. *et al*. (1988) Iron absorption in humans: bovine serum albumin compared with beef muscle and egg white. *Am J Clin Nutr* **47**: 102–107.

Hurrell, R.F., Lynch, S.R., Trinidad, T.P. *et al*. (1989) Iron absorption in humans as influenced by bovine milk proteins. *Am J Clin Nutr* **49**: 546–552.

Hurrell, R.F., Juillerat, M.A., Reddy, M.B. *et al*. (1992) Soy protein, phytate and iron absorption in humans. *Am J Clin Nutr* **56**: 573–578.

Idjradinata, P. and Pollitt, E. (1993) Reversal of developmental delays in iron deficient anaemic infants treated with iron. *Lancet* **341**: 1–4.

Idjradinata, P., Watkins, W.E., Pollitt, E. (1994) Adverse effect of iron supplementation on weight gain of iron-replete young children. *Lancet* **343**: 1252–1254.

Imlay, J.A., Chin, S.M., Linn, S. (1988) Toxic DNA damage by hydrogen peroxide through the Fenton reaction *in vivo* and *in vitro*. *Science* **240**: 640–642.

Ingram, D.M., Nottage, E., Roberts, R. (1991) The role of diet in the development of breast cancer: a case control study of patients with breast cancer, benign epithelial hyperplasia and fibrocystis disease of the breast. *Br J Cancer* **64**: 187–191.

Institute of Medicine (1990) *Nutrition during pregnancy. Part II: Nutrient Supplements*, National Academy Press, Washington DC: 272–298.

International Committee for Standardization in Haematology (ICSH) (1990) Revised recommendations for the measurements of the serum iron in human blood. *Brit J Haematol* **75**: 615–616.

Irigoyen, M., Davidson, L.L., Carriero, D., Seaman, C. (1991) Randomised, placebo controlled trial of iron supplementation in infants with low haemoglobin levels fed iron fortified formula. *Pediatrics* **88**: 320–326.

Irvin, T.T., Chattopadhyay, K., Smythe, A. (1978) Ascorbic acid requirements in postoperative patients. *Surg Gynecol Obstet* **147**: 49–55.

Ismail, M. and Brock, J.H. (1993) Binding of lactoferrin and transferrin to the human promonocytic cell line U937. Effect on iron uptake and release. *J Biol Chem* **268**: 21618–21625.

Iwamoto, N., Kawaguchi, T., Horikawa, K. *et al*. (1994) Haemolysis induced by ascorbic acid in paroxysmal nocturnal haemoglobinuria. *Lancet* **343**: 357.

Jackson, A.A. and Golden, M.H.N. (1983) Protein-energy malnutrition. In *The Oxford textbook of medicine* (eds D.J. Weatherall, J.G.G. Ledingham, D.A. Warrel), Oxford University Press, Oxford: 812–821.

Jacobs, A. and Miles, P.M. (1970) The formation of iron complexes with bile and bile constituents. *Gut* **11**: 732–734.

Jacobs, A. and Worwood, M. (1975) Ferritin in serum. Clinical and biochemical implications. *New Eng J Med* **292**: 951–956.

Jacobs, A. and Worwood, M. (1980) Iron metabolism, iron deficiency and iron overload. In *Blood and its Disorders* (eds R.M. Hardisty, D.J. Weatherall), Blackwell Scientific Publications, Oxford: 149–197.

Jacobs, A., Waters, W.E., Campbell, H., Barrow, A. (1969) A random sample from Wales. III. Serum iron, iron binding capacity and transferrin saturation. *Brit J Haematol* **17**: 581–587.

Jacobs, A., Greenman, D., Owen, E., Cavill, I. (1971) Ascorbic acid status in iron deficiency anemia. *J Clin Path* **24**: 694–697.

James, J. and Laing, G. (1994) Iron deficiency anaemia. *Current Paediatrics* **4**: 33–37.

James, J., Evans, J., Male, P. *et al*. (1988) Iron deficiency in inner-city preschool children: development of a general practice screening programme. *J Roy Coll Gen Prac* **38**: 250–252.

James, J., Lawson, P., Male, P., Oakhill, A. (1989) Preventing iron deficiency in preschool children by implementing an educational and screening programme in an inner city practice. *Br Med J* **299**: 838–840.

James, J., Bailward, T., Lawson, P., Laing, G. (1993) Treatment of iron deficiency anaemia with iron in children. *Lancet* **341**: 572.

James, S.L. and Hibbs, J.B. Jr (1990) The role of nitrogen oxides as effector molecules of parasite killing. *Parasitology Today* **6**: 303–305.

Janelle, K.C. and Barr, S. (1995) Nutrient intakes and eating behaviour scores of vegetarian and non-vegetarian women. *J Am Diet Assoc* **95**: 180–186.

Janghorbani, M., Ting, B.T.G., Fomon, S.J. (1986)

Erythrocyte incorporation of ingested stable isotope of iron ($^{58}$Fe). *Am J Haem* **21**: 277–288.

Jansen, C. and Harrill, I. (1977) Intakes and serum levels of protein and iron for 70 elderly women. *Am J Clin Nutr* **30**: 1414–1422.

Jarvis, J.H. and Jacobs, A. (1974) Morphological abnormalities in lymphocyte mitochondria associated with iron-deficiency anaemia. *J Clin Pathol* **27**: 973–979.

Jaz Winska, E.C., Lee, S.C., Webb, S.I. *et al.* (1993) Localization of the hemochromatosis gene close to D6S105. *Am J Hum Gen* **53**: 347–352.

Jellinger, K., Paulus, W., Grundke-Iqbal, I. *et al.* (1990) Brain iron and ferritin in Parkinson's and Alzheimer's diseases. *J Neural Transm* **2**: 327–340.

Jenner, P. (1993) Altered mitochondrial function, iron metabolism and glutathione levels in Parkinson's disease. *Acta Neurol Scand* **87**: 6–13.

Johnson, D.L. and McGowan, R.J. (1983) Anemia and infant behavior. *Nutr & Behav* **1**: 185–192.

Jones, H.R. and Hedley-Whyte, E.T. (1983) Idiopathic hemochromatosis (IHC): dementia and ataxia as presenting signs. *Neurology* **33**: 1479–1483.

Joosten, E., Pelemans, W., Hiele, M. *et al.* (1992) Prevalence and causes of anaemia in a geriatric hospitalized population. *Gerontol* **38**: 111–117.

Kahn, R., Romslo, I., Lamvik, J. (1990) Anemia in general practice. *Scand J Clin Lab Invest* **50**: 41–45.

Kannel, W.B., Gordon, T., Wolf, P.A., McNamara, P. (1972) Haemoglobin and the risk of cerebral infarction: The Framingham study. *Stroke* **3**: 409–420.

Kaplan, J., Sarnaik, S., Gitlin, J., Lusher, J. (1984) Diminished helper/suppressor lymphocyte ratios and natural killer activity in recipients of repeated blood transfusions. *Blood* **64**: 308–310.

Kaul, L., Heshmat, M.Y., Kovi, J. *et al.* (1987) The role of diet in prostate cancer. *Nutr Cancer* **9**: 123–128.

Kawamata, T., Tooyama, I., Yamada, T. *et al.* (1993) Lactotransferrin immunocytochemistry in Alzheimer and normal human brain. *Am J Pathol* **142**: 1574–1585.

Kelly, A.M., McDonald, D.J., MacDougal, A.N. (1978) Observations in maternal and foetal ferritin concentrations at term. *Br J Obstet Gynaecol* **85**: 338–343.

Kemp, J.D., Thorson, J.A., Stewart, B.C., Naumann, P.W. (1992) Inhibition of hematopoietic tumor growth by combined treatment with deferoxamine and an IgG monoclonal antibody against the transferrin receptor: evidence for a threshold model of iron deprivation toxicity. *Cancer Res* **52**: 4144–4148.

Keusch, G.T. and Farthing, M.J.G. (1986) Nutrition and infection. *Ann Rev Nutr* **6**: 131–154.

Kim, H.Y., Klausner, R.D., Rouault, T.A. (1995) Translational repressor activity is equivalent and is quantitatively predicted by in vitro RNA binding for two iron-responsive element-binding proteins, IRP1 and IRP2. *J Biol Chem* **270** (**10**): 4983–4986.

Kim, I., Yetley, E.A., Calvo, M.S. (1993) Variation in iron-status measures during the menstrual cycle. *Am J Clin Nutr* **58**: 705–709.

Kinlen, L.J. (1982) Meat and fat consumption and cancer mortality: a study of strict religious orders in Britain. *Lancet* **1**: 946–949.

Kivivuori, S.M., Anttila, R., Viinikka, L. *et al.* (1993) Serum transferrin receptor for assessment of iron status in healthy prepubertal and early pubertal boys. *Ped Res* **34**: 297–299.

Klausner, R.D., Rouault, T.A., Harford, J.B. (1993) Regulating the fate of mRNA: the controls of cellular iron metabolism. *Cell* **72**: 19–28.

Klingshirn, L.A., Pate, R.R., Bourque, S.P. *et al.* (1992) Effect of iron supplementation on endurance capacity in iron-depleted female runners. *Med Sci Sports Exerc* **24**: 819–824.

Kochanowski, B.A. and Sherman, A.R. (1985) Decreased antibody formation in iron-deficient rat pups – effect of iron repletion. *Am J Clin Nutr* **41**: 278–284.

Koerper, M.A. and Dallman, P.R. (1977) Serum iron concentration and transferrin saturation in the diagnosis of iron deficiency in children: normal developmental changes. *J Pediatr* **91**: 870–874.

Kohgo, Y., Niitsu, Y., Kondo, H. *et al.* (1987) Serum transferrin receptor as a new index of erythropoiesis. *Blood* **70**: 1955–1958.

Koj, A. (1985) Biological functions of acute phase proteins. In *The acute phase response to injury and infection* (eds A.H. Gordon, A. Koj), Elsevier, London: 145–160.

Konijn, A.M. and Hershko, C. (1977) Ferritin synthesis in inflammation 1. Pathogenesis of impaired iron release. *Brit J Haematol* **37**: 7–16.

Konijn, A.M., Carmel, M., Levy, R., Hershko, C. (1981) Ferritin synthesis in inflammation. II. Mechanism of increased ferritin synthesis. *Brit J Haematol* **49**: 361–370.

Krantz, S.B. (1991) Erythropoietin. *Blood* **77**: 419–434.

Kristal-Boneh, E., Froom, P., Harari, G. *et al.* (1993) Seasonal changes in red blood cell parameters. *Brit J Haematol* **85**: 603–607.

Kuhn, I.N., Monsen, E.R., Cook, J.D., Finch, C.A. (1968) Iron absorption in man. *J Lab Clin Med* **71**: 7I5–72I.

Kuhn, L.C. (1991) mRNA-protein interactions regulate critical pathways in cellular iron metabolism [Annotation]. *Brit J Haematol* **79**: 1–5.

Kuiper, M.A., Mulder, C., Van-Kamp, G.J. *et al.* (1994) Cerebrospinal fluid ferritin levels of patients with Parkinson's disease, Alzheimer's disease and multiple system atrophy. *J Neural Transm* **7**: 109–114.

Kulapongs, P., Vithayasai, V., Suskind, R.M., Olson, R.E. (1974) Cell-mediated immunity and phagocytosis and killing function in children with severe iron-deficiency anaemia. *Lancet* **ii**: 689–691.

Kuller, L.H., Eichner, J.E., Orchard, T.J. *et al.* (1991) The relation between serum albumin levels and risk of coronary heart disease in the Multiple Risk Factor Intervention Trial. *Am J Epidemiol* **134**: 1266–1277.

Kune, S., Kune, G.A., Watson, L.F. (1987) Case-control study of dietary etiological factors: the Melbourne colorectal study. *Nutr Cancer* **9**: 21–43.

Kutty, K., Kutty, G., Rodriguez, I.R. *et al.* (1994) Chromosomal localization of the human heme oxygenase genes: heme oxygenase-1 (HMOX1) maps to chromosome 22q12 and heme oxygenase-2 (HMOX2) maps to chromosome 16p13.3. *Genomics* **20**: 513–516.

Kuvibidila, S. and Sarpong, D. (1990) Mitogenic

response of lymph nodes and spleen lymphocytes from mice with moderate and severe iron deficiency anemia. *Nutr Res* **10**: 195–210.

Kuvibidila, S.R., Baliga, B.S., Suskind, R.M. (1990) The effect of iron-deficiency anemia on cytolytic activity of mice spleen and peritoneal cells against allogenic tumor cells. *Am J Clin Nutr* **38**: 238–244.

La Vecchia, C., Negri, E., Decarli, A. *et al.* (1988) A case-control study of diet and colo-rectal cancer in northern Italy. *Int J Cancer* **41**: 492–498.

Labbe, R.F. and Rettmer, R.L. (1989) Zinc protoporphyrin: a product of iron-deficient erythropoiesis. *Semin Hematol* **26**: 40–46.

Lacoste, H., Goyert, G.L., Goldman, L.S. *et al.* (1992) Acute iron intoxication in pregnancy: case report and review of the literature. *Obstet Gynecol* **80**: 500–501.

Lamanca, J.J. and Haymes, E.M. (1992) Effects of low ferritin concentration on endurance performance. *Int J Sport Nutr* **2**: 376–385.

Lamparelli, R.D., Bothwell, T.H., MacPhail, A.P. *et al.* (1988a) Nutritional anaemia in pregnant coloured women in Johannesburg. *S Afr Med J* **73**: 477–481.

Lamparelli, R.D., van der Westhuyzen, J., Bothwell, T.H. *et al.* (1988b) Anaemia in pregnant Indian women in Johannesburg. *S Afr Med J* **74**: 170–173.

Lancaster, J.R. and Hibbs, J.B. (1990) EPT demonstration of iron–nitrosyl complex formation by cytotoxic activated macrophages. *Proc Nat Acad Sci USA* **87** (**3**): 1223–1227.

Larkin, E.C. and Rao, G.A. (1990) Importance of fetal and neonatal iron: adequacy for normal development of the central nervous system. In *Brain Behaviour and Iron in the Infant Diet* (ed. J. Dobbing), Springer-Verlag, London: 43–57.

Larrick, J.W. and Cresswell, P. (1979) Modulation of cell surface iron transferrin receptors by cellular density and state of activation. *J Supramol Struct* **11**: 579–586.

Larsson, G., Milsom, I., Lindstedt, G., Rybo, G. (1992) The influence of a low-dose combined oral contraceptive on menstrual blood loss and iron status. *Contraception* **46**: 327–334.

Lauffer, R.B. (1991) Iron stores and the international variation in mortality from coronary artery disease. *Medical Hypotheses* **35**: 96–102.

Layrisse, M., Martinez-Torres, C., Leets, I. *et al.* (1984) Effect of histidine, cysteine, glutathione or beef on iron absorption in humans. *J Nutr* **114**: 217–223.

Lederman, M.H., Cohen, A., Lee, J.W.W. *et al.* (1984) Deferoxamine: a reversible S-phase inhibitor of human lymphocyte proliferation. *Blood* **64**: 748–753.

Lee, G.R. (1983) The anemia of chronic disease. *Semin Hematol* **20**: 61–80.

Lee, H.P. (1992) Risk factors for breast cancer by age and menopausal status: a case-control study in Singapore. *Cancer Causes Control* **3**: 313–322.

Lee, H.P., Gourley, L., Durry, S.W. *et al.* (1991) Dietary effects on breast-cancer risk in Singapore. *Lancet* **337**: 1197–1200.

Leggett, B.A., Brown, N.N., Bryant, S.J. *et al.* (1990) Factors affecting the concentrations of ferritin in serum in a healthy Australian population. *Clin Chem* **36/7**: 1350–1355.

Lemons, J.A., Moye, L., Hall, D., Simmons, M. (1982) Differences in the composition of preterm and term human milk during early lactation. *Pediatr Res* **16**: 145–151.

Letsky, E. (1991) The haematological system. In *Clinical Physiology in Obstetrics* (eds F. Hytten, G. Chamberlain), Blackwell Scientific Publications, Oxford: 39–82.

Levi, F., La Vecchia, C., Gulie, C., Negri, E. (1993) Dietary factors and breast cancer risk in Vaud, Switzerland. *Nutr Cancer* **19**: 327–335.

Levi, S., Yewdall, S.J., Harrison, P.M. *et al.* (1992) Evidence that H- and L-chains have co-operative roles in the iron-uptake mechanism of human ferritin. *Biochem J* **288**: 591–596.

Lieberman, E., Ryan, K.J., Monson, R.R., Schoenbaum, S.C. (1988) Association of maternal haematocrit with premature labour. *Am J Obstet Gynecol* **159**: 107–114.

Liew, F.Y. and Cox, E.G. (1991) Nonspecific defence mechanism: the role of nitric oxide. *Immunol Today* **12**: A17–A21.

Lindemann, R., Ekanger, R., Opstad, P.K. *et al.* (1978) Hematological changes in normal men during prolonged severe exercise. *Am Corr Ther J* **32**: 107–111.

Lipschitz, D.A. (1990) The anemia of chronic disease. *J Am Geriat Soc* **38**: 1258–1264.

Looker, A.C., Sempos, C.T., Liu, K. *et al.* (1990) Within-person variance in biochemical indicators of iron status: effects on prevalence estimates. *Am J Clin Nutr* **52**: 541–547.

Lozoff, B., Brittenham, G.M., Viteri, F.E. *et al.* (1982a) The effects of short-term iron therapy on developmental deficits in iron deficient anemic infants. *J Pediatr* **100**: 351–357.

Lozoff, B., Brittenham, G.M., Viteri, F.E. *et al* (1982b) Developmental deficits in iron-deficient infants: effects of age and severity in iron lack. *J Pediatr* **101**: 948–952.

Lozoff, B., Brittenham, G.M., Wolf, A.W. *et al.* (1987) Iron deficiency anemia and iron therapy. Effects on infant developmental test performance. *Pediatrics* **79**: 981–995.

Lozoff, B., Jimenez, E., Wolf, A.W. (1991) Long-term developmental outcome of infants with iron deficiency. *N Engl J Med* **325**: 687–694.

Lubin, J.H., Burns, P.E., Blot, W.J. *et al.* (1981) Dietary factors and breast cancer risk. *Int J Cancer* **28**: 685–689.

Lundberg, P.-A., Lindstedt, G., Andersson, T. *et al.* (1984) Increase in serum ferritin concentration induced by fasting. *Clin Chem* **30**: 161–163.

Lundstrom, U., Siimes, M.A., Dallman, P.R. (1977) At what age does iron supplementation become necessary in low birthweight infants? *J Pediatr* **91**: 878–883.

Lyle, R.M., Weaver, C.M., Sedlock, D.A. *et al.* (1992) Iron status in exercising women: the effect of oral iron therapy vs increased consumption of muscle foods. *Am J Clin Nutr* **56**: 1049–1055.

Lynch, S.R., Beard, J.L., Dassenko, S.A., Cook, J.D. (1984) Iron absorption from legumes in humans. *Am J Clin Nutr* **40**: 42–47.

Lynch, S.R., Dassenko, S.A., Morck, T.A. *et al.* (1985)

Soy protein products and heme iron absorption in humans. *Am J Clin Nutr* **41**: 13–20.

Lynch, S.R., Dassenko, S.A., Cook, J.D. *et al.* (1994) Inhibitory effect of a soybean protein-related moiety on iron absorption in humans. *Am J Clin Nutr* **60**: 567–572.

MacDermott, R.P. and Greenberger, N.J. (1969) Evidence for a humoral factor influencing iron absorption. *Gastroen* **57**: 117–125.

Macdougall, I.C., Cavill, I., Hulme, B. *et al.* (1992) Detection of functional iron deficiency during erythropoietin treatment: a new approach. *Br Med J* **304**: 225–226.

Macfarlane, B.J., Bezwoda, W.R., Bothwell, T.H. *et al.* (1988) Inhibitory effect of nuts on iron absorption. *Am J Clin Nutr* **47**: 270–274.

Macfarlane, B.J., van der Riet, W.B., Bothwell, T.H. *et al.* (1990) Effect of traditional oriental soy products on iron absorption. *Am J Clin Nutr* **51**: 873–880.

Mack, U., Storey, E.L., Powell, L.W., Halliday, J.W. (1985) Characterization of the binding of ferritin to the rat liver ferritin receptor. *Biochim Biophys Acta* **843**: 164–170.

Mackay, H.M. (1928) Anaemia in infancy; its prevalence and prevention. *Arch Dis Childh* **3**: 117–147.

Mackenzie, J.F. (1994) *The effect of iron fortification upon trace element metabolism in healthy term infants*. PhD thesis. University of Aberdeen.

Mackler, B., Person, R., Miller, L.R. *et al.* (1978) Iron deficiency in the rat: Biochemical studies of brain metabolism. *Pediatr Res* **12**: 217–220.

MacPhail, P. and Bothwell, T.H. (1992) The prevalence and causes of nutritional iron deficiency anaemia. In *Nutritional Anaemias* (eds S.J. Fomon, S. Zlotkin). Nestle Nutrition Workshop Series, vol. 30., Vevey/Raven Press, New York: 1–12.

MAFF (Ministry of Agriculture, Fisheries and Food) (1991) *Household food consumption and expenditure 1990*. MAFF, London.

MAFF (1992) *Household food consumption and expenditure 1991*. HMSO, London.

MAFF (1994) *National Food Survey 1993*. HMSO, London.

Magazanik, A., Weinstein, Y., Abarbanel, J. *et al.* (1991) Effect of an iron supplement on body iron status and aerobic capacity of young training women. *Eur J Appl Physiol* **62**: 317–323.

Magnussen, M.K., Sigfusson, N., Sigvaldason, H. *et al.* (1994) Low iron binding capacity as a risk factor for myocardial infarction. *Circulation* **89**: 102–8.

Magnusson, B., Bjorn-Rasmussen, E., Hallberg, L., Rossander, L. (1981) Iron absorption in relation to iron status. Model proposed to express results to food iron absorption measurements. *Scand J Haematol* **27**: 201–208.

Magnusson, B., Hallberg, L., Rossander, L., Swolin, B. (1984) Iron metabolism and 'sports anemia'. 1. A study of several iron parameters in elite runners with differences in iron status. *Acta Med Scand* **216**: 149–155.

Magnusson, G., Flodh, H., Malmfors, T. (1977) Oncological study in rats of Ferastral, an iron-poly-(sorbitol-gluconic acid) complex, after intramuscular administration. *Scand J Haematol* **32 (Suppl)**: 87–98.

Mahaffey, K.R. (1981) Nutritional factors in lead poisoning. *Nutr Rev* **39**: 353–362.

Mainou-Fowler, T. and Brock, J.H. (1985) Effect of iron deficiency on the response of mouse lymphocytes to concanavalin A: the importance of transferrin-bound iron. *Immunology* **54**: 325–332.

Mann, S., Wade, V.J., Dickson, D.P.E. *et al.* (1988) Structural specificity of haemosiderin iron cores in iron overload syndromes. *FEBS* **234**: 69–72.

Manousos, O., Day, N.E., Trichopoulos, D. *et al.* (1983) Diet and colorectal cancer: a case-control study in Greece. *Int J Cancer* **32**: 1–5.

Manzoni, O., Prezeau, L., Martin, P. *et al.* (1992) Nitric-oxide induced blockade of NMDA receptors. *Neuron* **8**: 653–662.

Marder, E., Nicoll, A., Polnay, L., Shulman, C.E. (1990) Discovering anaemia at child health clinics. *Arch Dis Child* **65**: 892–894.

Martinez, I., Torres, R., Frias, Z. *et al.* (1990) Factors associated with adenocarcinomas of the large bowel in Puerto Rico. *Revista Latinoamericana de oncologia*: 13–20.

Martinez, O.B. (1988) Indices of vitamin, iron and hematological status of a selected sample of elderly Canadians. *Nutr Res* **8**: 1345–1351.

Martinez-Torres, C., Romano, E., Layrisse, M. (1981) Effect of cysteine on iron absorption in man. *Am J Clin Nutr* **34**: 322–327.

Martinez-Torres, C., Cubeddu, L., Dillmann, E. *et al.* (1984) Effect of exposure to low temperatures on normal and iron-deficient subjects. *Am J Physiol* **246**: R380–R383.

Matzner, Y., Konijn, A.M., Shlomai, Z. *et al.* (1985) Differential effect of isolated placental isoferritins on *in vitro* T-lymphocyte function. *Br J Haematol* **59**: 443–448.

McCance, R.A. and Widdowson, E.M. (1937) Absorption and excretion of iron. *Lancet* **ii**: 680–684.

McFarlane, H., Reddy, S., Adcock, K.J. *et al.* (1970) Immunity, transferrin and survival in kwashiorkor. *Br Med J* **4**: 268–270.

McGregor, S.J., Naves, M.L., Birly, A.K. *et al.* (1991) Interaction of aluminium and gallium with human lymphocytes: the role of transferrin. *Biochim Biophys Acta* **1095**: 196–200.

McLachlan, D.R.C., Dalton, A.J., Kruck, T.P.A. *et al.* (1991) Intramuscular desferrioxamine in patients with Alzheimer's disease. *Lancet* **337**: 1304–1308.

McLane, J.A., Fell, R.D., McKay, R.H. *et al.* (1981) Physiological and biochemical effects of iron deficiency on rat skeletal muscle. *Am J Physiol* **241**: C47–C54.

McMahon, L.F., Ryan, M.J., Larson, D., Fisher, R.L. (1984) Occult gastrointestinal blood loss in marathon runners. *Ann Intern Med* **100**: 846–847.

Means, R.T. and Krantz, S.B. (1992) Progress in understanding the pathogenesis of the anemia of chronic disease. *Blood* **80**; 1639–1647.

Melhorn, D.K., Gross, S., Childers, G. (1979) Vitamin E dependent anemia in the premature infant. I. Effects of large doses of medicinal iron. *J Pediatr* **79**: 569–580.

Merhav, H., Amitai, Y., Palti, H., Godfrey, S. (1985) Tea drinking and microcytic anaemia in infants. *Am J Clin Nutr* **41**: 1201–1213.

Merlo, F., Filiberti, R., Reggiardo, G. *et al.* (1991) Regional and temporal differences in the Italian diet and their relationship with cancer risk. In *The Mediterranean diet and cancer prevention* (eds A. Giacosa, M.J. Hill), European Cancer Prevention Organization, Cosenza, Italy.

Milder, M.S., Cook, J.D., Finch, C.A. (1978) The influence of food iron absorption on the plasma iron level in idiopathic hemochromatosis. *Acta Haematol* **60**: 65–75.

Milledge, J.S., Bryson, E.I., Catley, D.M. *et al.* (1982) Sodium balance, fluid homeostasis and the renin–aldosterone system during the prolonged exercise of hill walking. *Clin Sci* **62**: 595–604.

Miller, A.B., Howe, G.R., Jain, M. *et al.* (1983) Food items and food groups as risk factors in a case-control study of diet and colorectal cancer. *Int J Cancer* **32**: 155–161.

Miller, C.B., Jones, R.J., Piantadosi, S. *et al.* (1990) Decreased erythropoietin response in patients with anaemia of cancer. *New Engl J Med* **322**: 1689–1692.

Miller, D.D. and Berner, L.A. (1989) Is solubility *in vitro* a reliable predictor of iron bioavailability? *Biol Tr Elm Res* **19**: 11–24.

Miller, D.D., Schricker, B.R., Rasmussen, R.R., Van Campen, D. (1981) An *in vitro* method for estimation of iron availability from meals. *Am J Clin Nutr* **34**: 2248–2256.

Mills, A. and Tyler, H. (1992) *Food and nutrition intakes of British infants aged 6–12 months*. HMSO, London.

Mills, A.F. (1990) Surveillance for anaemia: risk factors in pattern of milk intake. *Arch Dis Child* **65**: 428–432.

Mills, C.D., Shearer, J., Evans, R., Caldwell, M.D. (1992) Macrophage arginine metabolism and the inhibition or stimulation of cancer. *J Immunol* **149**: 2709–2714.

Mills, P.K., Beeson, W.L., Phillips, R.L., Fraser, G.E. (1989) Dietary habits and breast cancer incidence among seventh day adventists. *Cancer* **64**: 582–590.

Milman, N. and Ibsen, K.K. (1984) Serum ferritin in Danish children and adolescents. *Scand J Haematol* **33**: 260–266.

Milman, N. and Kirchhoff, M. (1991) Influence of blood donation and iron stores assessed by serum ferritin and haemoglobin in a population survey of 1433 Danish males. *Eur J Haematol* **47**: 134–139.

Milman, N., Andersen, H.C., Strandberg Pedersen, N. (1986) Serum ferritin and iron status in 'healthy' elderly individuals. *Scand J Clin Lab Invest* **46**: 19–26.

Milman, N., Ibsen, K.K., Christensen, J.M. (1987) Serum ferritin and iron status in mothers and newborn infants. *Acta Obstet Gynecol Scand* **66**: 205–211.

Milman, N., Agger, A.O., Nielsen, O.J. (1991) Iron supplementation during pregnancy. Effect on iron status markers, serum erythropoietin and human placental lactogen. A placebo controlled study in 207 Danish women. *Dan Med Bull* **38**: 471–476.

Milman, N., Kirchhoff, M., Jorgensen, T. (1992) Iron status markers, serum ferritin and hemoglobin in 1359 Danish women in relation to menstruation, hormonal contraception, parity, and postmenopausal hormone treatment. *Ann Hematol* **65**: 96–102.

Minihane, A.M., Fox, T.E., Fairweather-Tait, S.J. (1993) A continuous flow *in vitro* method to predict bioavailability of Fe from foods. *Proc Bio* **93 (vol. II)**: 175–179.

Mochizuki, D.Y., Eisenman, J.R., Conlon, P.J. *et al.* (1987) Interleukin 1 regulates hemopoietic activity, a role previously ascribed to hemopoietin 1. *Proc Natl Acad Sci USA* **84**: 5267–5271.

Mohler, D.H. and Wheby, M.S. (1986) Hemochromatosis heterozygotes may have a significant iron overload when they also have hereditary spherocytosis. *Am J Med Sci* **292**: 320–324.

Moison, R.M.W., Palinckx, J.J.S., Roest, M. *et al.* (1993) Induction of lipid peroxidation by pulmonary surfactant by plasma of preterm babies. *Lancet* **341**: 79–82.

Moldawar, L.L., Marano, M.A., Wei, H. *et al.* (1989) Cachetin/tumor necrosis factors alter red blood cell kinetics and induce anaemia *in vivo*. *FASEB J* **3**: 1637–1643.

Mollison, P.L., Engelfriet, C.P., Contreras, M. (1988) *Blood Transfusion in Clinical Medicine*, Blackwell Scientific Publications, Oxford.

Moncada, S., Palmer, R.M.J., Higgs, E.A. (1991) Nitric oxide physiology, pathophysiology and pharmacology. *Pharmacol Rev* **43**: 109–142.

Monsen, E.R. and Cook, J.D. (1976) Food iron absorption in human subjects. IV. The effect of calcium and phosphate salts on the absorption of non-heme iron. *Am J Clin Nutr* **29**: 1142–1148.

Monsen, E.R. and Cook JD (1979) Food iron absorption in human subjects. V. Effects of the major dietary constituents of a semisynthetic meal. *Am J Clin Nutr* **32**: 804–808.

Moore, C.A., Raha-Chowdhury, R., Fagan, D.G., Worwood, M. (1994) Liver iron concentrations in sudden infant death syndrome. *Arch Dis Childh* **70 (4)**: 295–298.

Moore, R.J., Friedl, K.E., Tulley, R.T., Askew, E.W. (1993) Maintenance of iron status in healthy men during an extended period of stress and physical activity. *Am J Clin Nutr* **58**: 923–927.

Morbidity and Mortality Weekly Report (1989) Anaemia during pregnancy in low income women. **38**: 400–404.

Morck, T.A., Lynch, S.R., Cook, J.D. (1983) Inhibition of food iron absorption by coffee. *Am J Clin Nutr* **37**: 4l6–420.

Moreau, M.C., Duval-Iflah, Y., Muller, M.C. *et al.* (1983) Effet de la lactoferrine bovine et des IgG bovines donnés per os sur l'implantation de *Escherichia coli* dans le tube digestif de souris gnotoxéniques et de nouveau nés humains. *Ann Microbiol (Inst Pasteur)* **134B**: 429–441.

Morgan, A.G., Kelleher, J., Walker, B.E. *et al.* (1973) A nutritional survey of the elderly: Haematological aspects. *Int J Vit Nutr Res* **43**: 461–471.

Morgan, E.H. (1961) Plasma iron and haemoglobin levels in pregnancy. The effect of oral iron. *Lancet* **1**: 9–12.

Morgan, E.H. (1980) The role of plasma transferrin in iron absorption in the rat. *Quart J Exp Physiol* **65**: 239–252.

Morgan, J.W., Fraser, G.E., Phillips, R.L., Andress, M.H. (1988) Dietary factors and colon cancer incidence among seven day adventists. *Am J Epidemiol* **128**: 918A.

Morris, C.M., Keith, A.B., Edwardson, J.A., Pullen, R.G.L. (1992a) Uptake and distribution of iron and transferrin in the adult rat brain. *J Neurochem* **59**: 300–306.

Morris, C.M., Keith, A.B., Edwardson, J.A., Pullen, R.G.L. (1992b) Brain iron uptake and brain iron levels in chronic iron overload. *Neurosci Res Comms* **10**: 45–51.

Morris, C.M., Candy, J.M., Bloxham, C.A., Edwardson, J.A. (1992c) Distribution of transferrin receptors in relation to cytochrome oxidase activity in the human spinal cord, lower brainstem and cerebellum. *J Neurol Sci* **111**: 158–172.

Morris, C.M., Candy, J.M., Edwardson, J.A. *et al* (1993) Evidence for the localization of haemopexin immunoreactivity in neurones in the human brain. *Neurosci Lett* **149**: 141–144.

Morris, C.M., Candy, J.M., Kerwin, J.M. *et al* (1994) Transferrin receptors in the normal human hippocampus and in Alzheimer's disease. *Neuropathol Appl Neurobiol* **20**: 473–477.

Morrison, H.I., Semenoiw, R.M., Mao, Y., Wigle, D.T. (1994) Serum iron and risk of fatal acute myocardial infarction. *Epidemiology* **5**: 243–6.

Morrow, J.J., Dagg, J.H., Goldberg, A.A. (1968) A controlled trial of iron therapy in sideropenia. *Scott Med J* **13**: 78–83.

Morton, R.E., Nysenbaum, A., Price, K. (1988) Iron status in the first year of life. *J Pediatr Gastroenterol Nutr* **7**: 707–712.

Moser, P.B., Reynolds, R.D., Acharya, S. *et al.* (1988) Copper, iron, zinc, and selenium dietary intake and status of Nepalese lactating women and their breast-fed infants. *Am J Clin Nutr* **47**: 729–734.

Moser, U. and Weber, F. (1984) Uptake of ascorbic acid by human granulocytes. *Int J Vit Nutr Res* **54**: 47–53.

Moss, D., Powell, L.W., Arosio, P., Halliday, J.W. (1992) Characterization of the ferritin receptors of human T lymphoid (MOLT-4) cells. *J Lab Clin Med* **119**: 273–279.

Moynihan, P.J., Anderson, C., Adamson, A.J. *et al.* (1994) Dietary sources of iron in English adolescents. *J Hum Nutr Dietet* **7**: 225–230.

Murakawa, H., Bland, C.E., Willis, W.T., Dallman, P.R. (1987) Iron deficiency and neutrophil function: different rates of the correction of the depressions in oxidative burst and myeloperoxidase activity after iron treatment. *Blood* **69**: 1464–1468.

Murphy, J.F., O'Riordan, J., Newcombe, R.G. *et al.* (1986) Relation of haemoglobin levels in first and second trimesters to outcome of pregnant women. *Lancet* **1**: 992–995.

Murray, M.J., Murray, A.B., Murray, N.J., Murray, M.B. (1975) Refeeding malaria and hyperferraemia. *Lancet* **i**: 653–654.

Murray, M.J., Murray, A.B., Murray, M.B., Murray, C.J. (1978) The adverse effect of iron repletion on the course of certain infections. *Br Med J* **2**: 1113–1115.

Murray, M.J., Murray, A., Murray, C.J. (1980) The salutory effect of milk on amoebiasis and its reversal by iron. *Br Med J* **2**: 1151–1152.

Narasinga Rao, B.S. and Prabhavathi, T. (1978) An *in vitro* method for predicting the bioavailability of iron from foods. *Am J Clin Nutr* **31**: l69–175.

Nathan, I., Hackett, A.F., Kirby, S. (1994) Vegetarianism and health: is a vegetarian diet adequate for the growing child. *Food Sci and Tech Today* **8(1)**: 13–15.

Neckers, L.M. and Cossman, J. (1983) Transferrin receptor induction in mitogen-stimulated human T lymphocytes is required for DNA synthesis and is regulated by interleukin 2. *Proc Nat Acad Sci USA* **89**: 3494–3498.

Neckers, L.M., Yenokida, G., James, S.P. (1984) The role of the transferrin receptor in human B-lymphocyte activation. *J Immunol* **133**: 2437–2441.

Negri, E., La Vecchia, C., Franceschi, S. *et al.* (1991) Vegetables and fruit consumption and cancer risk. *Int J Cancer* **47**: 350–354.

Neilands, J.B. (1982) Microbial envelope proteins related to iron. *Ann Rev Microbiol* **36**: 285–309.

Neilands, J.B. and Nakamura, K. (1985) Regulation of iron assimilation in microorganisms. *Nutr Rev* **43**: 193–203.

Nelson, M., Naismith, D.J., Burley, V. (1990) Nutrient intakes, vitamin–mineral supplementation, and intelligence in British schoolchildren. *Br J Nutr* **64**: 13–22.

Nelson, M., White, J., Rhodes, C. (1993) Haemoglobin, ferritin, and iron intakes in British children aged 12–14 years: a preliminary investigation. *Br J Nutr* **70**: 147–155.

Nelson, M., Bakaliou, F., Trivedi, A. (1994) Iron-deficiency anaemia and physical performance in adolescent girls from different ethnic backgrounds. *Br J Nutr* **72**: 427–433.

Nelson, S.E., Zeigler, E.E., Copeland, A.M. (1988) Lack of adverse reactions to iron fortified formulas. *Pediatrics* **81**: 360–364.

Nelson, W.E., Vaughan, V.C., McKay, R.J. (1969) *Textbook of Pediatrics*, W.B. Saunders, Philadelphia.

Newhouse, I.J. and Clement, D.B. (1988) Iron status in athletes. An update. *Sports Med* **5**: 337–352.

Newton, H.M.V., Schorah, C.J., Habibzadeh, N. *et al.* (1985) The cause and correction of low blood vitamin C concentrations in the elderly. *Am J Clin Nutr* **42**: 656–659.

Nickerson, H.J., Holubets, M., Tripp, A.D., Pierce, W.E. (1985) Decreased iron stores in high school female runners. *Am J Dis Child* **139**: 1115–1119.

Niederau, C., Fischer, R., Sonnenberg, A. *et al.* (1985) Survival and causes of death in cirrhotic and in non-cirrhotic patients with primary hemochromatosis. *N Engl J Med* **313**: 1256–1262.

Niki, E. (1991) Action of ascorbic acid as a scavenger of active and stable oxygen radicals. *Am J Clin Nutr* **54**: 1119S–1124S.

Nomura, A., Henderson, B.E., Lee, J. (1978) Breast cancer and diet among the Japanese in Hawaii. *Am J Clin Nutr* **31**: 2020–2025.

Nomura, A.M.Y., Hirrohata, T., Kolonel, L.N. *et al* (1985) Breast cancer in Caucasian and Japanese women in Hawaii. *Nat Cancer Inst Monogr* **69**: 191–196.

Northrop-Clewes, C.A., Lunn, P.G., Downes, R.M. (1994) Seasonal fluctuations in vitamin A status and health indicators in Gambian infants. *Proc Nutr Soc* **53**: 144A.

Nunez, M.T., Gaete, V., Watkins, J.A., Glass, J. (1990) Mobilization of iron from endocytic vesicles: the effects of acidification and reduction. *J Biol Chem* **265**: 6688–6692.

Nutrition Society (1993) *Review of Diet and Cancer*, Nutrition Society, London.

O'Connell, M., Halliwell, B., Moorhouse, C.P. *et al.* (1986) Formation of hydroxyl radicals in the presence of ferritin and haemosiderin. *Biochem J* **234**: 727–731.

O'Connell, M.J., Ward, R.J., Peters, T.J. (1987) Iron overload, lysosomes and free radicals. In *Cell Membranes and Diseases, Including Renal* (eds E. Reid, G.M.W. Cook, J.P. Luzio), Plenum Co., UK: 109–112.

O'Connor, D.L. (1991) Interaction of iron and folate during reproduction. *Prog Food Nutr Sci* **15**: 231–254.

Oettinger, L., Mills, W.B., Hahn, P.F. (1954) Iron absorption in premature and full term infants. *J Pediatr* **45**: 302–306.

Ohira, Y., Edgerton, V.R., Gardner, G.W. *et al.* (1979) Work capacity, heart rate and blood lactate responses in iron treatment. *Br J Haematol* **41**: 365–372.

Ohls, R.K. and Christensen, R.D. (1991) Recombinant erythropoietin compared with erythrocyte transfusion in the treatment of anaemia of prematurity. *J Pediatr* **119**: 781–788.

Okuda, M., Tokunaga, R., Taketani, S. (1992) Expression of haptoglobin receptors in human hepatoma cells. *Biochim Biophys Acta* **1136**: 143–149.

Olivares, M., Walter, T., Osorio, M. *et al.* (1989) Anaemia of a mild viral infection: the measles vaccine as a model. *Pediatrics* **84**: 851–855.

Olivares, M., Llaguno, S., Marin, V. *et al.* (1992) Iron status in low-birth-weight infants, small and appropriate for gestational age. *Acta Paediatr* **81**: 824–828.

Olivieri, N.F., Nathan, D.G., Macmillan, J.H. *et al.* (1994) Survival in medically treated patients with homozygous β-thalassemia. *N Engl J Med* **331**: 574–578.

O'Neil-Cutting, M.A. and Crosby, W.H. (1987) Blocking of iron absorption by a preliminary oral dose of iron. *Arch Intern Med* **147**: 489–49l.

Oppenheimer, S.J. (1989a) Iron and infection: the clinical evidence. *Acta Paed Scand* **361 (Suppl)**: 53–62.

Oppenheimer, S.J. (1989b) Iron and malaria. *Parasitology Today* **5**: 77–79.

Oppenheimer, S.J., Gibson, F.D., Macfarlane, S.B. *et al.* (1986) Iron supplementation increases prevalence and effects of malaria: report on clinical studies in Papua New Guinea. *Trans R Soc Trop Med Hyg* **80**: 603–612.

Oski, F.A. (1985) Iron requirements of the premature infant. In *Vitamin and Mineral Requirements in Preterm Infants* (ed. R.C. Tsang), Marcel Dekker, New York: 9–21.

Oski, F.A. (1993) Iron deficiency in infancy and childhood. *New Engl J Med* **329**: 190–3.

Oski, F.A. and Honig, A.S. (1978) The effects of therapy on the developmental scores of iron deficient infants. *J Pediatr* **92**: 21–25.

Oski, F.A. and Landaw, S.A. (1980) Inhibition of iron absorption from human milk by baby food. *Am J Dis Childh* **134**: 459–460.

Oski, F.A., Honig, A.S., Helu, B., Howanitz, P. (1983) Effect of iron therapy on behaviour performance in non-anaemic iron-deficient infants. *Pediatrics* **71**: 877–880.

Oster, O., Dahm, M., Oelert, H., Prellwitz, W. (1989) Concentrations of some trace elements (Se, Zn, Cu, Fe, Mg, K) in blood and heart tissue of patients with coronary heart disease. *Clin Chem* **35**: 851–856.

Otto, B.R., Verweij-van Vugt, A.M.J.J., MacLaren, D.M. (1992) Transferrin and heme-compounds as iron sources for pathogenic bacteria. *Crit Rev Microbiol* **18**: 217–233.

Owen, G.M. (1989) Iron nutrition: growth in infancy. In *Dietary Iron: birth to two years* (ed. L.J. Filer), Raven Press, New York: 103–117.

Palmer, A. and Burns, M.A. (1994) Selective increase in lipid peroxidation in the inferior temporal cortex in Alzheimer's disease. *Brain Res* **645**: 338–342.

Palmer, R.M.J., Ferrige, A.G., Moncada, S. (1987) Nitric oxide release accounts for the biological activity of endothelium-derived relaxing factor. *Nature* **327**: 524–526.

Palti, H., Pevsner, B., Adler, B. (1983) Does anemia in infancy affect achievement on developmental and intelligence tests? *Hum Biol* **55**: 189–194.

Pantopoulos, K., Weiss, G., Hentze, M.W. (1994) Nitric oxide and the post-transcriptional control of cellular iron traffic. *Trends Cell Bio* **4**: 82–86.

Parks, Y.A. and Wharton, B. (1989) Iron deficiency and the brain. In *Brain, Behaviour and Iron in Infant Diet* (ed. J. Dobbing), Springer-Verlag, London: 71–79.

Parks, Y.A., Aukett, M.A., Murray, J.A. *et al.* (1989) Mildly anaemic toddlers respond to iron. *Arch Dis Child* **64**: 400–406.

Pate, R.R., Miller, B.J., Davis, J.M. *et al.* (1993) Iron status of female runners. *Int J Sport Nutr* **3**: 222–231.

Payne, J.A. and Belton, N.R. (1992) Nutrient intake and growth in preschool children. II. Intake of minerals and vitamins. *J Hum Nutr & Dietet* **5**: 299–304.

Payne, S.M. (1989) Iron and virulence in *Shigella*. *Molec Microbiol* **3**: 1301–1306.

Perkkio, M.V., Jansson, L.T., Brooks, G.A. *et al.* (1985) Work performance in iron deficiency of increasing severity. *J Appl Physiol* **58**: 1477–1480.

Perry, G.S., Byers, T., Yip, R., Margen, S. (1992) Iron nutrition does not account for the hemoglobin differences between blacks and whites. *J Nutr* 1992; **122**: 1417–1424.

Persijn, J.-P., van der Slik, W., Riethorst, A. (1971) Determination of serum iron and latent iron-binding capacity (LIBC). *Clin Chim Acta* **35**: 91–98.

Peto, T.E.A. and Thompson, J.L. (1986) A reappraisal of the effects of iron and desferrioxamine on the growth of *Plasmodium falciparum* in vitro: the unimportance of serum iron. *Brit J Haematol* **63**: 273–280.

Peto, T.E.A., Pippard, M.J., Weatherall, D.J. (1983) Iron overload in mild sideroblastic anaemias. *Lancet* **2**: 375–378.

Petterson, T., Kivivuori, S.M., Siimes, M.A. (1994) Is serum transferrin receptor useful for detecting iron-deficiency in anaemic patients with chronic inflammatory diseases? *Brit J Rheumatol* **33**: 740–744.

Phillips, A., Shaper, A.G., Whincup, P.H. (1989) Association between serum albumin and mortality from cardiovascular disease, cancer and other causes. *Lancet* **ii**: 1434–1436.

Phillips, R.E., Looareesuwan, S., Warrell, D.A. *et al.* (1986) The importance of anaemia in cerebral and uncomplicated falciparum malaria: role of complications, dyserythropoiesis and iron sequestration. *Quart J Med* New Series **58 (227)**: 305–323.

Pietrangelo, A., Rocchi, E., Schiaffonati, L. *et al.* (1990) Liver gene expression during chronic dietary iron overload in rats. *Hepatology* **11**: 798–804.

Pietrangelo, A., Rocchi, E., Rigo, G. *et al.* (1992) Regulation of transferrin, transferrin receptor and ferritin gene expression in the duodenum of normal, anaemic and siderotic subjects. *Gastroent* **102**: 802–809.

Pilon, V.A., Howantitz, P.J., Howanitz, J.H., Domres, N. (1981) Day-to-day variation in serum ferritin concentration in healthy subjects. *Clin Chem* **27**: 78–82.

Pincus, T., Olsen, N.J., Russell, I.J. *et al.* (1990) Multicenter study of recombinant human erythropoietin in correction of anemia of rheumatoid arthritis. *Amer J Med* **89**: 161–168.

Pippard, M.J. (1989) Desferrioxamine-induced iron excretion in humans. *Bailliere's Clinical Haematology* **2**: 323–343.

Pippard, M.J. and Chanarin, I. (1988) Iron and folate supplements during pregnancy [Letter]. *Br Med J* **297**: 1611.

Pippard, M.J. and Wainscoat, J.S. (1987) Erythrokinetics and iron status in heterozygous β thalassaemia, and the effect of interaction with α thalassaemia. *Br J Haematol* **66**: 123–127.

Pollitt, E. (1993) Iron deficiency and cognitive function. *Annual Rev Nutr* **13**: 521–537.

Pollitt, E. and Kim, I. (1988) Learning and achievement among iron-deficient children. In *Brain Iron: Neurochemical and Behavioural Aspects* (ed. M.B.H. Youdin), Taylor and Francis, London.

Pollitt, E. and Metallinos-Katsaras, E. (1990) Iron deficiency and behaviour: Constructs, methods and validity of the findings. In *Nutrition and Brain* vol. 8 (eds R.J. Wurtman, J.J. Wurtman), Raven Press, New York.

Pollitt, E., Greenfield, D., Leibel, R.L. (1978) Significance of Bayley Scale score changes following iron therapy. *J Pediatr* **92**: 177–178.

Pollitt, E., Leibel, R.L., Greenfield, D. (1983) Iron deficiency and cognitive test performance in preschool children. *Nutr Behav* **1**: 137–146.

Pollitt, E., Soemantri, A.G., Yunis, F., Scrimshaw, N.S. (1985) Cognitive effects of iron deficiency anaemia [Letter]. *Lancet* **i**: 158.

Pollitt, E., Saco-Pollitt, C., Leibel, R.L., Viteri, F.E. (1986) Iron deficiency and behavioral development in infants and preschool children. *Am J Clin Nutr* **43**: 555–565.

Pollitt, E., Hathirat, P., Kotchabharkdi, N.J. *et al* (1989a) Iron deficiency and educational achievement in Thailand. *Am J Clin Nutr* **50**: 687–697.

Pollitt, E., Haas, J.J. and Levistky, D.A. (eds) (1989b) International conference on iron deficiency and behavioural development. *Am J Clin Nutr* **50 (Suppl)**: 565–705.

Pootrakul, P., Skikne, B.S., Cook, J.D. (1983) The use of capillary blood for measurements of circulating ferritin. *Am J Clin Nutr* **37**: 307–310.

Pootrakul, P., Wattanasaree, J., Anuwatanakulchai, M., Wasi, P. (1984) Increased red blood cell protoporphyrin in thalassemia: a result of relative iron deficiency. *Am J Clin Pathol* **82**: 289–293.

Pootrakul, P., Kitcharoen, K., Yansukon, P. *et al.* (1988) The effect of erythroid hyperplasia on iron balance. *Blood* **71**: 1124–1129.

Porter, A.M.W. (1983) Do some marathon runners bleed into the gut? *Br Med J* **287**: 1427.

Porter, J.B. and Huehns, E.R. (1989) The toxic effects of desferrioxamine. *Bailliere's Clinical Haematology* **2**: 459–474.

Powars, D.R. (1975) Natural history of sickle cell disease – the first ten years. *Semin Hematol* **12**; 267–285.

Powell, L.W., Halliday, J.W., Wilson, A. *et al.* (1993) *Proceedings of the 4th International Conference on Haemochromatosis and Clinical Problems in Iron Metabolism*, Jerusalem.

Powers, H.J., Bates, C.J., Prentice, A.M. *et al.* (1983) The relative effectiveness of iron and iron with riboflavin in correcting a microcytic anaemia in men and children in rural Gambia. *Hum Nut: Clin Nut* **37C**: 413–425.

Prema, K., Ramalakshmi, B.A., Madhavapeddi, R., Babu, S. (1982) Immune status of anaemic pregnant women. *Br J Obstet Gynaecol* **89**: 222–225.

Prieto, J., Barry, M., Sherlock, S. (1975) Serum ferritin in patients with iron overload and with acute and chronic liver diseases. *Gastroent* **68**: 525–533.

Propper, R.D., Shurin, S.B., Nathan, D.G. (1976) Reassessment of the use of desferrioxamine B in iron overload. *N Engl J Med* **294**: 1421–1423.

Prual, A., Galan, P., de Bernis, L., Hercberg, S. (1988) Evaluation of iron status in Chadian pregnant women: consequences of maternal iron deficiency on the haematopoietic status of newborns. *Trop Geogr Med* **40**: 1–6.

Pryor, W.A. (1988) Why is the hydroxyl radical the only radical that commonly adds to DNA? Hypothesis: it has a rare combination of electrophilicity, high thermochemical reactivity, and a mode of production that can occur near DNA. *Free Rad Biol Med* **4**: 219–223.

Pugh, L.G.C.E. (1969) Blood volume changes in outdoor exercise of 8–10 hour duration. *J Physiol* **200**: 345–351.

Puschmann, M. and Ganzoni, A.M. (1977) Increased resistance of iron-deficient mice to *Salmonella* infection. *Infect Immunity* **17**: 663–664.

Quarterman, J. (1987) Metal absorption and the intestinal mucus layer. *Digestion* **37**: l–9.

Raftos, J., Schuller, M., Lovric, V.A. (1983) Iron stores assessed in blood donors by hematofluorometry. *Transfusion* **23**: 226–228.

Raja, K.B., Simpson, R.J., Pippard, M.J., Peters, T.J. (1988) *In vivo* studies on the relationship between intestinal iron ($Fe^{3+}$) absorption, hypoxia and erythropoiesis in the mouse. *Br J Haematol* **68**: 373–378.

Read, M.S. (1975) Anaemia and behaviour: Nutrition, growth and development. *Mod Probl Pediatr* **14**: 189–202.

Rector, W.G. Jr, Fortuin, N.J., Conley, C.L. (1982) Non-hematologic effects of chronic iron deficiency. A study of patients with polycythaemia vera treated solely with venesections. *Medicine* **61**: 382–389.

Reddy, S. and Sanders, T.A.B. (1990) Haematological

studies on pre-menopausal Indian and Caucasian vegetarians compared with Caucasian omnivores. *Br J Nutr* **64**: 331–338.

Reeves, J.D., Yip, R., Kiley, V.A., Dallman, P.R. (1984) Iron deficiency in infants: the influence of mild antecedent infection. *J Pediatr* **105**: 874–879.

Reibnegger, G., Fuchs, D., Hausen, A. (1987) The dependence of cell-mediated immune activation in malaria on age and endemicity. *Trans R Soc Trop Med Hyg* **81**: 729–733.

Reif, D.W. (1992) Ferritin as a source of iron for oxidative damage. *Free Rad Biol Med* **12**: 417–427.

Reif, D.W. and Simmons, R.D. (1990) Nitric oxide mediates iron release from ferritin. *Arch Biochem Biophys* **283**: 537–541.

Riboli, E., Cornée, J., Macquart-Moulin, G. *et al.* (1991) Cancer and polyps of the colorectum and lifetime consumption of beer and other alcoholic beverages. *Am J Epidemiol* **134 (2)**: 157–165.

Richmond, H.G. (1959) Induction of sarcoma in the rat by iron-dextran complex. *Brit Med J* **1**: 947–949.

Richter, G.W. (1984) Studies of iron overload. Rat liver siderosome ferritin. *Lab Invest* **50**: 26–35.

Riederer, P., Sofic, E., Rausch, W.-D. *et al.* (1989) Transition metals, ferritin, glutathione and ascorbic acid in Parkinsonian brains. *J Neurochem* **52**: 515–520.

Riemersma, R.A., Wood, D.A., MacIntyre, C.C.A. *et al.* (1991) Anti-oxidants and pro-oxidants in coronary heart disease. *Lancet* **337**: 677.

Rios, E., Hunter, R.E., Cook, J.D. *et al* (1975) The absorption of iron as supplements in infant cereal and infant formulas. *Pediatrics* **55**: 686–693.

Roberts, A.K., Chierici, R., Sawatzki, G. *et al.* (1992) Supplementation of an adapted formula with bovine lactoferrin: 1. Effect on the infant faecal flora. *Acta Paediatr* **81**: 119–124.

Roberts, F.D., Charalambous, P., Fletcher, L. *et al.* (1993) Effect of chronic iron overload on procollagen gene expression. *Hepatology* **18**: 590–595.

Robontham, J.L. and Lietman, P.S. (1980) Acute iron poisoning. A review. *Am J Dis Child* **134**: 875–879.

Rodger, R.S.C., Fletcher, K., Fail, B.J. *et al.* (1987) Factors influencing haematological measurements in healthy adults. *J Chron Dis* **40**: 943–947.

Rogers, J.T., Bridges, K.R., Durmowiez, G.P. *et al.* (1990) Translational control during the acute phase response. *J Biol Chem* **265**: 14572–14578.

Romslo, I. and Talstad, I. (1988) Day-to-day variations in serum iron, serum iron binding capacity, serum ferritin and erythrocyte protoporphyrin concentrations in anaemic subjects. *Eur J Haematol* **40**: 79–82.

Romslo, I., Haram, K., Sagan, N., Augensen, K. (1983) Iron requirements in normal pregnancy as assessed by serum ferritin, serum transferrin saturation and erythrocyte protoporphyrin determinations. *Br J Obstet Gynaecol* **90**: 101–107.

Roskams, A.J. and Connor, J.R. (1990) Aluminum access to the brain: a role for transferrin and its receptor. *Proc Natl Acad Sci USA* **87**: 9024–9027.

Rossander, L., Hallberg, L., Bjorn-Rasmussen, E. (1979) Absorption of iron from breakfast meals. *Am J Clin Nutr* **32**: 2484–2489.

Rossander-Hulthen, L., Gleerup, A., Hallberg, L. (1990) Inhibitory effect of oat products on non-haem iron absorption in man. *Eur J Clin Invest* **44**: 783–791.

Rutledge, J.N., Hilal, S.K., Silver, A.J. *et al.* (1987) Study of movement disorders and brain iron by MR. *Amer J Radiol* **149**: 365–379.

Saarinen, U.M. and Siimes, M.A. (1977a) Developmental changes in serum iron, total iron-binding capacity, and transferrin saturation in infancy. *J Pediatr* **91**: 875–877.

Saarinen, U.M. and Siimes, M.A. (1977b) Iron absorption from infant milk formula and the optimal level of iron supplementation. *Acta Paediatr Scand* **66**: 719–722.

Saarinen, U.M. and Siimes, M.A. (1978) Serum ferritin in assessment of iron nutrition in healthy infants. *Acta Paediatr Scand* **67**: 741–751.

Saarinen, U.M. and Siimes, M.A. (1979) Iron absorption from breast milk, cows milk and iron supplemented formula: an opportunistic use of changes in total body iron determined by haemoglobin, ferritin and body weight in 132 infants. *Pediatr Res* **13**: 173–175.

Saarinen, U.M., Siimes, M.A., Dallman, P.R. (1977) Iron absorption in infants: high bioavailability of breast milk iron as indicated by the extrinsic tag method of iron absorption and by the concentration of serum ferritin. *J Pediatr* **91**: 36–39.

Sadrzadeh, S.M.H., Anderson, D.K., Panter, S.S. *et al.* (1987) Hemoglobin potentiates central nervous system damage. *J Clin Invest* **79**: 662–664.

Salonen, J.T., Nyyssonen, K., Korpela, H. *et al.* (1992a) High stored iron levels are associated with excess risk of myocardial infarction in Eastern Finnish men. *Circulation* **86**: 803–811.

Salonen, J.T., Salonen, R., Nyyssonen, K., Korpela, H. (1992b) Iron sufficiency is associated with hypertension and excess risk of myocardial infarction. The Kuopio ischaemic heart disease risk factor study. *Circulation* **85**: 864.

Samokyszyn, V.M., Miller, D.M., Reif, D.W., Aust, S.D. (1989) Inhibition of superoxide and ferritin-dependent lipid peroxidation by ceruloplasmin. *J Biol Chem* **264**: 21–26.

Sanders, T.A.B. and Manning, J. (1992) The growth and development of vegan children. *J Hum Nutr & Dietet* **5**: 11–21.

Sandstrom, B., Fairweather-Tait, S., Hurrell, R., Van Dokkum, W. (1993) Methods for studying mineral and trace element absorption in humans using stable isotopes. *Nutr Res Rev* **6**: 71–95.

Santos, P.C. and Falcão, P.R. (1990) Decreased lymphocyte subsets and K-cell activity in iron deficiency anemia. *Acta Hematol* **84**: 118–121.

Schacter, B.A. (1988) Heme catabolism by heme oxygenase: physiology, regulation, and mechanism of action. *Sem Hematol* **25**: 349–369.

Schade, A.L. and Caroline, L. (1946) An iron binding component in human blood plasma. *Science* **104**: 340–341.

Schelp, F.P., Pongpaew, P., Sutjahjo, S.R. *et al.* (1981) Proteinase inhibitors and other biochemical criteria in infants and primary schoolchildren from urban and rural environments. *Br J Nutr* **45**: 451–459.

Schifman, R.B. and Rivers, S.L. (1987) Red blood cell

zinc protoporphyrin to evaluate anemia risk in deferred blood donors. *Am J Clin Pathol* **87**: 511–514.

Schoene, R.B., Escourrou, P., Robertson, H.T. *et al* (1983) Iron repletion decreases maximal exercise lactate concentrations in female athletes with minimal iron-deficiency anemia. *J Lab Clin Med* **102**: 306–312.

Schofield, C., Stewart, J., Wheeler, E. (1989) The diets of pregnant and post-pregnant women in different social groups in London and Edinburgh: calcium, iron, retinol, ascorbic acid and folic acid. *Br J Nutr* **62**: 363–377.

Scholl, T.O., Hediger, M.L., Fischer, R.L., Shearer, J.W. (1992) Anaemia vs iron deficiency: increased risk of pre-term delivery in a prospective study. *Am J Clin Nutr* **55**: 985–988.

Schultink, W., van der Ree, M., Matulessi, P., Gross, R. (1993) Low compliance with an iron-supplementation program: a study among pregnant women in Jakarta, Indonesia. *Am J Clin Nutr* **57**: 135–139.

Scott, E. (1960) Prevalence of pernicious anaemia in Great Britain. *J Coll Gen Pract* **3**: 80–84.

Scott, J.M., Goldie, H., Hay, S.H. (1975) Anaemia of pregnancy: the changing postwar pattern. *Br Med J* **1**: 259–261.

Scott, R.B. (1993) Common blood disorders: a primary care approach. *Geriat* **48**: 72–80.

Selby, J.V. and Friedman, G.D. (1988) Epidemiologic evidence of an association between body iron stores and risk of cancer. *Int J Cancer* **41**: 677–682.

Sempos, C.T., Looker, A.C., Gillum, R.F., Makin, D.M. (1994) Body iron stores and the risk of coronary heart disease. *New Eng J Med* **330**: 1119–1124.

Seshadri, S. and Gopaldas, T. (1989) Impact of iron supplementation on cognitive functions in preschool and school aged children: the Indian experience. *Am J Clin Nutr* **50 (suppl)**: 675–696.

Seymour, C.A. and Peters, T.J. (1978) Organelle pathology in primary and secondary haemochromatosis with special reference to lysosomal changes. *Br J Haem* **40**: 239–253.

Shannon, K.M., Mentzer, W.C., Abels, R.I. (1991) Recombinant human erythropoietin in the anaemia of prematurity: results of a placebo controlled pilot study. *J Pediatr* **118**: 949–955.

Shultink, W., Gross, R., Gliwitzki, M. *et al.* (1995) Effect of daily vs twice weekly iron supplementation in Indonesian preschool children with low iron status. *Am J Clin Nutr* **61**: 111–115.

Siegenberg, D., Baynes, R.D., Bothwell, T.H. *et al.* (1991) Ascorbic acid prevents the dose-dependent inhibitory effects of polyphenols and phytates on non-heme-iron absorption. *Am J Clin Nutr* **53**: 537–541.

Siimes, M.A. and Salmenpera, L. (1989) The weanling: iron for all or one? *Acta Paed Scand* **361 (Suppl)**: 103–108.

Siimes, M.A., Addiego, J.E., Dallman, P.R. (1974) Ferritin in serum: diagnosis of iron deficiency and iron overload in infants and children. *Blood* **43**: 581–590.

Siimes, M.A., Vuori, E., Kuitunen, P. (1979) Breast milk iron – a declining concentration during the course of lactation. *Acta Paedr Scand* **68**: 29–31.

Siimes, M.A., Salmenpera, L., Perheentupa, J. (1984) Exclusive breast-feeding for 9 months: risk of iron deficiency. *J Pediatr* **104**: 196–199.

Simeon, D.T. and Grantham-McGregor, S.M. (1990) Nutritional deficiencies and children's behaviour and mental development. *Nutr Res Rev* **3**: 1–24.

Simmer, K., Iles, C.A., James, C., Thompson, R.P. (1987) Are iron-folate supplements harmful? *Am J Clin Nutr* **45**: 122–125.

Simon, M., Bourel, M., Fauchet, R., Genetet, B. (1976) Association of HLA-A3 and HLA-B14 antigens with idiopathic hemochromatosis. *Gut* **17**: 332–334.

Simpson, R.J. and Peters, T.J. (1987) Iron-binding lipids of rabbit duodenal brush-border membrane. *Biochim Biophys Acta* **898**: 181–186.

Sinclair, A.J., Girling, A.J., Gray, L. *et al.* (1991) Disturbed handling of ascorbic acid in diabetic patients with and without microangiopathy during high dose ascorbate supplementation. *Diabetol* **34**: 171–175.

Singh, S., Hider, R.C., Porter, J.B. (1990) A direct method for quantification of non-transferrin-bound iron. *Analyt Biochem* **186**: 320–323.

Singla, P.N., Chand, S., Khanna, S., Agarwal, K.N. (1978) Effect of maternal anaemia on the placenta and newborn infant. *Acta Paed Scand* **67**: 645–648.

Singla, P.N., Gupta, V.K., Agarwal, K.N. (1985) Storage iron in human foetal organs. *Acta Paediatr Scand* **74**: 701–706.

Sirota, L., Kupfer, B., Moroz, C. (1989) Placental isoferritin as a physiological downregulator of cellular immunoreactivity in pregnancy. *Clin Exp Immunol* **77**: 257–62.

Skikne, B. and Baynes, R.D. (1994) Iron absorption. In *Iron Metabolism in Health and Disease* (eds J.H. Brock, J.W. Halliday, M.J. Pippard, L.W. Powell), W.B. Saunders, London: 177.

Skikne, R.S. and Cook, J.D. (1992) Effect of enhanced erythropoiesis on iron absorption. *J Lab Clin Med* **20**: 746–751.

Skikne, B., Lynch, S., Borek, D., Cook, J. (1984) Iron and blood donation. *Clinics in Haematology* **13**: 271–287.

Skikne, B.S., Flowers, C.H., Cook, J.D. (1990) Serum transferrin receptor: a quantitative measure of tissue iron deficiency. *Blood* **75**: 1870–1876.

Smith, A., Farooqui, S.M., Morgan, W.T. (1991) The murine haemopexin receptor. *Biochem J* **276**: 417–425.

Smith, A.W., Hendrickse, R.G., Harrison, C. *et al.* (1989) The effects on malaria of treatment of iron-deficiency anaemia with oral iron in Gambian children. *Ann Trop Paed* **9**: 17–23.

Snow, R.W., Byass, P., Shenton, F.C., Greenwood, B.M. (1991) The relationship between anthropometric measurements and measurements of iron status and susceptibility to malaria in Gambian children. *Trans Royal Soc Trop Med Hyg* **85**: 584–589.

Soemantri, A.G., Pollitt, E., Kim, I. (1985) Iron deficiency anaemia and educational achievement. *Am J Clin Nutr* **42**: 1221–1228.

Soewondo, S., Hussaini, M., Pollitt, E. (1989) effects of iron deficiency on attention and learning processes in preschool children: Bandung, Indonesia. *Am J Clin Nutr* **50 (Suppl)**: 667–674.

Sokoll, L.J. and Dawson-Hughes, B. (1992) Calcium supplementation and plasma ferritin concentrations in premenopausal women. *Am J Clin Nutr* **56**: 1045–1048.

Sonderer, B., Wild, P., Wyler, R. *et al.* (1987) Murine glia cells in culture can be stimulated to generate reactive oxygen. *J Leukocyte Biol* **42**: 463–473.

Southon, S., Wright, A.J.A., Finglas, P.M. *et al* (1992) Micronutrient intake and psychological performance of schoolchildren: consideration of the value of calculated nutrient intakes for the assessment of micronutrient status in children. *Proc Nutr Soc* **51**: 315–324.

Southon, S., Wright, A.J.A., Finglas, P.M. *et al.* (1994) Dietary intake and micronutrient status of adolescents: effect of vitamin and trace element supplementation on indices of status and performance in tests of verbal and non-verbal intelligence. *Br J Nutr* **71**: 897–918.

Srai, S.K.S., Epstein, O., Denham, E.S., McIntyre, N. (1984) The ontogeny of iron absorption and its possible relationship to pathogenesis of haemochromatosis. *Hepatology* **4**: 1033.

Srikantia, S.G. (1958) Ferritin in nutritional oedema. *Lancet* **i**: 667–668.

Srikantia, S.G., Prasad, J.S., Bhaskaram, C., Krishnamachari, K.A.V.R. (1976) Anaemia and the immune response. *Lancet* **i**: 1307–1309.

Stadtman, E.R. (1991) Ascorbic acid and oxidative inactivation of proteins. *Am J Clin Nutr* **54**: 1125S–1128S.

Stal, P., Glaumann, H., Hultcrantz, R. (1990) Liver cell damage and lysosomal iron storage in patients with idiopathic haemochromatosis. A light and electron microscopic study. *J Hepatology* **11**: 172–180.

Stampfer, M.J., Grodstein, F., Rosenberg, I. *et al.* (1993) A prospective study of plasma ferritin and risk of myocardial infarction in US physicians [Abstract]. *Circulation* **87**: 688.

Stark, D.D. (1991) Hepatic iron overload: paramagnetic pathology. *Radiology* **179**: 333–335.

Statland, B.E. and Winkel, P. (1977) Relationship of day-to-day variation of serum iron concentrations to iron-binding capacity in healthy young women. *Am J Clin Pathol* **67**: 84–90.

Statland, B.E., Winkel, P., Bokieland, H. (1976) Variation of serum iron concentration in healthy young men: within day and day-to-day changes. *Clin Biochem* **9**: 26–29.

Statland, B.E., Winkel, P., Harris, S.C. *et al.* (1977) Evaluation of biologic sources of variation of leucocyte counts and other hematologic quantities using very precise automated analyzers. *Am J Clin Pathol* **69**: 48–54.

Statutory Instrument (1984) *The Bread and Flour Regulations*, SI No 1304, HMSO, London.

Statutory Instrument (1994) *The Food Labelling (Amendment) Regulations*, SI No 804, HMSO, London.

Statutory Instrument (1995) *The Infant Formula and Follow-on Formula Regulations*, SI No 77, HMSO, London.

Steer, P., Alain, M.A., Wadsworth, J., Welch, A. (1995) Relation between maternal haemoglobin concentration and birth weight in different ethnic groups. *Br Med J* **310**: 489–491.

Stevens, D. (1991) Epidemiology of hypochromic anaemia in young children. *Arch Dis Child* **66**: 886–889.

Stevens, R.G. (1993) Iron and cancer. In *Iron and Human Disease* (ed. R.B. Lauffer), CRC Press, Boca Raton: 333–348.

Stevens, R.G., Jones, D.Y., Mocozzi, M.S., Taylor, P.R. (1988) Body iron stores and the risk of cancer. *New Engl J Med* **319**: 1047–1052.

Stockman, J.A. (1988) Erythropoietin: off again, on again. *J Pediatr* **112**: 906–908.

Stocks, J., Gutteridge, J.M.C., Sharp, R.J., Dormandy, T.L. (1974) Assay using brain homogenate for measuring the antioxidant activity of biological fluids. *Clin Sci Mol Med* **47**: 215–222.

Stokes, J. (1962) Haematological factors as related to the sex difference in coronary-artery disease. *Lancet* **ii**: 25.

Strain, J.J., Thompson, K.A., Barker, M.E., Carville, D.G.M. (1990) Iron sufficiency in the population of Northern Ireland: estimates from blood measurements. *Br J Nutr* **64**: 219–224.

Sturrock, A., Alexander, J., Lamb, J. *et al* (1990) Characterization of a transferrin-independent uptake system for iron in HeLa cells. *J Biol Chem* **65**: 3139–3145.

Sullivan, J.L. (1981) Iron and the sex difference in heart disease risk. *Lancet* **i**: 1293–1294.

Sullivan, J.L. (1992a) Stored iron and ischemic-heart-disease. Empirical support for a new paradigm. *Circulation* **86**: 1036–1037.

Sullivan, J.L. (1992b) Stored iron as a risk factor for ischaemic heart disease. In *Iron and Human Disease* (ed. R.B. Lauffer), CRC Press, London.

Sullivan, J.L. (1995) Hemochromatosis and coronary heart disease [Letter]. *JAMA* **273 (1)**: 25–26.

Sullivan, J.R., Till, G.O., Ward, P.A., Newton, R.B. (1989) Nutritional iron restriction diminishes acute complement-dependent lung injury. *Nutr Res* **9**: 625–634.

Sulzer, J.L., Wesley, H.H., Leonig, F. (1973) Nutrition and behaviour in Headstart children: results from the Tulane study. In *Nutrition, Development and Social Behaviour* (ed. D.J. Kallen), DHEW Publications (NIH), Washington DC.

Sussman, H.H. (1992) Iron in cancer. *Pathobiology* **60**: 2–9.

Svanberg, B., Arvidsson, B., Bjorn-Rasmussen, E. *et al.* (1975) Dietary iron absorption in pregnancy. A longitudinal study with repeated measurements of non-haem iron absorption from a whole diet. *Acta Obstet Gynecol Scand* **48 (Suppl)**: 43–68.

Swaiman, K.F. (1991) Hallervorden-Spatz syndrome and brain iron metabolism. *Arch Neurol* **48**: 1285–1293.

Symes, A.L., Missala, K., Sourkes, T.L. (1971) Iron and riboflavin dependent metabolism of a monoamine in the rat in vivo. *Science* **174**: 153–155.

Taetle, R., Rhyner, K., Castagnola, J. *et al.* (1985) Role of transferrin, Fe, and transferrin receptors in myeloid leukemia cell growth. *J Cell Invest* 1985; **75**: 1061–1067.

Takkunen, H., Reunanen, A., Knekt, P., Aromaa, A. (1989) Body iron stores and the risk of cancer [Letter]. *N Engl J Med* **320**: 1013–1014.

Talamini, R., La Vecchia, C., Decarli, A. *et al.* (1984) Social factors, diet and breast cancer in a northern Italian population. *Br J Cancer* **49**: 723–729.

Taylor, C., Rogers, G., Goodman, C. *et al.* (1987) Hematologic, iron related, and acute-phase protein responses to sustained strenuous exercise. *J Appl Physiol* **62**: 464–469.

Taylor, E.M., Crowe, A., Morgan, E.H. (1991) Transferrin and iron intake by the brain: effects of altered iron status. *J Neurochem* **57**: 1584–1592.

Taylor, P.G., Martinez-Torres, C., Romano, E.L., Layrisse, M. (1986) The effect of cysteine-containing peptides released during meat digestion on iron absorption in humans. *Am J Clin Nutr* **43**: 68–71.

Teichmann, R. and Sremmel, W. (1990) Iron uptake by human upper small intestine microvillous membrane vesicles: indication for a facilitated transport mechanism mediated by a membrane iron-binding protein. *J Clin Invest* **86**: 2145–2153.

Temlett, J.A., Landsberg, J.P., Watt, F., Grime, G.W. (1994) Increased iron in the substantia nigra compacta of the MPTP-lesioned hemiparkinsonian African Green Monkey: evidence from proton microprobe elemental microanalysis. *J Neurochem* **62**: 134–146.

Tershakovec, A.M. and Weller, S.C. (1991) Iron status of inner-city elementary school children: lack of correlation between anemia and iron deficiency. *Am J Clin Nutr* 1991; **54**: 1071–1076.

Theil, E.C. (1990) The ferritin family of iron storage proteins. *Adv Enzymol Related Areas of Mol Biol* 1990; **63**: 421–449.

Thomas, D.J., Marshall, J., Russell, R.R.W. *et al.* (1977) Effect of haematocrit on cerebral blood flow in man. *Lancet* **2**: 941–943.

Thomsen, J.K., Prien Larsen, J.C., Devantier, A., Fogh Andersen, N. (1993) Low dose iron supplementation does not cover the need for iron during pregnancy. *Acta Obstet Gynecol Scand* **72**: 93–98.

Thun, M.J., Calle, E.E., Namboodiri, M.M. *et al.* (1992) Risk factors for fatal colon cancer in a large prospective study. *JNCI* **84**: 1491–1500.

Thurnham, D.I. (1990a) Anti-oxidant vitamins and cancer prevention. *J Micronut Anal* **7**: 279–299.

Thurnham, D.I. (1990b) Antioxidants and prooxidants in malnourished populations. *Proc Nutr Soc* **48**: 247–259.

Thurnham, D.I. (1992a) Functionally important antioxidants and free radical scavengers in foods. *Food Sci Tech Today* **6**: 42–46.

Thurnham, D.I. (1992b) Micronutrients, how important in old age? *Eur J Clin Nutr* **46**: S29–S37.

Thurnham, D.I. (1993) Chemical aspects and biological mechanisms of anticancer nutrients in plant foods. In *Food and Cancer Prevention* (eds K.W. Waldron, I.T. Johnson, G.R. Fenwick), Royal Society of Chemistry, London: 109–118.

Thurnham, D.I. (1994) β-Carotene, are we misreading the signals in risk groups? Some analogies with vitamin C. *Proc Nutr Soc* **53**: 557–569.

Thurnham, D.I. (1995) Micronutrient levels following surgery or during disease: A protective response or increased requirements? *J Irish Coll Phys Surg* **24**: 31–35.

Tietz, N.W. and Rinker, A.D. (1994) Should accuracy of iron measurements be improved? *Clin Chem* **40**: 1347–1348.

Tomkins, A. and Watson, F. (1989) *Malnutrition and infection – a review.* Lavenham, Suffolk: The Lavenham Press Ltd: 1–136.

Torrance, J.D. and Bothwell, T.H. (1980) Tissue iron stores. In *Iron. Methods in Hematology*, vol. I (ed. J.D. Cook), Churchill Livingstone, New York: 90–115.

Touitou, Y., Proust, J., Carayon, A. *et al.* (1985) Plasma ferritin in old age. Influence of biological and pathological factors in a large elderly population. *Clin Chim Acta* **149**: 37–45.

Trowbridge, I.S. and Schackelford, D.A. (1986) Structure and function of transferrin receptors and their relationship to cell growth. *Biochem Soc Symp* **51**: 117–129.

Tunessen, W.W. and Oski, F.A. (1987) Consequences of starting whole cow milk at 6 months of age. *J Pediatr* **111**: 813–816.

Tuntawiroon, M., Sritongkul, N., Rossander-Hulten, L. *et al.* (1990) Rice and iron absorption in man. *Eur J Clin Nutr* **44**: 489–497.

Tuntawiroon, M., Sritongkul, N., Brune, M. *et al.* (1991) Dose-dependent inhibitory effect of phenolic compounds in foods on nonheme-iron absorption in men. *Am J Clin Nutr* **53**: 554–557.

Turnlund, J.R., Smith, R.G., Kretsch, M.J. *et al.* (1990) Milk's effect on the bioavailability of iron from cereal-based diets in young women by use of *in vitro and in vivo* methods. *Am J Clin Nutr* **52**: 373–378.

Tuyns, A.J., Kaaks, R., Haelterman, M. (1988) Colorectal cancer and the consumption of foods: a case-control study in Belgium. *Nutr Cancer* 1988; **11**: 189–204.

Ueda, F., Raja, K.B., Simpson, R.J. *et al.* (1993) Rate of $^{59}$Fe uptake into brain and cerebrospinal fluid and the influence thereon of antibodies against the transferrin receptor. *J Neurochem* **60**: 106–113.

Uphadyay, S.K., Agarwal, D.K., Tripathi, A.M., Agarwarl, K.N. (1987) *Nutritional status, physical work capacity and mental function in school children*, Scientific Report Series No 6, Nutrition Foundation of India: 40–41.

US Preventive Services Task Force (1993) Routine iron supplements during pregnancy. *JAMA* **270**: 2846–2854.

Valberg, L.S., Sorbie, J., Ludwig, T., Pelletier, D. (1976) Serum ferritin and the iron status of Canadians. *Can Med Assoc J* **114**: 417–421.

Valberg, L.S., Ghent, C.N., Lloyd, D.A. *et al.* (1978) Diagnostic efficacy of tests for the detection of iron overload in chronic liver disease. *Can Med Assoc J* **119**: 229–236.

Valberg, L.S., Flanagan, P.R., Kertesz, A., Ebers, G.C. (1989) Abnormalities in iron metabolism in multiple sclerosis. *Can J Neurol Sci* **16**: 184–186.

Valdez, D.H., Gee, J.M., Fairweather-Tait, S.J., Johnson, I.T. (1992) A comparison of methods for the *in vitro* determination of the effects of tea on iron availability from foods. *Food Chem* **44**: 331–335.

Vallance, S. (1986) Platelets, leucocytes and buffy layer vitamin C after surgery. *Hum Nutr: Clin Nutr* **40C**: 35–41.

Van Asbeck, B.S., Verbrugh, H.A., van Oost, B.A. *et al.* (1982) *Listeria monocytogenes* meningitis and decreased phagocytosis associated with iron overload. *Br Med J* **284**: 542–544.

Van Asbeck, B.S., Marx, J.J.M., Struyvenberg, J., Verhoef, J. (1984) Functional defects in phagocytic cells from patients with iron overload. *J Infect* **8**: 232–240.

Van Dijk, J.P. (1988) Regulatory aspects of placental iron transfer – a comparative study. *Placenta* **9**: 215–226.

Van Heerden, C., Oosthuizen, K., Van Wyk, H. *et al.* (1983) Evaluation of neutrophil and lymphocyte function in subjects with iron deficiency. *S Afr Med J* **24**: 111–113.

Van Rensburg, S.J., Carstens, M.E., Potocnik, F.C.V. *et al.* (1993) Increased frequency of the transferrin C2 subtype in Alzheimer's disease. *NeuroReport* **4**: 1269–1271.

Van Snick, J.L., Masson, P.L., Heremans, J.F. (1974) The involvement of lactoferrin in the hyposideremia of inflammation. *J Exp Med* **140**: 1068–1084.

Vatten, L.J., Solvoll, K., Loken, E.B. (1990) Frequency of meat and fish intake and the risk of breast cancer in a prospective study of 14,500 Norwegian women. *Int J Cancer* **46**: 12–15.

Vidnes, A. and Opstad, P.K. (1981) Serum ferritin in young men during prolonged heavy physical exercise. *Scand J Haematol* **27**: 165–170.

Vlajinac, H., Adanja, B., Jarebinski, M. (1987) Case-control study of the relationship of diet and colon cancer. *Arch Geschwulstforsh* **57**: S493–S498.

Voorhes, M.L., Stuart, M.J., Stockman, J.A., Oski, F.A. (1975) Iron deficiency anaemia and increased urinary norepinephrine excretion. *J Pediatr* **86**: 542–547.

Vostrejs, M., Moran, P.L., Seligman, P.A. (1988) Transferrin synthesis by small cell lung cancer cells acts as an autocrine regulator of cellular proliferation. *J Clin Invest* **82**: 331–339.

Vuori, E., Makinen, S.M., Kara, R. *et al.* (1980) The effects of the dietary intakes of copper, iron, manganese and zinc on the trace element content of milk. *Am J Clin Nutr* **33**: 227–231.

Wagstaff, M., Worwood, M., Jacobs, A. (1982) Iron and isoferritins in iron overload. *Clin Sci* **62**: 529–540.

Walter, T. (1989) Infancy: mental and motor development. *Am J Clin Nutr* **50 (Suppl)**: 655–661.

Walter, T., Kovalskys, J., Stekel, A. (1983) Effect of mild iron deficiency on infant mental development scores. *J Pediatr* **102**: 519–522.

Walter, T., Arredondo, S., Arévalo, M., Stekel, A. (1986) Effect of iron therapy on phagocytosis and bactericidal activity in neutrophils of iron deficient infants. *Am J Clin Nutr* **44**: 877–882.

Walter, T., De Andraca, I., Chadud, P., Perales, C.G. (1989a) Iron deficiency anemia: adverse effects on infant psychomotor development. *Pediatr* **84**: 7–17.

Walter, T., Olivares, M., Pizarro, F. (1989b) Iron and infection. In *Dietary iron: birth to two years* (ed. L.J. Filer), Raven Press, New York.

Walters, G.O., Miller, F.M., Worwood, M. (1973) Serum ferritin concentration and iron stores in normal subjects. *J Clin Pathol* **26**: 770–772.

Walters, G.O., Jacobs, A., Worwood, M. *et al.* (1975) Iron absorption in normal subjects and patients with idiopathic haemochromatosis: relationship with serum ferritin concentration. *Gut* **16**: 188–192.

Ward, N.I. and Mason, J.A. (1987) Neutron activation analysis techniques for identifying elemental status in Alzheimer's disease. *J Radioanal Nucl Chem* **113**: 515–526.

Ward, R.J., Florence, A.L., Baldwin, D. *et al.* (1991) Biochemical and biophysical investigations of the ferrocene-loaded rat: an animal model of primary haemochromatosis. *Eur J Biochem* **202**: 405–410.

Ward, R.J., Ramsey, M.H., Dickson, D.P.E. *et al.* (1992) Chemical and structural characterisation of iron cores of haemosiderin isolated from different sources. *Eur J Biochem* **209**: 847–850.

Ward, R.J., Ramsey, M.H., Dickson, D.P.E. *et al.* (1995a) Further characterisation of forms of haemosiderin in iron loaded tissues. *Eur J Biochem* (in press).

Ward, R.J., Dexter, D.T., Florence, A. *et al.* (1995b) Brain iron in the ferrocene loaded rat. Its chelation and influence on dopamine metabolism. *Biochem Pharmacol* (in press).

Ward, R.J., Zhang, Y., Crichton, R.R. *et al.* (1995c) Hepatic iron toxicity: a free radical mediated process, fact or artifact. *Arch Phys Biochem* **103**: B30.

Wasserman, G., Graziano, J.H., Factor-Litvak, P. *et al.* (1992) Independent effects of lead exposure and iron deficiency anemia on developmental outcome at age 2 years. *J Pediatr* **121**: 695–703.

Waterlow, J.C. (1947) *Fatty liver disease in infants in the British West Indies*. Medical Research Council Special Report Series No. 263, Medical Research Council, London: 22–25.

Weatherall, D.J., Clegg, J.B., Higgs, D.R., Wood, W.G. (1989) The hemoglobinopathies. In *The Metabolic Bases of Inherited Disease*, 6th edn (eds C.R. Scriver, A.L. Beaudet, W.S. Sly, D. Valle), McGraw-Hill, New York: 2281–2339.

Weaver, J. and Pollack, S. (1989) Low M iron isolated from guinea pig reticulocytes as AMP-Fe and ATP-Fe complexes. *Biochem J* **261**: 787–792.

Weber, J., Werre, J.M., Julius, H.W. and Marx, J.J.M. (1988) Decreased iron absorption in patients with active rheumatoid arthritis, with and without iron deficiency. *Ann Rheum Dis* **47**: 404–409.

Webb, T.E. and Oski, F.A. (1973a) The effects of iron deficiency anemia on scholastic achievement, behavioural stability and perceptual sensitivity of adolescents [Abstract]. *Pediatr Res* **7**: 294.

Webb, T.E. and Oski, F.A. (1973b) Iron deficiency anemia and scholastic achievement in young adolescents. *J Pediatr Res* **30**: 47–63.

Webb, T.E. and Oski, F.A. (1974) Behavioral status of young adolescents with iron deficiency anemia. *J Spec Educ* **8**: 153–156.

Weight, L.M., Darge, B.L., Jacobs, P. (1991) Athletes' pseudoanaemia. *Eur J Appl Physiol* **62**: 358–362.

Weight, L.M., Jacobs, P., Noakes, T.D. (1992a) Dietary iron deficiency and sports anaemia. *Br J Nutr* **68**: 253–260.

Weight, L.M., Klein, M., Noakes, T.D., Jacobs, P. (1992b) 'Sports anemia' – a real or apparent phenomenon in endurance-trained athletes? *Int J Sports Med* **13**: 344–347.

Weinberg, E.D. (1983) Iron in neoplastic disease. *Nutr Cancer* **4**: 223–233.

Weinberg, E.D. (1984) Iron withholding: a defense against infection and neoplasia. *Physiol Rev* **64**: 65–102.

Weinberg, E.D. (1992) Roles of iron in neoplasia: promotion, prevention and therapy. *Biol Trace Element Res* **34**: 123–140.

Weiss, G., Fuchs, D., Hausen, A. *et al.* (1992) Iron modulates interferon-gamma effects in the human myelomonocytic cell line THP-1. *Exp Hematol* **20**: 605–610.

Weiss, G., Fuchs, D., Wachter, H. (1993a) High stored iron levels and the risk of myocardial infarction. *Circulation* **87**: 1425.

Weiss, G., Goosen, B., Doppler, W. *et al.* (1993b) Translation regulation via iron-responsive elements by the nitric oxide/NO-synthase pathway. *EMBO J* **12**: 3651–3657.

Weiss, S.J. (1989) Tissue destruction by neutrophils. *New Engl J Med* **320**: 365–376.

Werkman, S., Shifman, L., Shelly, T. (1964) Psychosocial correlates of iron deficiency in early childhood. *Psychosom Med* **26**: 125.

Wharton, B. (1992a) Trace element deficiency in man: classifications and methods of assessment. *Food Chemistry* **43**: 219–224.

Wharton, B. (1992b) Which milk for normal infants? *Eur J Clin Nutr* **46**: S27–S32.

Wharton, B.A. (1989) Iron nutrition in childhood: the interplay of genes, development and environment. *Acta Paed Scand* **361 (Suppl)**: 5–11.

Wharton, B.A. and Clarke, B. (1990) Childhood nutrition. *Med Int*: 3375–3381.

Wharton, M., Granger, D.L., Durack, D.T. (1988) Mitochondrial iron loss from leukemia cells injured by macrophages. A possible mechanism for electron transport chain defects. *J Immunol* **141**: 1311–1317.

Wharton, P.A., Eaton, P.M., Wharton, B.A. (1984) Subethnic variation in the diets of Moslem, Sikh and Hindu pregnant women at Sorrento Maternity Hospital, Birmingham. *Br J Nutr* **52**: 469–476.

Wheby, M.S. (1980) Effect of iron therapy on serum ferritin levels in iron-deficiency anemia. *Blood* **56**: 138–140.

Wheby, M.S., Conrad, M.E., Hedberg, S.E., Crosby, W.H. (1962) The role of bile in the control of iron absorption. *Gastroent* **42**: 319–324.

Whitcombe, D.M., Albertson, D.G., Cox, T.M. (1994) Molecular analysis of functional and nonfunctional genes for human ferrochelatase: isolation and characterization of a FECH pseudogene and its sublocalization on chromosome 3. *Genomics* **20**: 482–486.

White, A., Nicolaas, G., Foster, K. *et al.* (1993) *Health Survey for England 1991*, HMSO, London.

Whittaker, P.G., Lind, T., Williams, J.G. (1991) Iron absorption during normal human pregnancy: a study using stable isotopes. *Br J Nutr* **65**: 457–463.

WHO (World Health Organization) (1968) *Nutritional Anaemias*. Report of a WHO Scientific Group. Technical Report Series No. 405, WHO, Geneva.

WHO (World Health Organization) Group of Experts (1972) *Nutritional Anaemias*. WHO Technical Report Series No 503, WHO, Geneva.

Widdowson, E.M. and Spray, C.M. (1951) Chemical development in utero. *Arch Dis Childh* **26**: 205–214.

Wiggers, P., Dalhoj, J., Hyltoft Peterson, P. *et al.* (1991) Screening for haemochromatosis: influence of analytical imprecision, diagnostic limit and prevalence on test validity. *Scand J Clin Lab Investig* **51**: 143–148.

Willett, W.C., Stampfer, M.J., Colditz, G.A. *et al.* (1990) Relation of meat, fat and fibre intake to the risk of colon cancer in a prospective study among women. *N Engl J Med* **323**: 1664–1672.

Willett, W.C., Stampfer, M.J., Colditz, G.A. *et al.* (1992) Relation of meat, fat, and fiber intake to the risk of colon cancer in women [Correspondence]. *N Engl J Med* **326**: 201–202.

Williams, R., Manenti, F., Williams, H.S., Pitcher, C.S. (1966) Iron absorption in idiopathic haemochromatosis before, during and after venesection therapy. *Br Med J* **ii**: 78–8l.

Williams, W.J. (1990) Clinical evaluation of the patient. In *Hematology*, 4th edn (eds W.J. Williams, A.J. Erslev, M.A. Lichtman), McGraw-Hill, New York: 9–23.

Willis, W.T., Gohil, K., Brooks, G.A., Dallman, P.R. (1990) Iron deficiency: improved exercise performance within 15 hours of iron treatment in rats. *J Nutr* **120**: 909–916.

Witte, D.L. (1991) Can serum ferritin be effectively interpreted in the presence of the acute-phase response? *Clin Chem* **37**: 484–485.

Witte, D.L., Kraemer, D.F., Johnson, G.F. *et al.* (1986) Prediction of bone marrow iron findings from tests performed on peripheral blood. *Am J Clin Pathol* **85**: 202–206.

Wong, C.T. and Saha, N. (1990) Inter-relationships of storage iron in the mother, the placenta and the newborn. *Acta Obstet Gynecol Scand* **69**: 613–616.

Wood, M.M. and Elwood, P.C. (1966) Symptoms of iron deficiency anaemia. A community survey. *Br J Prev Soc Med* **20**: 117–121.

Woodson, R.D. (1984) Hemoglobin concentration and exercise capacity. *Am Rev Respir Dis* **129 (Suppl)**: S72–S75.

Worwood, M. (1980a) Serum ferritin. In *Methods in Hematology*, vol. I (ed. J.D. Cook), Churchill-Livingstone, New York: 59–89.

Worwood, M. (1980b) Serum ferritin. In *Iron in Biochemistry and Medicine* II (ed. A. Jacobs, M. Worwood), Academic Press, London and New York: 203–244.

Worwood, M. (1982) Ferritin in human tissues and serum. *Clinics in Haematology* **11**: 275–307.

Worwood, M. (1986) Serum ferritin. *Clin Sci* **70**: 215–220.

Worwood, M. (1990) Ferritin. *Blood Reviews* **4**: 259–269.

Worwood, M. and Darke, C. (1993) Serum ferritin, blood donation, iron stores and haemochromatosis. *Transfus Med* **2**: 21–28.

Worwood, M., Cragg, S.J., Jacobs, A. *et al.* (1980) Binding of serum ferritin to concanavalin A: patients with homozygous β-thalassaemia and transfusional iron overload. *Brit J Haematol* **46**: 409–416.

Worwood, M., Cragg, S.J., Wagstaff, M., Jacobs, A. (1979) Binding of human serum ferritin to concanavalin A. *Clin Sci* **56**: 83–87.

Worwood, M., Darke, C., Trenchard, P. (1991a) Hereditary haemochromatosis and blood donation. *Br Med J* **302**: 59.

Worwood, M., Thorpe, S.J., Heath, A. *et al.* (1991b) Stable lyophilised reagents for the serum ferritin assay. *Clin Lab Haematol* **13**: 297–305.

Wright, A.J. and Southon, S. (1990) The effectiveness of various iron supplement regimens in improving the iron status of anaemic rats. *Brit J Nutr* **63**: 579–585.

Wright, A.J.A., Southon, S., Bailey, A.L. *et al.* (1995) Intake and status of non-institutionalised elderly subjects in Norwich: Comparison with younger adults and adolescents from the same general community. *Br J Nutr* (in press).

Yagi, K. (1987) Lipid peroxides and human diseases. *Chem Phys Lipid* **45**: 337–351.

Yarnell, J.W.G., Baker, I.A., Sweetnam, P.M. *et al.* (1991) Fibrinogen, viscosity, and white blood cell count are major risk factors for ischaemic heart disease. The Caerphilly and Speedwell collaborative heart disease studies. *Circulation* **83**: 836–844.

Yasui, M., Ota, K., Garruto, R.M. (1993) Concentrations of zinc and iron in the brains of Guamanian patients with amyotrophic lateral sclerosis and Parkinsonism-dementia. *Neurotoxicol* **14**: 445–450.

Yehuda, S. (1990) Neurochemical basis of behavioural effects of brain iron deficiency in animals. In *Brain Behaviour and Iron in the Infant Diet* (ed. J. Dobbing), Springer-Verlag, London: 63–76.

Yim, C.-Y., Bastian, N.R., Smith, J.C. *et al.* (1993) Macrophage nitric oxide synthesis delays progression of ultraviolet light-induced murine skin cancers. *Cancer Res* **53**: 5507–5511.

Yip, R. (1994) Changes in iron metabolism with age. In *Iron Metabolism in Health and Disease* (eds J.H. Brock, J.W. Halliday, M.J. Pippard, L.W. Powell), W.R. Saunders, London: 427–448.

Yip, R. and Dallman, P.R. (1988) The roles of inflammation and iron deficiency as causes of anemia. *Am J Clin Nutr* **48**: 1295–1300.

Yip, R. and Williamson, D.F. (1989) Body iron stores and risk of cancer [Correspondence]. *N Engl J Med* **320**: 1012.

Yip, R., Binkin, N.J., Fleshood, L., Trowbridge, F.L. (1987a) Declining prevalence of anemia among low income children in the United States. *J Amer Med Assoc* **258**: 1619–1623.

Yip, R., Walsh, K.M., Goldfarb, M.G., Binkin, N.J. (1987b) Declining prevalence of anemia in a middle class society. A pediatric success story? *Pediatrics* **80**: 330–334.

Yoshioka, M., Matsushita, T., Chuman, Y. (1984) Inverse association of serum ascorbic acid level and blood pressure or rate of hypertension in male adults aged 30–39 years. *Int J Vit Nutr Res* **54**: 343–347.

Youdim, M.B.H. (1990) Neuropharmacological and neurobiochemical aspects of iron deficiency. In *Brain Behaviour and Iron in the Infant Diet* (ed. J. Dobbing), Springer-Verlag, London: 83–99.

Youdim, M.B.H. and Green, A.R. (1977) Biogenic monoamine metabolism and functional activity in iron deficient rats: behavioral correlates. In *Iron Metabolism* (eds R. Porter, W. Fitzimmons), Ciba Foundation Symposium 51, New York, Elsevier, North Holland.

Youdim, M.B.H., Grahame-Smith, D.G., Woods, H.F. (1975) Some properties of human platelet monoamine oxidase in iron deficiency anaemia. *Clin Sci Mol Med* **50**: 479–485.

Youdim, M.B.H., Green, A.R., Bloomfield, M.R. *et al.* (1980) The effects of iron deficiency on brain biogenic monamine biochemistry and function in rats. *Neuropharmacol* **19**: 259–267.

Youdim, M.B.H., Yehuda, S., Ben-Uriah, Y. (1981) Iron-deficiency-induced circadian rhythm reversal of dopaminergic-mediated behaviors and thermoregulation in rats. *Eur J Pharmacol* **74**: 295–301.

Young, S.P. and Aisen, P. (1988) The liver and iron. In *The liver: biology and pathobiology*, 2nd edn (eds I.M. Arias, W.B. Jakoby, H. Popper, D. Schachter, D.A. Shafritz), Raven Press, New York: 535–550.

Young, T.B. and Wolf, D.A. (1988) Case-control study of proximal and distal colon cancer and diet in Wisconsin. *Int J Cancer* **42**: 167–175.

Yu, G.S.M., Steinkirchner, T.M., Rao, G.A., Larkin, E.C. (1986) Effect of prenatal iron deficiency on myelination in rat pups. *Am J Pathol* **125**: 620–624.

Zanella, A., Gridelli, L., Berzuini, A. *et al.* (1989) Sensitivity and predictive value of serum ferritin and free erythrocyte protoporphyrin. *J Lab & Clin Med* **113**: 73–78.

Zeigler, E.E., Fomon, S.J., Nelson, S.E. (1990) Cow milk feeding in infancy; further observations on blood loss from the gastrointestinal tract. *J Pediatr* **116**: 11–18.

Ziegler, R.G., Morris, L.E., Blot, W.J. *et al.* (1981) Esophageal cancer among black men in Washington, DC. II. Role of nutrition. *JNCI* **67**: 1199–1206.

Zipursky, A., Brown, E.J., Watts, J. (1987) Oral vitamin E supplementation for the prevention of anaemia in premature infants: a controlled trial. *Pediatrics* **79**: 61–68.

Zoli, A., Altomonte, L., Mirone, L. *et al.*, (1994) Serum transferrin receptors in rheumatoid arthritis. *Ann Rheum Dis* **53**: 699–670.

Zurlo, M.G., De Stefano, P., Borgna-Pignatti, C. *et al.* (1989) Survival and causes of death in thalassaemia major. *Lancet* **2**: 27–29.

# INDEX

Page references in **bold** refer to figures and page references in *italic* refer to tables